Handbook of
Power Plant Engineering

Handbook of Power Plant Engineering

James Taylor

Oxford Book Company
Jaipur ▪ New Delhi

Published By

Oxford Book Company

Regd. Off. "Bony Residency", Gate No. 2
Opp. Tilak Public School, Vishweswariya Nagar,
Gopalpura Road, Jaipur - 302018, India
Ph: 0141 - 2761280, Telefax: 2761381
email - oxfordbookjpr@gmail.com
website - oxfordbookcompany.com

Delhi Office: 102, 1st Floor, Satyam House,
4327/3, Ansari Road, Darya Ganj,
New Delhi - 110002, India
Ph: 011-45652440

Distributed by
KEYA INTERNATIONAL PRESS LLP
E-mail: keyainternationalpressllp@gmail.com

First published in India, 2024

ISBN : 9789355241078

Typeset by Shivangi Computers, Jaipur

Printed and bound in India by Replika Press Pvt. Ltd.

Contents

Preface

Electricity became a subject of scientific interest in the late 17th century with the work of William Gilbert. Over the next two centuries a number of important discoveries were made including the incandescent lightbulb and the voltaic pile. Probably the greatest discovery with respect to power engineering came from Michael Faraday who in 1831 discovered that a change in magnetic flux induces an electromotive force in a loop of wire—a principle known as electromagnetic induction that helps explain how generators and transformers work. In 1881 two electricians built the world's first power station at Godalming in England. The station employed two water wheels to produce an alternating current that was used to supply seven Siemens arc lamps at 250 volts and thirty-four incandescent lamps at 40 volts. However supply was intermittent and in 1882 Thomas Edison and his company, The Edison Electric Light Company, developed the first steam-powered electric power station on Pearl Street in New York City. The Pearl Street Station consisted of several generators and initially powered around 3,000 lamps for 59 customers.

The power station used direct current and operated at a single voltage. Since the direct current power could not be easily transformed to the higher voltages necessary to minimise power loss during transmission, the possible distance between the generators and load was limited to around half-a-mile (800 m). That same year in London Lucien Gaulard and John Dixon Gibbs demonstrated the first transformer suitable for use in a real power system. The practical value of Gaulard and Gibbs' transformer was demonstrated in 1884 at Turin where the transformer was used to light up forty kilometres (25 miles) of railway from a single alternating current generator. Despite the success of the system, the pair made some fundamental mistakes. Perhaps the most serious was connecting the primaries of the transformers in series so that switching one lamp on or off would affect other lamps further down the line. Following the demonstration George Westing house, an American entrepreneur, imported a number of the transformers along with a Siemens generator and set his engineers to experimenting with them in the hopes of improving them for use in a commercial power system.

One of Westing house's engineers, William Stanley, recognised the problem with connecting transformers in series as opposed to parallel and also realised that making the iron core of a transformer a fully enclosed loop would improve the voltage regulation of the secondary winding. Using this knowledge he built a much improved alternating current power system at Great Barrington, Massachusetts in 1886. Then in 1887 and 1888 another engineer called Nikola Tesla filed a range of patents related to power systems including one for a two-phase induction motor. Although Tesla cannot necessarily be attributed with building the first induction motor, his design, unlike others, was practical for industrial use. By 1890 the power industry had flourished and power companies had built literally thousands of power systems in the United States and Europe - these networks were effectively dedicated to providing electric lighting. During this time a fierce rivalry known as the "War of Currents" emerged between Edison, Westing house and Tesla over which form of transmission (direct or alternating current) was superior. In 1891, Westing house installed the first major power system that was designed to drive an electric motor and not just provide electric lighting. The installation powered a 100 horsepower (75 kW) synchronous motor at Telluride, Colorado with the motor being started by a Tesla induction motor. On the other side of the Atlantic, Oskar von Miller built a 20 kV 176 km three-phase transmission line from Lauffen am Neckar to Frankfurt am Main for the Electrical Engineering Exhibition in Frankfurt. In 1895, after a protracted decision-making process, the Adams No. 1 generating station at Niagara Falls began transmitting three-phase alternating current power to Buffalo at 11 kV. Following completion of the Niagara Falls project, new power systems increasingly chose alternating current as opposed to direct current for electrical transmission.

This Handbook covers Plant Design Engineering, Process Plant Engineering, Power Plant Engineering, Piping Engineering, Piping Stress Analysis, Plant Design Management System, Electrical Engineering, Instrumentation Design, Mechanical Equipments, Civil & Structural Engineering, Project Management, Industrial Process Measurement, Programmable Logic Controllers (PLCs), Variable Speed Drives, Supervisory Control and Data Acquisition System (SCADA), Distributed Control System (DCS), AC/DC Drives and MMI.

— James Taylor

Chapter 1

Introduction

Pondage

Pondage usually refers to the comparably small water storage behind the weir of a run-of-the-river hydroelectric power plant. It is considerably less storage than the reservoirs of large dams and conventional hydroelectric stations which can store water for long periods such as a dry season or year. It is usually stored during periods of low electricity demand and days when the power plant is inactive, enabling its use as a peaking power plant in dry seasons and a base load power plant during wet seasons. Ample pondage allows a power plant meet hourly load fluctuations for a period of a week or more.

Calculation

Because of the possibility of limited pondage, calculating its effect on power generation is important in determining how often the plant can be operated. The pondage factor is a rough index of the amount pondage needed when the stream flow is constant and the plant is operational during specific periods. It is calculated by the ratio of total inflow hours per week to the hours the power plant will be operated in that same period. For example, if a river has an inflow for seven days a week and the plant operates for five of those days for eight hours:

$$\frac{7\times24}{5\times8}=4.2$$

24-hour inflows and plant operation would have a pondage factor of one.

If power is used for twelve hours per day and during the inactive hours, relatively all inflows can be stored, then power can be doubled during active hours. In 12 hours there are 43,200 seconds and in

1 acre ft (1,233 m^3) there are 43,560 cu ft (1,233 m^3). This gives an estimation and guide that for twelve hours of pondage, there must be as much storage available in the pond as cubic feet per section received.

Power Engineering

Power engineering, also called power systems engineering, is a subfield of engineering that deals with the generation, transmission and distribution of electric power as well as the electrical devices connected to such systems including generators, motors and transformers. Although much of the field is concerned with the problems of three-phase AC power - the standard for large-scale power transmission and distribution across the modern world - a significant fraction of the field is concerned with the conversion between AC and DC power as well as the development of specialised power systems such as those used in aircraft or for electric railway networks.

History

Figure 1: A sketch of the Pearl Street Station

Electricity became a subject of scientific interest in the late 17th century with the work of William Gilbert. Over the next two centuries a number of important discoveries were made including the incandescent lightbulb and the voltaic pile. Probably the greatest discovery with respect to power engineering came from Michael Faraday who in 1831 discovered that a change in magnetic flux induces an electromotive force in a loop of wire—a principle known as electromagnetic induction that helps explain how generators and transformers work.

In 1881 two electricians built the world's first power station at Godalming in England. The station employed two waterwheels to produce an alternating current that was used to supply seven Siemens arc lamps at 250 volts and thirty-four incandescent lamps at 40 volts. However supply was intermittent and in 1882 Thomas Edison and his company, The Edison Electric Light Company, developed the first steam-powered electric power station on Pearl Street in New York City. The Pearl Street Station consisted of several generators and initially powered around 3,000 lamps for 59 customers. The power station used direct current and operated at a single voltage. Since the direct current power could not be easily transformed to the higher voltages necessary to minimise power loss during transmission, the possible distance between the generators and load was limited to around half-a-mile (800 m).

That same year in London Lucien Gaulard and John Dixon Gibbs demonstrated the first transformer suitable for use in a real power system. The practical value of Gaulard and Gibbs' transformer was demonstrated in 1884 at Turin where the transformer was used to light up forty kilometres (25 miles) of railway from a single alternating current generator. Despite the success of the system, the pair made some fundamental mistakes. Perhaps the most serious was connecting the primaries of the transformers in series so that switching one lamp on or off would affect other lamps further down the line. Following the demonstration George Westinghouse, an American entrepreneur, imported a number of the transformers along with a Siemens generator and set his engineers to experimenting with them in the hopes of improving them for use in a commercial power system.

One of Westinghouse's engineers, William Stanley, recognised the problem with connecting transformers in series as opposed to parallel and also realised that making the iron core of a transformer a fully enclosed loop would improve the voltage regulation of the secondary winding. Using this knowledge he built a much improved alternating current power system at Great Barrington, Massachusetts in 1886. Then in 1887 and 1888 another engineer called Nikola Tesla filed a range of patents related to power systems including one for a two-phase induction motor. Although Tesla cannot necessarily be attributed with building the first induction motor, his design, unlike others, was practical for industrial use.

By 1890 the power industry had flourished and power companies had built literally thousands of power systems (both direct and alternating current) in the United States and Europe - these networks were effectively dedicated to providing electric lighting. During this time a fierce rivalry known as the "War of Currents" emerged between Edison, Westinghouse and Tesla over which form of transmission (direct or alternating current) was superior. In 1891, Westinghouse installed the first major power system that was designed to drive an electric motor and not just provide electric lighting.

The installation powered a 100 horsepower (75 kW) synchronous motor at Telluride, Colorado with the motor being started by a Tesla induction motor. On the other side of the Atlantic, Oskar von Miller built a 20 kV 176 km three-phase transmission line from Lauffen am Neckar to Frankfurt am Main for the Electrical Engineering Exhibition in Frankfurt. In 1895, after a protracted decision-making process, the Adams No. 1 generating station at Niagara Falls began transmitting three-phase alternating current power to Buffalo at 11 kV. Following completion of the Niagara Falls project, new power systems increasingly chose alternating current as opposed to direct current for electrical transmission.

Although the 1880s and 1890s were seminal decades in the field, developments in power engineering continued throughout the 20th and 21st century. In 1936 the first commercial HVDC (high voltage direct current) line using Mercury arc valves was built between Schenectady and Mechanicville, New York. HVDC had previously been achieved by installing direct current generators in series (a system known as the Thury system) although this suffered from serious reliability issues. In 1957 Siemens demonstrated the first solid-state rectifier (solid-state rectifiers are now the standard for HVDC systems) however it was not until the early 1970s that this technology was used in commercial power systems.

In 1959 Westinghouse demonstrated the first circuit breaker that used SF_6 as the interrupting medium. SF_6 is a far superior dielectric to air and, in recent times, its use has been extended to produce far more compact switching equipment (known as switchgear) and transformers. Many important developments also came from extending innovations in the information technology and telecommunications field to the power engineering field. For example, the development of computers meant load flow studies could be run more efficiently allowing for much better planning of power systems. Advances in information technology and telecommunication also allowed for much better remote control of the power system's switchgear and generators.

Basics of Electric Power

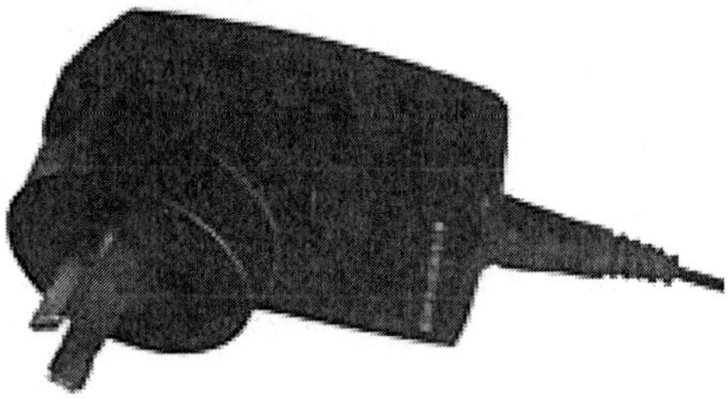

Figure 2: An external AC to DC power adapter used for household appliances

Electric power is the mathematical product of two quantities: current and voltage. These two quantities can vary with respect to time (AC power) or can be kept at constant levels (DC power).

Most refrigerators, air conditioners, pumps and industrial machinery use AC power whereas most computers and digital equipment use DC power (the digital devices you plug into the mains typically have an internal or external power adapter to convert from AC to DC power). AC power has the advantage of being easy to transform between voltages and is able to be generated and utilised by brushless machinery. DC power remains the only practical choice in digital systems and can be more economical to transmit over long distances at very high voltages.

The ability to easily transform the voltage of AC power is important for two reasons: Firstly, power can be transmitted over long distances with less loss at higher voltages. So in power networks where generation is distant from the load, it is desirable to step-up the voltage of power at the generation point and then step-down the voltage near the load. Secondly, it is often more economical to install turbines that produce higher voltages than would be used by most appliances, so the ability to easily transform voltages means this mismatch between voltages can be easily managed.

Solid state devices, which are products of the semiconductor revolution, make it possible to transform DC power to different voltages, build brushless DC machines and convert between AC and DC power. Nevertheless devices utilising solid state technology are often more expensive than their traditional counterparts, so AC power remains in widespread use.

Power

Figure 3: Transmission lines transmit power across the grid.

Power Engineering deals with the generation, transmission and distribution of electricity as well as the design of a range of related devices. These include transformers, electric generators, electric motors and power electronics.

The power grid is an electrical network that connects a variety of electric generators to the users of electric power. Users purchase electricity from the grid avoiding the costly exercise of having to generate their own. Power engineers may work on the design and maintenance of the power grid as well as the power systems that connect to it. Such systems are called on-grid power systems and may supply the grid with additional power, draw power from the grid or do both.

Power engineers may also work on systems that do not connect to the grid. These systems are called off-grid power systems and may be used in preference to on-grid systems for a variety of reasons. For example, in remote locations it may be cheaper for a mine to generate its own power rather than pay for connection to the grid and in most mobile applications connection to the grid is simply not practical.

Today, most grids adopt three-phase electric power with alternating current. This choice can be partly attributed to the ease with which this type of power can be generated, transformed and used. Often (especially in the USA), the power is split before it reaches residential customers whose low-power appliances rely upon single-phase electric power. However, many larger industries and organisations still prefer to receive the three-phase power directly because it can be used to drive highly efficient electric motors such as three-phase induction motors. Transformers play an important role in power transmission because they allow power to be converted to and from higher voltages. This is important because higher voltages suffer less power loss during transmission. This is because higher voltages allow for lower current to deliver the same amount of power, as power is the product of the two. Thus, as the voltage steps up, the current steps down. It is the current flowing through the components that result in both the losses and the subsequent heating. These losses, appearing in the form of heat, are equal to the current squared times the electrical resistance through which the current flows, so as the voltage goes up the losses are dramatically reduced.

For these reasons, electrical substations exist throughout power grids to convert power to higher voltages before transmission and to lower voltages suitable for appliances after transmission.

Components

Power engineering is a network of interconnected components which convert different forms of energy to electrical energy. Modern power engineering consists of three main subsystems: the generation subsystem, the transmission subsystem, and the distribution subsystem. In the generation subsystem, the power plant produces the electricity. The transmission subsystem transmits the electricity to the load centres. The distribution subsystem continues to transmit the power to the customers.

Generation

Generation of electrical power is a process whereby energy is transformed into an electrical form. There are several different transformation processes, among which are chemical, photo-voltaic, and electromechanical. Electromechanical energy conversion is used in converting energy from coal, petroleum, natural gas, uranium into electrical energy. Of these, all except the wind energy conversion process take advantage of the synchronous AC generator coupled to a steam, gas or hydro turbine such that the turbine converts steam, gas, or water flow into rotational energy, and the synchronous generator then converts the rotational energy of the turbine into electrical energy. It is the turbine-generator conversion process that is by far most economical and consequently most common in the industry today.

The AC synchronous machine is the most common technology for generating electrical energy. It is called synchronous because the composite magnetic field produced by the three stator windings rotate at the same speed as the magnetic field produced by the field winding on the rotor. A simplified circuit model is used to analyse steady-state operating conditions for a synchronous machine. The phasor diagram is an effective tool for visualising the relationships between internal voltage, armature current, and terminal voltage.

The excitation control system is used on synchronous machines to regulate terminal voltage, and the turbine-governor system is used to regulate the speed of the machine. However, in highly interconnected systems, such as the "Western system", the "Texas system" and the "Eastern system", *one* machine will usually be assigned as the so-called "swing machine", and which generation may be increased or decreased to compensate for small changes in load, thereby maintaining the system frequency at precisely 60 Hz. Should the load dramatically change, as which happens with a system separation, then a combination of "spinning reserve" and the "swing machine" may be used by the system's load dispatcher.

The operating costs of generating electrical energy is determined by the fuel cost and the efficiency of the power station. The efficiency depends on generation level and can be obtained from the heat rate curve. We may also obtain the incremental cost curve from the heat rate curve. *Economic dispatch* is the process of allocating the required load demand between the available generation units such that the cost of operation is minimised. *Emission dispatch* is the process of allocating the required load demand between the available generation units such that air pollution occurring from operation is minimised. In large systems, particularly in the West, a combination of economic and emission dispatch may be used.

Transmission

The electricity is transported to load locations from a power station to a transmission subsystem. Therefore we may think of the transmission system as providing the medium of transportation for electric energy. The transmission system may be subdivided into the bulk transmission system and the sub-transmission system. The functions of the bulk transmission are to interconnect generators, to interconnect various areas of the network, and to transfer electrical energy from the generators to the major load centres. This portion of the system is called "bulk" because it delivers energy only to so-called bulk loads such as the distribution system of a town, city, or large industrial plant. The function of the sub-transmission system is to interconnect the bulk power system with the distribution system. Transmission circuits may be built either underground or overhead. Underground cables are used predominantly in urban areas where acquisition of overhead rights of way are costly or not possible. They are also used for transmission under rivers, lakes and bays. Overhead transmission is used otherwise because, for a given voltage level, overhead conductors are much less expensive than underground cables.

The transmission system is a highly integrated system. It is referred to the substation equipment and transmission lines. The substation equipment contain the transformers, relays, and circuit breakers. Transformers are important static devices which transfer electrical energy from one circuit with another in the transmission subsystem. Transformers are used to step up the voltage on the transmission line to reduce the power loss which is dissipated on the way. A relay is functionally a level-detector; they perform a switching action when the input voltage (or current) meets or exceeds a specific and adjustable value. A circuit breaker is an automatically operated electrical switch designed to protect an electrical circuit from damage caused by overload or short circuit.

A change in the status of any one component can significantly affect the operation of the entire system. There are three possible causes for power flow limitations to a transmission line. These causes are thermal overload, voltage instability, and rotor angle instability. Thermal overload is caused by excessive current flow in a circuit causing overheating. Voltage instability is said to occur when the power required to maintain voltages at or above acceptable levels exceeds the available power. Rotor angle instability is a dynamic problem that may occur following faults, such as short circuit, in the transmission system. It may also occur tens of seconds after a fault due to poorly damped or undamped oscillatory response of the rotor motion. As long as the *equal area criteria* is maintained, the interconnected system will remain stable. Should the *equal area criteria* be violated, it becomes necessary to separate the unstable component from the remainder of the system.

Distribution

The distribution system transports the power from the transmission system to the customer. The distribution systems are typically radial because networked systems are more expensive. The equipment associated with the distribution system includes the substation transformers connected to the transmission systems, the distribution lines from the transformers to the customers and the protection and control equipment between the transformer and the customer. The protection equipment includes lightning protectors, circuit breakers, disconnectors and fuses. The control equipment includes voltage regulators, capacitors, relays and demand side management equipment.

Electric power, energy generated through the conversion of other forms of energy, such as mechanical, thermal, or chemical energy. Electric energy is unrivalled for many uses, as for lighting, computer operation, motive power, and entertainment applications. For other uses it is competitive, as for many industrial heating applications, cooking, space heating, and railway traction.

Electric power is characterised by current or the flow of electric charge and voltage or the potential of charge to deliver energy. A given value of power can be produced by any combination of current and voltage values. If the current is direct, electronic charge progresses always in the same direction through the device receiving power. If the current is alternating, electronic charge moves back and forth in the device and in the wires connected to it.

For many applications either type of current is suitable, but alternating current (AC) is most widely available because of the greater efficiency with which it can be generated and distributed. A direct current (DC) is required for certain industrial applications, such as electroplating and electrometallurgical processes and for most electronic devices.

The wide-scale production and distribution of electric power was made possible by the development of the electric generator, a device that operates on the basis of the induction principle formulated in 1831 by the English scientist Michael Faraday and independently by the American scientist Joseph Henry. The first public power station employing an electric generator began operation in London in January 1882. A second such station opened later that same year in New York City. Both used DC systems, which proved inefficient for long-distance power transmission. By the early 1890s the first practical AC generator was built at the Lauffen power station in Germany, and service to Frankfurt am Main was initiated in 1891.

There are two primary sources for driving generators—hydro and thermal. Hydroelectric power is derived from generators and turbines driven by falling water. Most other electric energy is obtained from generators coupled to turbines driven by steam produced either by a nuclear reactor or by burning fossil fuels—namely, coal, oil, and natural gas.

Until the 1930s, hydroelectric-power plants equipped with water-turbine generating units produced the largest percentage of electric energy because they were less expensive to operate than thermal-power plants using steam-turbine units. Since that time, major technological advances have reduced the cost of thermal-power generation, while the cost of developing more remote hydroelectric sites has increased. By 1990, hydroelectric-power production<script src=”http://adserver.adtechus.com/addyn/3.0/5308.1/1371256/0/170/ADTECH;target=_blank;grp=692;key=science technology; kvqsegs = D; kvsource = science; kvtopicid = 182711; kvchannel = SCIENCE; misc = 1322794880513" ></script> constituted only 18 percent of global electric energy output. Thermal plants using nuclear energy or gas turbines to run steam-electric units are among these technological advances. Alternative electric energy sources include solar cells, wind turbines, fuel cells, and geothermal-power stations.

Electric energy generated at a central power station is transmitted to bulk delivery points, or substations, from which it is distributed to consumers. Transmission is accomplished by an extensive network of high-voltage power lines, including overhead wires and underground and submarine cables. Voltages higher than those suitable for power plant generators are required when transmitting alternating current over long distances in order to reduce the power losses that result from the resistance of transmission lines. Step-up transformers are employed at the generating station to increase the transmission voltage. At the substations other transformers step down the voltage to levels suitable for distribution systems.

Chapter 2

Nuclear Power Plant

Figure 1: A nuclear power station. The nuclear reactor is contained inside the cylindrical containment buildings to the right - left is a cooling tower venting water vapour from the non-radioactive side of the plant.

A nuclear power plant (NPP) is a thermal power station in which the heat source is one or more nuclear reactors. As in a conventional thermal power station the heat is used to generate steam which drives a steam turbine connected to a generator which produces electricity.

Nuclear power plants are usually considered to be base load stations, which are best suited to constant power output.

History

Figure 2: The control room at a U.S. nuclear power plant.

Electricity was generated for the first time ever by a nuclear reactor on December 20, 1951 at the EBR-I experimental station near Arco, Idaho in the United States. On June 27, 1954, the world's first nuclear power plant to generate electricity for a power grid started operations at Obninsk, USSR. The world's first commercial scale power station, Calder Hall in England opened in October 17, 1956.

Systems

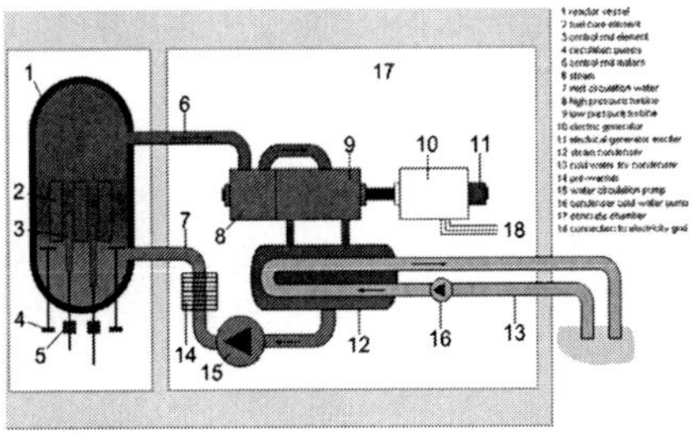

Figure 3: BWR schematic.

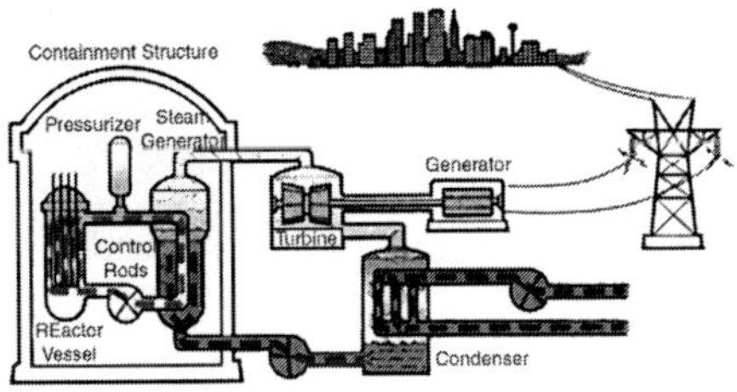

Figure 4: Pressurised water reactor

The conversion to electrical energy takes place indirectly, as in conventional thermal power plants: The heat is produced by fission in a nuclear reactor (in light water reactor). Directly or indirectly water vapour-steam is produced. The pressurised steam is then usually fed to a multi-stage steam turbine. Steam turbines in Western nuclear power plants are among the largest steam turbines ever. After the steam turbine has expanded and partially condensed the steam, the remaining vapour is condensed in a condenser. The condenser is a heat exchanger which is connected to secondary side such as a river or a cooling tower. The water then pumped back into the nuclear reactor and the cycle begins again. The water-steam cycle corresponds to the Rankine cycle.

Nuclear Reactors

A nuclear reactor is a device to initiate and control a sustained nuclear chain reaction. The most common use of nuclear reactors is for the generation of electric energy and for the propulsion of ships.

The nuclear reactor is the heart of the plant. In its central part, the reactor core's heat is generated by controlled nuclear fission. With this heat, a coolant is heated as it is pumped through the reactor and thereby removes the energy from the reactor. Heat from nuclear fission is used to raise steam, which runs through turbines, which in turn powers either ship's propellers or electrical generators.

Since nuclear fission creates radioactivity, the reactor core is surrounded by a protective shield. This containment absorbs radiation and prevents radioactive material from being released into the environment. In addition, many reactors are equipped with a dome of concrete to protect the reactor against external impacts.

In nuclear power plants, different types of reactors, nuclear fuels, and cooling circuits and moderators are sometimes used.

Steam Turbine

The object of the steam turbine is to convert the heat contained in steam into mechanical energy. The engine house with the steam turbine is usually structurally separated from the main reactor building. It is aligned to prevent debris from the destruction of a turbine in operation from flying towards the reactor.

In the case of a pressurised water reactor, the steam turbine hermetically separated from the nuclear system. To detect a leak in the steam generator and thus the passage of radioactive water at an early stage is the outlet steam of the steam generator mounted an activity metre. In contrast, boiling water reactors and the steam turbine with radioactive water applied and therefore part of the control area of the nuclear power plant.

Generator

The generator converts kinetic energy supplied by the turbine into electrical energy. Low-pole AC synchronous generators of high rated power are used.

Cooling System

A cooling system removes heat from the reactor core and transports it to another area of the plant, where the thermal energy can be harnessed to produce electricity or to do other useful work. Typically the hot coolant is used as a heat source for a boiler, and the pressurised steam from that boiler powers one or more steam turbine driven electrical generators.

The reactor coolant pump in the case of the DWR has the task to circulate the coolant between the reactor and steam generators. In western nuclear power plants, the nuclear reactor is fed with four redundant pumps (loops), each separated by Redundancy structurally accommodated in the reactor building. The design of the pump corresponds to a Centrifugal with a one-piece forged body. The throughput is up to 10,000 l / s at a pressure of 175 bar and a maximum allowable temperature of 350 °C. There is an increase in pressure through the main coolant pump when DWR indicates pressure loss in the reactor, steam generators and piping system. Even after the failure of the main coolant pumps (RESA is the result of) the circulation and thus the heat dissipation is by so-called Natural circulation guaranteed.

In the case of boiling water reactor, the reactor pressure vessel force circulation pumps to avoid core wings attached to their interpretation is approximately equal to those in a PWR. You are responsible for the safety of the plant is not absolutely necessary.

Besides these main coolant pump of a nuclear power plant has usually still have several emergency supplies at different pressure levels, the case of disturbances maintain the cooling of the reactor core.

Safety Valves

In the event of an emergency, two independent safety valves can be used to prevent pipes from bursting or the reactor from exploding. The valves are designed so that they can derive all of the supplied flow rates with little increase in pressure. In the case of the BWR, the steam is directed into the condensate chamber and condenses there. The chambers on a heat exchanger are connected to the intermediate cooling circuit.

Feedwater Pump

The water level in the steam generator and nuclear reactor is controlled using the feedwater system. The feedwater pump has the task of taking the water from the feedwater tank up to the vapour pressure in the reactor and steam generator at rates of 2200 kg/s. The power required is about 20 MW per pump.

Emergency Power Supply

The emergency power supplies of a nuclear power plant are built up by several layers of redundancy, such as diesel generators, gas turbine generators and battery buffers. The battery backup provides uninterrupted coupling of the diesel/gas turbine units to the power supply network. If necessary, the emergency power supply allows the safe shut down of the nuclear reactor. Less important auxiliary systems such as, for example, heat tracing of pipelines are not supplied by these back ups. The majority of the required power is used to supply the feed pumps in order cool reactor and remove the decay heat after shut down.

People in a Nuclear Power Plant

Nuclear power plants typically employ just under a thousand people per reactor (including security guards and engineers associated with the plant but possibly working elsewhere).

- Nuclear engineers
- Reactor operators
- Health physicists
- Emergency response team personnel
- Nuclear Regulatory Commission Resident Inspectors

In the United States and Canada, workers except for management, professional (such as engineers) and security personnel are likely to be members of either the International Brotherhood of Electrical Workers (IBEW) or the Utility Workers Union of America (UWUA).

Economics

The economics of new nuclear power plants is a controversial subject, since there are diverging views on this topic, and multi-billion dollar investments ride on the choice of an energy source. Nuclear power plants typically have high capital costs for building the plant, but low direct fuel costs (with much of the costs of fuel extraction, processing, use and long term storage externalised). Therefore, comparison with other power generation methods is strongly dependent on assumptions about construction timescales and capital financing for nuclear plants. Cost estimates also need to take into account plant decommissioning and nuclear waste storage costs. On the other hand measures to mitigate global warming, such as a carbon tax or carbon emissions trading, may favour the economics of nuclear power.

In recent years there has been a slowdown of electricity demand growth and financing has become more difficult, which has an impact on large projects such as nuclear reactors, with very large upfront costs and long project cycles which carry a large variety of risks. In Eastern Europe, a number of long-established projects are struggling to find finance, notably Belene in Bulgaria and the additional reactors at Cernavoda in Romania, and some potential backers have pulled out. Where cheap gas is available and its future supply relatively secure, this also poses a major problem for nuclear projects.

Analysis of the economics of nuclear power must take into account who bears the risks of future uncertainties. To date all operating nuclear power plants were developed by state-owned or regulated utility monopolies where many of the risks associated with construction costs, operating performance, fuel price, and other factors were borne by consumers rather than suppliers. Many countries have now liberalised the electricity market where these risks, and the risk of cheaper competitors emerging before capital costs are recovered, are borne by plant suppliers and operators rather than consumers, which leads to a significantly different evaluation of the economics of new nuclear power plants. Following the 2011 Fukushima I nuclear accidents, costs are likely to go up for currently operating and new nuclear power plants, due to increased requirements for on-site spent fuel management and elevated design basis threats.

Safety

There are trades to be made between safety, economic and technical properties of different reactor designs for particular applications. Historically these decisions were often made in private by scientists, regulators and engineers, but this may be considered problematic, and since Chernobyl and Three Mile Island, many involved now consider informed consent and morality should be primary considerations.

Complexity

Nuclear power plants are some of the most sophisticated and complex energy systems ever designed. Any complex system, no matter how well it is designed and engineered, cannot be deemed failure-proof. Stephanie Cooke has said that:

> *The reactors themselves were enormously complex machines with an incalculable number of things that could go wrong. When that happened at Three Mile Island in 1979, another fault line in the nuclear world was exposed. One malfunction led to another, and then to a series of others, until the core of the reactor itself began to melt, and even the world's most highly trained nuclear engineers did not know how to respond. The accident revealed serious deficiencies in a system that was meant to protect public health and safety.*

A fundamental issue related to complexity is that nuclear power systems have exceedingly long lifetimes. The timeframe involved from the start of construction of a commercial nuclear power station, through to the safe disposal of its last radioactive waste, may be 100 to 150 years.

Failure Modes of Nuclear Power Plants

There are concerns that a combination of human and mechanical error at a nuclear facility could result in significant harm to people and the environment: Operating nuclear reactors contain large amounts of radioactive fission products which, if dispersed, can pose a direct radiation hazard, contaminate soil and vegetation, and be ingested by humans and animals. Human exposure at high enough levels can cause both short-term illness and death and longer-term death by cancer and other diseases.

It is impossible for a commercial nuclear reactor to explode like a nuclear bomb since the fuel is never sufficiently enriched for this to occur. Nuclear reactors can fail in a variety of ways. Should the instability of the nuclear material generate unexpected behaviour, it may result in an uncontrolled power excursion. Normally, the cooling system in a reactor is designed to be able to handle the excess heat this causes; however, should the reactor also experience a loss-of-coolant accident, then the fuel may melt or cause the vessel it is contained in to overheat and melt. This event is called a nuclear meltdown. After shutting down, for some time the reactor still needs external energy to power its cooling systems. Normally this energy is provided by the power grid to that the plant is connected, or by emergency diesel generators. Failure to provide power for the cooling systems, as happened in Fukushima I, can cause serious accidents.

Nuclear safety rules in the United States "do not adequately weigh the risk of a single event that would knock out electricity from the grid and from emergency generators, as a quake and tsunami recently did in Japan", Nuclear Regulatory Commission officials said in June 2011. Intentional cause of such failures may be the result of nuclear terrorism.

Vulnerability of Nuclear Plants to Attack

Nuclear reactors become preferred targets during military conflict and, over the past three decades, have been repeatedly attacked during military air strikes, occupations, invasions and campaigns:

- In September 1980, Iran bombed the Al Tuwaitha nuclear complex in Iraq.
- In June 1981, an Israeli air strike completely destroyed Iraq's Osirak nuclear research facility.

- Between 1984 and 1987, Iraq bombed Iran's Bushehr nuclear plant six times.
- In Iraq in 1991, the U.S. bombed three nuclear reactors and an enrichment pilot facility.
- In 1991, Iraq launched Scud missiles at Israel's Dimona nuclear power plant.
- In September 2007, Israel bombed a Syrian reactor under construction.

In the U.S., plants are surrounded by a double row of tall fences which are electronically monitored. The plant grounds are patrolled by a sizeable force of armed guards. The NRC's "Design Basis Threat" criteria for plants is a secret, and so what size of attacking force the plants are able to protect against is unknown. However, to scram (make an emergency shutdown) a plant takes fewer than 5 seconds while unimpeded restart takes hours, severely hampering a terrorist force in a goal to release radioactivity.

Attack from the air is an issue that has been highlighted since the September 11 attacks in the U.S. However, it was in 1972 when three hijackers took control of a domestic passenger flight along the east coast of the U.S. and threatened to crash the plane into a U.S. nuclear weapons plant in Oak Ridge, Tennessee. The plane got as close as 8,000 feet above the site before the hijackers' demands were met. The most important barrier against the release of radioactivity in the event of an aircraft strike on a nuclear power plant is the containment building and its missile shield. Current NRC Chairman Dale Klein has said "Nuclear power plants are inherently robust structures that our studies show provide adequate protection in a hypothetical attack by an aeroplane. The NRC has also taken actions that require nuclear power plant operators to be able to manage large fires or explosions—no matter what has caused them."

In addition, supporters point to large studies carried out by the U.S. Electric Power Research Institute that tested the robustness of both reactor and waste fuel storage and found that they should be able to sustain a terrorist attack comparable to the September 11 terrorist attacks in the U.S. Spent fuel is usually housed inside the plant's "protected zone" or a spent nuclear fuel shipping cask; stealing it for use in a "dirty bomb" is extremely difficult. Exposure to the intense radiation would almost certainly quickly incapacitate or kill anyone who attempts to do so. In September 2010, analysis of the Stuxnet computer worm suggested that it was designed to sabotage a nuclear power plant. Such a cyber attack would bypass the physical safeguards in place and so the exploit demonstrates an important new vulnerability.

Plant Location

Figure 5: Fort Calhoun Nuclear Generating Station surrounded by the 2011 Missouri River Floods on June 16, 2011

In many countries, plants are often located on the coast, in order to provide a ready source of cooling water for the essential service water system. As a consequence the design needs to take the risk of flooding and tsunamis into account. The World Energy Council (WEC) argues disaster risks are changing and increasing the likelihood of disasters such as earthquakes, cyclones, hurricanes, typhoons, flooding.

Climate change and increased temperatures, lower precipitation levels and an increase in the frequency and severity of droughts may lead to fresh water shortages. Seawater is corrosive and so nuclear energy supply is likely to be negatively affected by the fresh water shortage. This generic problem may become increasingly significant over time. Failure to calculate the risk of flooding correctly lead to a Level 2 event on the International Nuclear Event Scale during the 1999 Blayais Nuclear Power Plant flood, while flooding caused by the 2011 Tôhoku earthquake and tsunami lead to the Fukushima I nuclear accidents.

The design of plants located in seismically active zones also requires the risk of earthquakes and tsunamis to be taken into account. Japan, India, China and the USA are among the countries to have plants in earthquake-prone regions. Damage caused to Japan's Kashiwazaki-Kariwa Nuclear Power Plant during the 2007 Chûetsu offshore earthquake underlined concerns expressed by experts in Japan prior to the Fukushima accidents, who have warned of a *genpatsu-shinsai* (domino-effect nuclear power plant earthquake disaster).

Nuclear Safety Systems

The three primary objectives of nuclear safety systems as defined by the Nuclear Regulatory Commission are to shut down the reactor, maintain it in a shutdown condition, and prevent the release of radioactive material during events and accidents. These objectives are accomplished using a variety of equipment, which is part of different systems, of which each performs specific functions.

Controversy

Figure 6: The abandoned city of Prypiat, Ukraine, following the Chernobyl disaster. The Chernobyl nuclear power plant is in the background.

The nuclear power debate is about the controversy which has surrounded the deployment and use of nuclear fission reactors to generate electricity from nuclear fuel for civilian purposes. The debate about nuclear power peaked during the 1970s and 1980s, when it "reached an intensity unprecedented in the history of technology controversies", in some countries.

Proponents argue that nuclear power is a sustainable energy source which reduces carbon emissions and can increase energy security if its use supplants a dependence on imported fuels. Proponents advance the notion that nuclear power produces virtually no air pollution, in contrast to the chief viable alternative of fossil fuel. Proponents also believe that nuclear power is the only viable course to achieve energy independence for most Western countries. They emphasize that the risks of storing waste are small and can be further reduced by using the latest technology in newer reactors, and the operational safety record in the Western world is excellent when compared to the other major kinds of power plants. Opponents say that nuclear power poses many threats to people and the environment. These threats include health risks and environmental damage from uranium mining, processing and transport, the risk of nuclear weapons proliferation or sabotage, and the unsolved problem of radioactive nuclear waste. They also contend that reactors themselves are enormously complex machines where many things can and do go wrong, and there have been many serious nuclear accidents. Critics do not believe that these risks can be reduced through new technology. They argue that when all the energy-intensive stages of the nuclear fuel chain are considered, from uranium mining to nuclear decommissioning, nuclear power is not a low-carbon electricity source.

Reprocessing

Nuclear reprocessing technology was developed to chemically separate and recover fissionable plutonium from irradiated nuclear fuel. Reprocessing serves multiple purposes, whose relative importance has changed over time. Originally reprocessing was used solely to extract plutonium for producing nuclear weapons. With the commercialisation of nuclear power, the reprocessed plutonium was recycled back into MOX nuclear fuel for thermal reactors. The reprocessed uranium, which constitutes the bulk of the spent fuel material, can in principle also be re-used as fuel, but that is only economic when uranium prices are high. Finally, the breeder reactor can employ not only the recycled plutonium and uranium in spent fuel, but all the actinides, closing the nuclear fuel cycle and potentially multiplying the energy extracted from natural uranium by more than 60 times.

Nuclear reprocessing reduces the volume of high-level waste, but by itself does not reduce radioactivity or heat generation and therefore does not eliminate the need for a geological waste repository. Reprocessing has been politically controversial because of the potential to contribute to nuclear proliferation, the potential vulnerability to nuclear terrorism, the political challenges of repository siting (a problem that applies equally to direct disposal of spent fuel), and because of its high cost compared to the once-through fuel cycle. The Obama administration stepped back from President Bush's plans for commercial-scale reprocessing and reverted to a program focused on reprocessing-related scientific research.

Accident Indemnification

The Vienna Convention on Civil Liability for Nuclear Damage puts in place an international framework for nuclear liability. However states with a majority of the world's nuclear power plants, including the U.S., Russia, China and Japan, are not party to international nuclear liability conventions. In the U.S., insurance for nuclear or radiological incidents is covered (for facilities licensed through 2025) by the Price-Anderson Nuclear Industries Indemnity Act.

Under the Energy policy of the United Kingdom through its Nuclear Installations Act of 1965, liability is governed for nuclear damage for which a UK nuclear licensee is responsible. The Act requires compensation to be paid for damage up to a limit of £150 million by the liable operator for ten years after the incident. Between ten and thirty years afterwards, the Government meets this obligation. The Government is also liable for additional limited cross-border liability (about £300 million) under international conventions (Paris Convention on Third Party Liability in the Field of Nuclear Energy and Brussels Convention supplementary to the Paris Convention).

Decommissioning

Nuclear decommissioning is the dismantling of a nuclear power plant and decontamination of the site to a state no longer requiring protection from radiation for the general public. The main difference from the dismantling of other power plants is the presence of radioactive material that requires special precautions. Generally speaking, nuclear plants were designed for a life of about 30 years. Newer plants are designed for a 40 to 60-year operating life. Decommissioning involves many administrative and technical actions. It includes all clean-up of radioactivity and progressive demolition of the plant. Once a facility is decommissioned, there should no longer be any danger of a radioactive accident or to any persons visiting it. After a facility has been completely decommissioned it is released from regulatory control, and the licensee of the plant no longer has responsibility for its nuclear safety.

Historic Accidents

Figure 7: The 2011 Fukushima Daiichi nuclear disaster in Japan, the worst nuclear accident in 25 years, displaced 50,000 households after radiation leaked into the air, soil and sea. Radiation checks led to bans of some shipments of vegetables and fish.

The nuclear industry says that new technology and oversight have made nuclear plants much safer, but 57 accidents have occurred since the Chernobyl disaster in 1986. Two thirds of these mishaps occurred in the US. The French Atomic Energy Agency (CEA) has concluded that technical innovation cannot eliminate the risk of human errors in nuclear plant operation. An interdisciplinary team from MIT have estimated that given the expected growth of nuclear power from 2005–2055, at least four serious nuclear power accidents would be expected in that period.

Flexibility of Nuclear Power Plants

It is often claimed that nuclear stations are inflexible in their output, implying that other forms of energy would be required to meet peak demand. While that is true for the vast majority of reactors, this is no longer true of at least some modern designs. Nuclear plants are routinely used in load following mode on a large scale in France. Unit A at the German Biblis Nuclear Power Plant is designed to in- and decrease his output 15 % per minute between 40 and 100 % of its nominal power. Boiling water reactors normally have load-following capability, implemented by varying the recirculation water flow.

Future Power Plants

A number of new designs for nuclear power generation, collectively known as the Generation IV reactors, are the subject of active research and may be used for practical power generation in the future. Many of these new designs specifically attempt to make fission reactors cleaner, safer and/or less of a risk to the proliferation of nuclear weapons. Passively safe plants (such as the ESBWR) are available to be built and other designs that are believed to be nearly fool-proof are being pursued. Fusion reactors, which may be viable in the future, diminish or eliminate many of the risks associated with nuclear fission.

The 1600 MWe European Pressurised Reactor reactor is being built in Olkiluoto, Finland. A joint effort of French AREVA and German Siemens AG, it will be the largest reactor in the world. In December 2006 construction was about 18 months behind schedule so completion was expected 2010-2011.

As of March, 2007, there are seven nuclear power plants under construction in India, and five in China.

In November 2011 Gulf Power stated that by the end of 2012 it hopes to finish buying off 4000 acres of land north of Pensacola, Florida in order to build a posable nuclear power plant.

Russia has begun building the world's first floating nuclear power plant. The £100 million vessel, the *Lomonosov*, is the first of seven plants that Moscow says will bring vital energy resources to remote Russian regions.

By 2025, Southeast Asia nations would have a total of 29 nuclear power plants, Indonesia will have 4 nuclear power plants, Malaysia 4, Thailand 5 and Vietnam 16 from nothing at all in 2011.

Containment Building

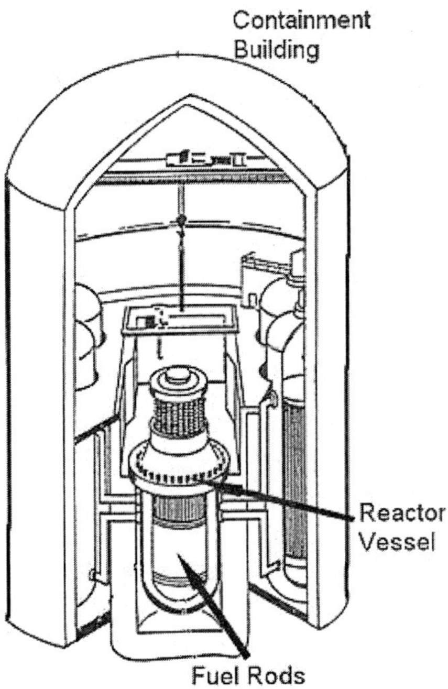

Figure 8: NRC drawing of containment building.

A containment building, in its most common usage, is a steel or reinforced concrete structure enclosing a nuclear reactor. It is designed, in any emergency, to contain the escape of radiation to a maximum pressure in the range of 60 to 200 psi (410 to 1400 kPa). The containment is the fourth and final barrier to radioactive release (part of a nuclear reactor's defence in depth strategy), the first being the fuel ceramic itself, the second being the metal fuel cladding tubes, the third being the reactor vessel and coolant system.

Each nuclear plant in the US is designed to withstand certain conditions which are spelled out as "Design Basis Accidents" in the

Final Safety Analysis Report (FSAR). The FSAR is available for public viewing, usually at a public library near the nuclear plant.

The containment building itself is typically an airtight steel structure enclosing the reactor normally sealed off from the outside atmosphere. The steel is either free-standing or attached to the concrete missile shield. In the United States, the design and thickness of the containment and the missile shield are governed by federal regulations (10 CFR 50.55a), and must be strong enough to withstand the impact of a fully loaded passenger airliner without rupture.

While the containment plays a critical role in the most severe nuclear reactor accidents, it is only designed to contain or condense steam in the short term (for large break accidents) and long term heat removal still must be provided by other systems. In the Three Mile Island accident the containment pressure boundary was maintained, but due to insufficient cooling, some time after the accident, radioactive gas was intentionally let from containment by operators to prevent over pressurisation. This, combined with further failures caused the release of minimal amounts of radioactive gas to atmosphere during the accident.

Information is still being studied of the failures at Fukushima. While the plant had operated safely since 1971, an earthquake and tsunami well beyond the design basis resulted in failure of AC power, backup generators and batteries which defeated all safety systems. This resulted in partial or complete meltdown of fuel rods, damage to fuel storage pools and buildings, significant release of radioactive debris to surrounding area, air and sea, and resorting to the use of fire engines and concrete pumps to deliver cooling water to spent fuel pools and containments.

Types

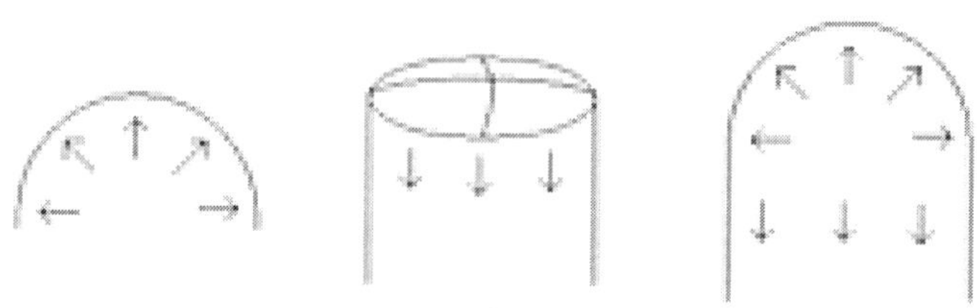

Figure 9: If the outward pressure from steam in a limiting accident is the dominant force, containments tend towards a spherical design, whereas if weight of the structure is the dominant force, designs tend towards a can design. Modern designs tend towards a combination.

Containment systems for nuclear power reactors are distinguished by size, shape, materials used, and suppression systems. The kind of containment used is determined by the type of reactor, generation of the reactor, and the specific plant needs.

Suppression systems are critical to safety analysis and greatly affect the size of containment. Suppression refers to condensing the steam after a major break has released it from the cooling system. Because decay heat doesn't go away quickly, there must be some long term method of suppression, but this may simply be heat exchange with the ambient air on the surface of containment. There are several common designs, but for safety-analysis purposes containments are categorised as either "large-dry," "sub-atmospheric," or "ice-condenser."

Pressurised Water Reactors

For a pressurised water reactor, the containment also encloses the steam generators and the pressuriser, and is the entire reactor building. The missile shield around it is typically a tall cylindrical or domed building. PWR containments are typically large (up to 10 times larger than a BWR) because the containment strategy during the leakage design basis accident entails providing adequate volume for the steam/air mixture that results from a loss-of-coolant-accident to expand into, limiting the ultimate pressure (driving force for leakage) reached in the containment building.

Early designs including Siemens, Westinghouse, and Combustion Engineering had a mostly can-like shape built with reinforced concrete. As concrete has a very good compression strength compared to tensile, this is a logical design for the building materials since the extremely heavy top part of containment exerts a large downward force that prevents some tensile stress if containment pressure were to suddenly go up. As reactor designs have evolved, many nearly spherical containment designs for PWRs have also been constructed. Depending on the material used, this is the most apparently logical design because a sphere is the best structure for simply containing a large pressure. Most current PWR designs involve some combination of the two, with a cylindrical lower part and a half-spherical top.

The Spent Fuel Pool is Outside of the Containment Building in Most PWR Designs.

Modern designs have also shifted more towards using steel containment structures. In some cases steel is used to line the inside of the concrete, which contributes strength from both materials in the hypothetical case that containment becomes highly pressurised. Yet other newer designs call for both a steel and concrete containment, notably the AP1000 and the European Pressurised Reactor plan to use both, which gives missile protection by the outer concrete and pressurising ability by the inner steel structure. The AP1000 has planned vents at the bottom of the concrete structure surrounding the steel structure under the logic that it would help move air over the steel structure and cool containment in the event of a major accident (in a similar way to how a cooling tower works).

Figure 10: Three Mile Island was an early PWR design by Babcock and Wilcox, and shows a 'can' containment design that is common to all of its generations

Figure 11: A more detailed image for the 'can' type containment from the French Brennilis Nuclear Power Plant

Figure 12: The twin PWR reactor containments at the Cook Nuclear Plant in Michigan

Figure 13: A German plant exhibiting a nearly completely spherical containment design, which is very common for German PWRs

The Russian VVER design is mostly the same as Western PWRs in regards to containment, as it is a PWR itself.

Old RBMK designs, however, do not use containments, which was one of many technical oversights of the Soviet Union that contributed to the Chernobyl accident in 1986.

Boiling Water Reactors

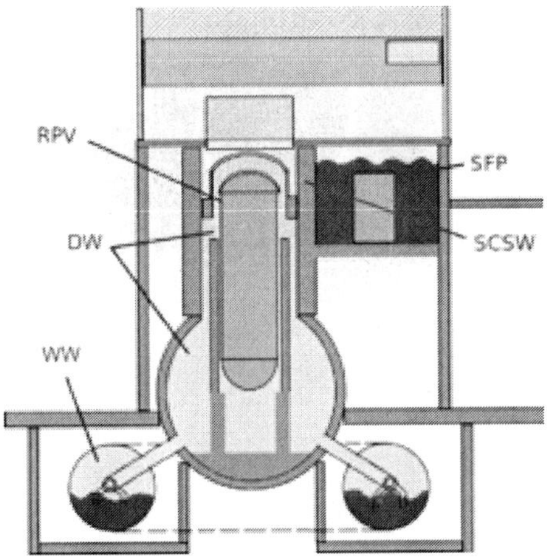

Figure 14: Cross-section sketch of a typical BWR Mark I containment. DW = drywell, WW = wetwell, SF = spent fuel area

In a BWR, the containment strategy is a bit different. A BWR's containment consists of a drywell where the reactor and associated cooling equipment is located and a wetwell. The drywell is much smaller than a PWR containment and plays a larger role. During the theoretical leakage design basis accident the reactor coolant flashes to steam in the drywell, pressurising it rapidly. Vent pipes or tubes from the drywell direct the steam below the water level maintained in the wetwell (also known as a torus or suppression pool), condensing the steam, limiting the pressure ultimately reached. Both the drywell and the wetwell are enclosed by a secondary containment building, maintained at a slight sub-atmospheric or negative pressure during normal operation and refuelling operations.

The containment designs are referred to by the names Mark I, Mark II, and Mark III. The Mark I is the oldest, distinguished by a drywell containment which resembles an inverted lightbulb above the wetwell which is a steel torus containing water. The Mark II was used with late BWR-4 and BWR-5 reactors. It is called an "over-under" configuration with the drywell forming a truncated cone on a concrete slab. Below is a cylindrical suppression chamber made of concrete rather than just sheet metal. Both use a lighweight steel or concrete "secondary containment" over the top floor which is kept at a slight negative pressure so that air can be filtered. The top level is a large open space with an overhead crane suspended between the two long walls for moving heavy fuel caskets from the ground floor, and removing / replacing hardware from the reactor and reactor well.

The reactor well can be flooded and is straddled by pools separated by gates on either side for storing reactor hardware normally placed above the fuel rods, and for fuel storage. A refuelling platform has a specialised telescoping mast for lifting and lowering fuel rod assemblies with precision through the "cattle chute" to the reactor core area. The Mark III uses a concrete dome, somewhat like PWRs, and has a separate building for storing used fuel rods on a different floor level. All three types house also use the large body of water in the suppression pools to quench steam released from the reactor system during transients.

The Mark I containment was used in those reactors at the Fukushima I Nuclear Power Plant which were involved in the Fukushima I nuclear accidents. The site suffered from a combination of two beyond design-basis events, a powerful earthquake which may have damaged reactor plumbing and structures and 15 metre tsunami which destroyed fuel tanks, generators and wiring causing back up generators to fail, and battery-powered pumps also eventually failed. Insufficient cooling and failure of pumps needed to restore water lost to boiling off led to partial or possible complete meltdowns of fuel rods which were completely uncovered by water. This led to releases of significant amounts of radioactive material to the air and sea, and hydrogen explosions.

However PWR reactors also require years of powered pumped cooling water. The thin secondary containments were not designed to withstand hydrogen explosions, and suffered blown out or destroyed roofs and walls, and destruction of all equipment on the refuelling floor including cranes and refuelling platform. Unit 3 suffered a particularly spectacular explosion which created a plume of debris over 300m high which resulted in a collapse of the north end of the top floor, and buckled concrete columns on its west side as can be seen by aerial photographs. Although they were fitted with modified hardened vent systems to vent hydrogen into exhaust stacks, they may have not been effective without power. Unit 2 had a large panel removed to vent gases, but suffered an explosion to the lower suppression area. Even before the Fukushima incident, Mark I containment had been criticized as being more likely to fail during a blackout.

From a distance, the BWR design looks very different from PWR designs because usually a square building is used for containment. Also, because there is only one loop through the turbines and reactor, and the steam going through the turbines is also slightly radioactive, the turbine building has to be considerably shielded as well. This leads to two buildings of similar construction with the taller one housing the reactor and the short long one housing the turbine hall and supporting structures.

Figure 15: A representative one-unit German BWR showing containment around both the turbine and reactor buildings

Figure 16: A typical two-unit BWR at the Brunswick Nuclear Generating Station

Figure 17: Modern plants have tended towards a design that is not completely cylindrical or spherical, like this painted containment at the Clinton Nuclear Generating Station

CANDU Plants

CANDU power stations, named after Canadian-invented Deuterium-Uranium design, make use of a wider variety of containment designs and suppression systems than other plant designs. Due to the nature of the core design, the size of containment for the same power rating is often larger than for a typical PWR, but many innovations have reduced this requirement. Many multiunit CANDU stations utilise a water spray equipped vacuum building. All individual Candu units on site are connected to this Vacuum building by a large pressure relief duct which is also part of containment. The Vacuum building rapidly draws in and condenses any steam from a postulated break, allowing the reactor building pressure to return to subatmospheric conditions. This minimises any possible fission product release to the environment. Additionally, there have been similar designs that use double containment, in which containment from two units are connected allowing a larger containment volume in the case of any major incident. This has been pioneered by one Indian HWR design where a double unit and suppression pool was implemented.

The most recent Candu designs, however, call for a single conventional dry containment for each unit.

Figure 18: The Bruce B Generating Station, showing a large vacuum building serving 4 separate units that have a BWR-like shielding around them individually

Figure 19: The Qinshan Nuclear Power Plant is two-unit site where the containment system is autonomous for each unit

Figure 20: A single unit of the Pickering Nuclear Generating Station, showing a slightly different shape from a typical PWR containment, which is mostly due to the larger footprint required by the Candu design

Design and Testing Requirements

Figure 21: NRC image of Containment area inside a Containment building.

Title 10 of the Code of Federal Regulations, Part 50, Appendix A, General Design Criteria (GDC 54-57) or some other design basis provides the basic design criteria for isolation of lines penetrating the containment wall. Each large pipe penetrating the containment, such as the steam lines, has isolation valves on it, configured as allowed by Appendix A; generally two valves. For smaller lines, one on the inside and one on the outside. For large, high-pressure lines, space for relief valves and maintenance considerations cause the designers to install the isolation valves near to where the lines exit containment. In the event of a leak in the high-pressure piping that carries the reactor coolant, these valves rapidly close to prevent radioactivity from escaping the containment. Valves on lines for standby systems penetrating containment are normally closed. The containment isolation valves may also close on a variety of other signals such as the containment high pressure experienced during a high-energy line break (e.g. main steam or feedwater lines). The containment building serves to contain the steam/resultant pressure, but there is typically no radiological consequences associated with such a break at a pressurised water reactor.

During normal operation, the containment is air-tight and access is only through marine style airlocks. High air temperature and radiation from the core limit the time, measured in minutes, people can spend inside containment while the plant is operating at full power. In the event of a worst-case emergency, called a "design basis accident" in NRC regulations, the containment is designed to seal off and contain a meltdown. Redundant systems are installed to prevent a meltdown, but as a matter of policy, one is assumed to occur and thus the requirement for a containment building. For design purposes, the reactor vessel's piping is assumed to be breached, causing a "LOCA" (loss Of coolant accident) where the water in the reactor vessel is released to the atmosphere inside the containment and flashes into steam.

The resulting pressure increase inside the containment, which is designed to withstand the pressure, triggers containment sprays ("dousing sprays") to turn on to condense the steam and thus reduce the pressure. A SCRAM ("neutronic trip") initiates very shortly after the break occurs. The safety systems close non-essential lines into the air-tight containment by shutting the isolation valves. Emergency Core Cooling Systems are quickly turned on to cool the fuel and prevent it from melting.

Containment buildings in the U.S. are subjected to mandatory testing of the containment and containment isolation provisions under 10 CFR Part 50, Appendix J. Containment Integrated Leakage Rate Tests (Type "A" tests or CILRTs) are performed on a 15 year basis. Local Leakage Rate Tests (Type B or Type C testing, or LLRTs) are performed much more frequently both to identify the possible leakage in an accident and to locate and fix leakage paths. LLRTs are performed on containment isolation valves, hatches and other appurtenances penetrating the containment. A nuclear plant is required by its operating license to prove containment integrity prior to restarting the reactor after each shutdown. The requirement can be met with satisfactory local or integrated test results (or a combination of both when an ILRT is performed).

In 1988, Sandia National Laboratories conducted a test of slamming a jet fighter into a large concrete block at 481 miles per hour (775 km/h). The aeroplane left only a 2.5-inch-deep (64 mm) gouge in the concrete. Although the block was not constructed like a containment building missile shield, it was not anchored, etc., the results were considered indicative. A subsequent study by EPRI, the Electric Power Research Institute, concluded that commercial airliners did not pose a danger.

The Turkey Point Nuclear Generating Station was hit directly by Hurricane Andrew in 1992. Turkey Point has two fossil fuel units and two nuclear units. Over $90 million of damage was done, largely to a water tank and to a smokestack of one of the fossil-fuelled units on-site, but the containment buildings were undamaged.

Nuclear Fuel Cycle

The nuclear fuel cycle, also called nuclear fuel chain, is the progression of nuclear fuel through a series of differing stages. It consists of steps in the *front end*, which are the preparation of the fuel, steps in the *service period* in which the fuel is used during reactor operation, and steps in the *back end*, which are necessary to safely manage, contain, and either reprocess or dispose of spent nuclear fuel. If spent fuel is not reprocessed, the fuel cycle is referred to as an *open fuel cycle* (or a *once-through fuel cycle*); if the spent fuel is reprocessed, it is referred to as a *closed fuel cycle.*

Basic Concepts

Nuclear power relies on fissionable material that can sustain a chain reaction with neutrons. Examples of such materials include uranium and plutonium. Most nuclear reactors use a moderator to lower the kinetic energy of the neutrons and increase the probability that fission will occur. This allows reactors to use material with far lower concentration of fissile isotopes than nuclear weapons. Graphite and heavy water are the most effective moderators, because they slow the neutrons through collisions without absorbing them. Reactors using heavy water or graphite as the moderator can operate using natural uranium.

A Light water reactor (LWR) uses water in the form that occurs in nature, and require fuel that is enriched in fissile isotopes, typically uranium enriched to 3-5% in the less common isotope U-235, the only fissile isotope that is found in significant quantity in nature. One alternative to this low-enriched uranium (LEU) fuel are Mixed Oxide (MOX) fuels produced by blending plutonium with natural or depleted uranium, and these fuels provide an avenue to utilise surplus weapons-grade plutonium. Another type of MOX fuel involves mixing LEU with thorium, which generates the fissile isotope U-233. Both plutonium and U-233 are produced from the absorption of neutrons by irradiating fertile materials in a reactor, in particular the common uranium isotope U-238 and thorium, respectively, and can be separated from spent uranium and thorium fuels in reprocessing plants.

Some reactors do not use moderators to slow the neutrons. Like nuclear weapons, which also use unmoderated or "fast" neutrons, these Fast-neutron reactors require much higher concentrations of fissile isotopes in order to sustain a chain reaction. They are also capable of breeding fissile isotopes from fertile materials; a Breeder reactor is one that generates more fissile material in this way than it consumes.

During the nuclear reaction inside a reactor, the fissile isotopes in nuclear fuel are consumed, producing more and more fission products, most of which are considered radioactive waste. The buildup of fission products and consumption of fissile isotopes eventually stop the nuclear reaction, causing the fuel to become a spent nuclear fuel. When 3% enriched LEU fuel is used, the spent fuel typically consists of roughly 1% U-235, 95% U-238, 1% plutonium and 3% fission products. Spent fuel and other high-level radioactive waste is extremely hazardous, although nuclear reactors produce relatively small volumes of waste compared to other power plants because of the high energy density of nuclear fuel. Safe management of these byproducts of nuclear power, including their storage and disposal, is a difficult problem for any country using nuclear power.

Front End

Figure 22: Uranium ore - the principal raw material of nuclear fuel

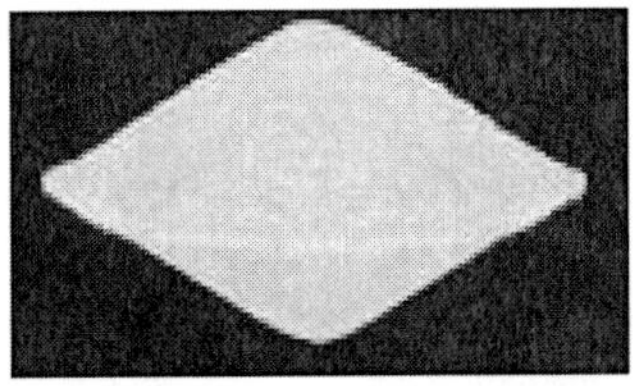

Figure 23: Yellowcake - the form in which uranium is transported to a conversion plant

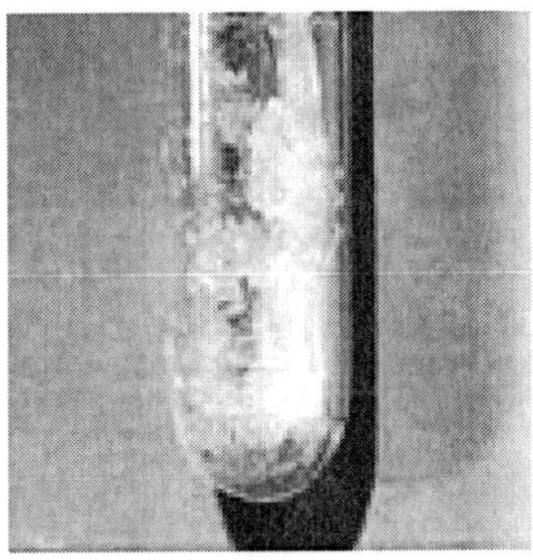

Figure 24: UF_6 - used in enrichment

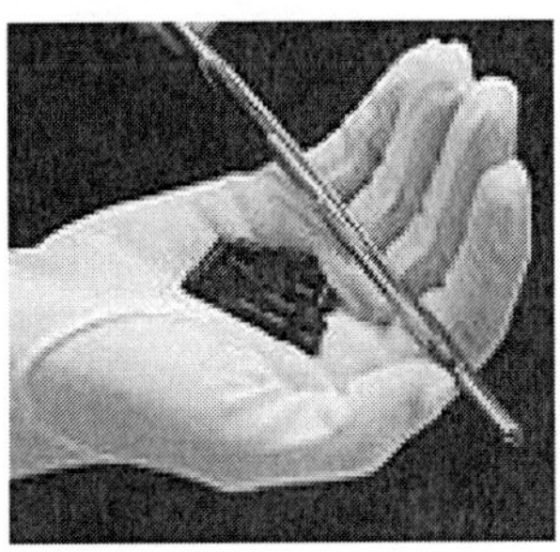

Figure 25: Nuclear fuel - a compact, inert, insoluble solid

Exploration

A deposit of uranium, such as uraninite, discovered by geophysical techniques, is evaluated and sampled to determine the amounts of uranium materials that are extractable at specified costs from the deposit. Uranium reserves are the amounts of ore that are estimated to be recoverable at stated costs. Uranium in nature consists primarily of two isotopes, U-238 and U-235. The numbers refer to the atomic mass number for each isotope, or the number of protons and neutrons in the atomic nucleus. Naturally occurring uranium consists of approximately 99.28% U-238 and 0.71% U-235. The atomic nucleus of U-235 will nearly always fission when struck by a free neutron, and the isotope is therefore said to be a "fissile" isotope. The nucleus of a U-238 atom on the other hand, rather than undergoing fission when struck by a free neutron, will nearly always absorb the neutron and yield an atom of the isotope U-239.

This isotope then undergoes natural radioactive decay to yield Pu-239, which, like U-235, is a fissile isotope. The atoms of U-238 are said to be fertile, because, through neutron irradiation in the core, some eventually yield atoms of fissile Pu-239.

Mining

Uranium ore can be extracted through conventional mining in open pit and underground methods similar to those used for mining other metals. In-situ leach mining methods also are used to mine uranium in the United States. In this technology, uranium is leached from the in-place ore through an array of regularly spaced wells and is then recovered from the leach solution at a surface plant. Uranium ores in the United States typically range from about 0.05 to 0.3% uranium oxide (U_3O_8).

Some uranium deposits developed in other countries are of higher grade and are also larger than deposits mined in the United States. Uranium is also present in very low-grade amounts (50 to 200 parts per million) in some domestic phosphate-bearing deposits of marine origin. Because very large quantities of phosphate-bearing rock are mined for the production of wet-process phosphoric acid used in high analysis fertilizers and other phosphate chemicals, at some phosphate processing plants the uranium, although present in very low concentrations, can be economically recovered from the process stream.

Milling

Mined uranium ores normally are processed by grinding the ore materials to a uniform particle size and then treating the ore to extract the uranium by chemical leaching. The milling process commonly yields dry powder-form material consisting of natural uranium, "yellowcake", which is sold on the uranium market as U_3O_8.

Uranium Conversion

Milled uranium oxide, U_3O_8, must be converted to uranium hexafluoride, UF_6, which is the form required by most commercial uranium enrichment facilities currently in use. A solid at room temperature, uranium hexafluoride can be changed to a gaseous form at moderately higher temperature of 57 °C (134 °F). The uranium hexafluoride conversion product contains only natural, not enriched, uranium.

Triuranium octaoxide (U_3O_8) is also converted directly to ceramic grade uranium dioxide (UO_2) for use in reactors not requiring enriched fuel, such as CANDU. The volumes of material converted directly to UO_2 are typically quite small compared to the amounts converted to UF_6.

Enrichment

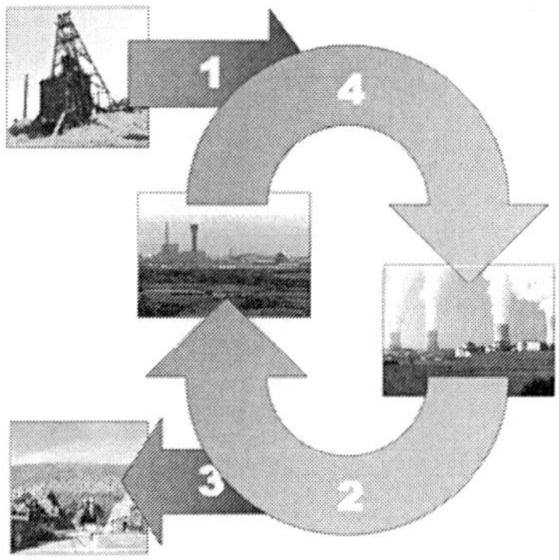

Figure 26: Nuclear fuel cycle begins when uranium is mined, enriched and manufactured to nuclear fuel (1) which is delivered to a nuclear power plant. After usage in the power plant the spent fuel is delivered to a reprocessing plant (if fuel is recycled) (2) or to a final repository (if no recycling is done) (3) for geological disposition. In reprocessing 95% of spent fuel can be recycled to be returned to usage in a nuclear power plant (4).

The concentration of the fissionable isotope, U-235 (0.71% in natural uranium) is less than that required to sustain a nuclear chain reaction in light water reactor cores. Natural UF_6 thus must be enriched in the fissionable isotope for it to be used as nuclear fuel. The different levels of enrichment required for a particular nuclear fuel application are specified by the customer: light-water reactor fuel normally is enriched to 3.5% U-235, but uranium enriched to lower concentrations is also required. Enrichment is accomplished using one or more methods of isotope separation. Gaseous diffusion and gas centrifuge are the commonly used uranium enrichment technologies, but new enrichment technologies are currently being developed.

The bulk (96%) of the byproduct from enrichment is depleted uranium (DU), which can be used for armour, kinetic energy penetrators, radiation shielding and ballast. Still, there are vast quantities of depleted uranium in storage. The United States Department of Energy alone has 470,000 tonnes. About 95% of depleted uranium is stored as uranium hexafluoride (UF_6).

Fabrication

For use as nuclear fuel, enriched uranium hexafluoride is converted into uranium dioxide (UO_2) powder that is then processed into pellet form. The pellets are then fired in a high temperature sintering furnace to create hard, ceramic pellets of enriched uranium. The cylindrical pellets then undergo a grinding process to achieve a uniform pellet size. The pellets are stacked, according to each nuclear reactor core's design specifications, into tubes of corrosion-resistant metal alloy. The tubes are sealed to contain the fuel pellets: these tubes are called fuel rods. The finished fuel rods are grouped in special fuel assemblies that are then used to build up the nuclear fuel core of a power reactor.

The metal used for the tubes depends on the design of the reactor. Stainless steel was used in the past, but most reactors now use zirconium. For the most common types of reactors, boiling water reactors (BWR) and pressurised water reactors (PWR), the tubes are assembled into bundles with the tubes spaced precise distances apart. These bundles are then given a unique identification number, which enables them to be tracked from manufacture through use and into disposal.

Service Period

Transport of Radioactive Materials: Transport is an integral part of the nuclear fuel cycle. There are nuclear power reactors in operation in several countries but uranium mining is viable in only a few areas. Also, in the course of over forty years of operation by the nuclear industry, a number of specialised facilities have been developed in various locations around the world to provide fuel cycle services and there is a need to transport nuclear materials to and from these facilities. Most transports of nuclear fuel material occur between different stages of the cycle, but occasionally a material may be transported between similar facilities. With some exceptions, nuclear fuel cycle materials are transported in solid form, the exception being uranium hexafluoride (UF_6) which is considered a gas. Most of the material used in nuclear fuel is transported several times during the cycle. Transports are frequently international, and are often over large distances. Nuclear materials are generally transported by specialised transport companies.

Since nuclear materials are radioactive, it is important to ensure that radiation exposure of both those involved in the transport of such materials and the general public along transport routes is limited.

Packaging for nuclear materials includes, where appropriate, shielding to reduce potential radiation exposures. In the case of some materials, such as fresh uranium fuel assemblies, the radiation levels are negligible and no shielding is required.

Other materials, such as spent fuel and high-level waste, are highly radioactive and require special handling. To limit the risk in transporting highly radioactive materials, containers known as spent nuclear fuel shipping casks are used which are designed to maintain integrity under normal transportation conditions and during hypothetical accident conditions.

In-core Fuel Management

A nuclear reactor core is composed of a few hundred "assemblies", arranged in a regular array of cells, each cell being formed by a fuel or control rod surrounded, in most designs, by a moderator and coolant, which is water in most reactors.

Because of the fission process that consumes the fuels, the old fuel rods must be changed periodically to fresh ones (this period is called a cycle). However, only a part of the assemblies (typically one-third) are removed since the fuel depletion is not spatially uniform. Furthermore, it is not a good policy, for efficiency reasons, to put the new assemblies exactly at the location of the removed ones. Even bundles of the same age may have different burn-up levels, which depends on their previous positions in the core.

Thus the available bundles must be arranged in such a way that the yield is maximised, while safety limitations and operational constraints are satisfied. Consequently reactor operators are faced with the so-called optimal fuel reloading problem, which consists in optimising the rearrangement of all the assemblies, the old and fresh ones, while still maximising the reactivity of the reactor core so as to maximise fuel burn-up and minimise fuel-cycle costs.

This is a discrete optimisation problem, and computationally infeasible by current combinatorial methods, due to the huge number of permutations and the complexity of each computation. Many numerical methods have been proposed for solving it and many commercial software packages have been written to support fuel management.

This is an on-going issue in reactor operations as no definitive solution to this problem has been found and operators use a combination of computational and empirical techniques to manage this problem.

The Study of Used Fuel

Used nuclear fuel is studied in Post irradiation examination, where used fuel is examined to know more about the processes that occur in fuel during use, and how these might alter the outcome of an accident. For example, during normal use, the fuel expands due to thermal expansion, which can cause cracking.

Most nuclear fuel is uranium dioxide, which is a cubic solid with a structure similar to that of calcium fluoride. In used fuel the solid state structure of most of the solid remains the same as that of pure cubic uranium dioxide. SIMFUEL is the name given to the simulated spent fuel which is made by mixing finely ground metal oxides, grinding as a slurry, spray drying it before heating in hydrogen/argon to 1700 °C.

In SIMFUEL, 4.1% of the volume of the solid was in the form of metal nanoparticles which are made of molybdenum, ruthenium, rhodium and palladium. Most of these metal particles are of the a phase (hexagonal) of Mo-Ru-Rh-Pd alloy, while smaller amounts of the α (cubic) and o (tetragonal) phases of these metals were found in the SIMFUEL. Also present within the SIMFUEL was a cubic perovskite phase which is a barium strontium zirconate ($Ba_xSr_{1-x}ZrO_3$).

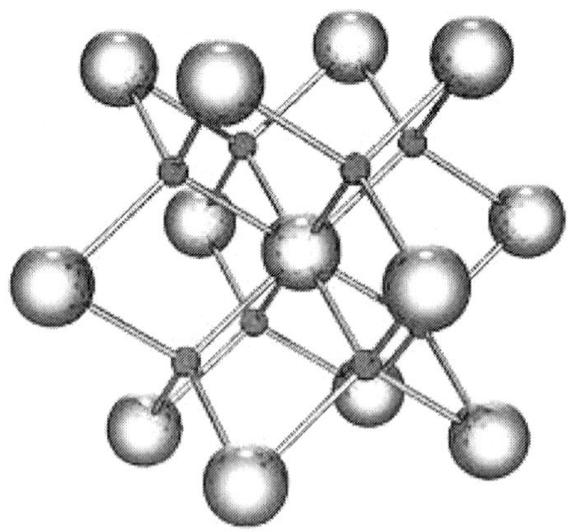

Figure 27: The solid state structure of uranium dioxide, the oxygen atoms are in green and the uranium atoms in red

Uranium dioxide is very insoluble in water, but after oxidation it can be converted to uranium trioxide or another uranium(VI) compound which is much more soluble. Uranium dioxide (UO_2) can be oxidised to an oxygen rich hyperstoichiometric oxide (UO_{2+x}) which can be further oxidised to U_4O_9, U_3O_7, U_3O_8 and $UO_3.2H_2O$.

Because used fuel contains alpha emitters (plutonium and the minor actinides), the effect of adding an alpha emitter (^{238}Pu) to uranium dioxide on the leaching rate of the oxide has been investigated. For the crushed oxide, adding ^{238}Pu tended to increase the rate of leaching, but the difference in the leaching rate between 0.1 and 10% ^{238}Pu was very small.

The concentration of carbonate in the water which is in contact with the used fuel has a considerable effect on the rate of corrosion, because uranium(VI) forms soluble anionic carbonate complexes such as $[UO_2(CO_3)_2]^{2-}$ and $[UO_2(CO_3)_3]^{4-}$. When carbonate ions are absent, and the water is not strongly acidic, the hexavalent uranium compounds which form on oxidation of uranium dioxide often form insoluble hydrated uranium trioxide phases.

By 'sputtering', using uranium metal and an argon/oxygen gas mixture, thin films of uranium dioxide can be deposited upon gold surfaces. These gold surfaces modified with uranium dioxide have been used for both cyclic voltammetry and AC impedance experiments, and these offer an insight into the likely leaching behaviour of uranium dioxide.

Fuel Cladding Interactions

The study of the nuclear fuel cycle includes the study of the behaviour of nuclear materials both under normal conditions and under accident conditions. For example, there has been much work on how uranium dioxide based fuel interacts with the zirconium alloy tubing used to cover it. During use, the fuel swells due to thermal expansion and then starts to react with the surface of the zirconium alloy, forming a new layer which contains both fuel and zirconium (from the cladding). Then, on the fuel side of this mixed layer, there is a layer of fuel which has a higher caesium to uranium ratio than most of the fuel.

This is because xenon isotopes are formed as fission products that diffuse out of the lattice of the fuel into voids such as the narrow gap between the fuel and the cladding. After diffusing into these voids, it decays to caesium isotopes. Because of the thermal gradient which exists in the fuel during use, the volatile fission products tend to be driven from the centre of the pellet to the rim area. Below is a graph of the temperature of uranium metal, uranium nitride and uranium dioxide as a function of distance from the centre of a 20 mm diametre pellet with a rim temperature of 200 °C. The uranium dioxide (because of its poor thermal conductivity) will overheat at the centre of the pellet, while the other more thermally conductive forms of uranium remain below their melting points.

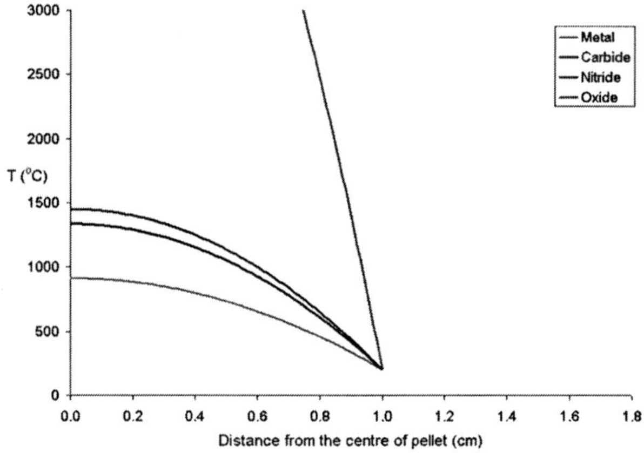

Figure 28: Temperature profile for a 20 mm diametre fuel pellet with a power density of 1000 W per cubic metre. The fuels other than uranium dioxide are not compromised.

Normal and Abnormal Conditions

The nuclear chemistry associated with the nuclear fuel cycle can be divided into two main areas, one area is concerned with operation under the intended conditions while the other area is concerned with maloperation conditions where some alteration from the normal operating conditions has occurred or (*more rarely*) an accident is occurring.

The releases of radioactivity from normal operations are the small planned releases from uranium ore processing, enrichment, power reactors, reprocessing plants and waste stores. These can be in a different chemical/physical form to the releases which could occur under accident conditions. In addition the isotope signature of a hypothetical accident may be very different to that of a planned normal operational discharge of radioactivity to the environment.

Just because a radioisotope is released it does not mean it will enter a human and then cause harm. For instance the migration of radioactivity can be altered by the binding of the radioisotope to the surfaces of soil particles.

For example caesium binds tightly to clay minerals such as illite and montmorillonite hence it remains in the upper layers of soil where it can be accessed by plants with shallow roots (such as grass). Hence grass and mushrooms can carry a considerable amount of ^{137}Cs which can be transferred to humans through the food chain. But ^{137}Cs is not able to migrate quickly through most soils and thus is unlikely to contaminate well water. Colloids of soil minerals can migrate through soil so simple binding of a metal to the surfaces of soil particles does not fix the metal totally.

According to Jiøí Hála's text book the distribution coefficient K_d is the ratio of the soil's radioactivity (Bq g^{-1}) to that of the soil water (Bq ml^{-1}). If the radioisotope is tightly bound to the minerals in the soil then less radioactivity can be absorbed by crops and grass growing on the soil.

- Cs-137 K_d = 1000
- Pu-239 K_d = 10000 to 100000
- Sr-90 K_d = 80 to 150
- I-131 K_d = 0.007 to 50

One of the best countermeasures in dairy farming against ^{137}Cs is to mix up the soil by deeply ploughing the soil. This has the effect of putting the ^{137}Cs out of reach of the shallow roots of the grass, hence the level of radioactivity in the grass will be lowered. Also after a nuclear war or serious accident the removal of top few cm of soil and its burial in a shallow trench will reduce the long term gamma dose to humans due to ^{137}Cs as the gamma photons will be attenuated by their passage through the soil.

Even after the radioactive element arrives at the roots of the plant, the metal may be rejected by the biochemistry of the plant. The details of the uptake of ^{90}Sr and ^{137}Cs into sunflowers grown under hydroponic conditions has been reported. The caesium was found in the leaf veins, in the stem and in the apical leaves. It was found that 12% of the caesium entered the plant, and 20% of the strontium. This paper also reports details of the effect of potassium, ammonium and calcium ions on the uptake of the radioisotopes.

In livestock farming an important countermeasure against ^{137}Cs is to feed animals a small amount of prussian blue. This iron potassium cyanide compound acts as a ion-exchanger. The cyanide is so tightly bonded to the iron that it is safe for a human to eat several grams of prussian blue per day.

The prussian blue reduces the biological half-life (different from the nuclear half-life) of the caesium. The physical or nuclear half-life of ^{137}Cs is about 30 years. This is a constant which can not be changed but the biological half-life is not a constant. It will change according to the nature and habits of the organism for which it is expressed. Caesium in humans normally has a biological half-life of between one and four months. An added advantage of the prussian blue is that the caesium which is stripped from the animal in the droppings is in a form which is not available to plants. Hence it prevents the caesium from being recycled.

The form of prussian blue required for the treatment of humans or animals is a special grade. Attempts to use the pigment grade used in paints have not been successful. Note that a good source of data on the subject of caesium in Chernobyl fallout exists at (*Ukrainian Research Institute for Agricultural Radiology*).

Release of Radioactivity from Fuel During Normal Use and Accidents

The IAEA assume that under normal operation the coolant of a water-cooled reactor will contain some radioactivity but during a reactor accident the coolant radioactivity level may rise.

The IAEA state that under a series of different conditions different amounts of the core inventory can be released from the fuel, the four conditions the IAEA consider are *normal operation*, a spike in coolant activity due to a sudden shutdown/loss of pressure (core remains covered with water), a cladding failure resulting in the release of the activity in the fuel/cladding gap (this could be due to the fuel being uncovered by the loss of water for 15–30 minutes where the cladding reached a temperature of 650-1250 °C) or a melting of the core (the fuel will have to be uncovered for at least 30 minutes, and the cladding would reach a temperature in excess of 1650 °C).

Based upon the assumption that a PWR contains 300 tons of water, and that the activity of the fuel of a 1 GWe reactor is as the IAEA predict, then the coolant activity after an accident such as the three mile island accident (where a core is uncovered and then recovered with water) can be predicted.

Releases from Reprocessing Under Normal Conditions

It is normal to allow used fuel to stand after the irradiation to allow the short-lived and radiotoxic iodine isotopes to decay away. In one experiment in the USA fresh fuel which had not been allowed to decay was reprocessed (the Green run) to investigate the effects of a large iodine release from the reprocessing of short cooled fuel. It is normal in reprocessing plants to scrub the off gases from the dissolver to prevent the emission of iodine. In addition to the emission of iodine the noble gases and tritium are released from the fuel when it is dissolved. It has been proposed that by voloxidation (heating the fuel in a furnace under oxidising conditions) the majority of the tritium can be recovered from the fuel.

A paper was written on the radioactivity in oysters found in the Irish Sea. These were found by gamma spectroscopy to contain ^{141}Ce, ^{144}Ce, ^{103}Ru, ^{106}Ru, ^{137}Cs, ^{95}Zr and ^{95}Nb. Additionally, a zinc activation product (^{65}Zn) was found, which is thought to be due to the corrosion of magnox fuel cladding in cooling ponds. It is likely that the modern releases of all these isotopes from Windscale is smaller.

On-load Reactors

Some reactor designs, such as RBMKs or CANDU reactors, can be refuelled without being shut down. This is achieved through the use of many small pressure tubes to contain the fuel and coolant, as opposed to one large pressure vessel as in pressurised water reactor (PWR) or boiling water reactor (BWR) designs. Each tube can be individually isolated and refuelled by an operator-controlled fuelling machine, typically at a rate of up to 8 channels per day out of roughly 400 in CANDU reactors.

On-load refuelling allows for the optimal fuel reloading problem to be dealt with continuously, leading to more efficient use of fuel. This increase in efficiency is partially offset by the added complexity of having hundreds of pressure tubes and the fuelling machines to service them.

Interim Storage

After its operating cycle, the reactor is shut down for refuelling. The fuel discharged at that time (spent fuel) is stored either at the reactor site (commonly in a spent fuel pool) or potentially in a common facility away from reactor sites. If on-site pool storage capacity is exceeded, it may be desirable to store the now cooled aged fuel in modular dry storage facilities known as Independent Spent Fuel Storage Installations (ISFSI) at the reactor site or at a facility away from the site.

The spent fuel rods are usually stored in water or boric acid, which provides both cooling (the spent fuel continues to generate decay heat as a result of residual radioactive decay) and shielding to protect the environment from residual ionising radiation, although after at least a year of cooling they may be moved to dry cask storage.

Transportation—Reprocessing

Figure 29: The Sellafield reprocessing plant

Spent fuel discharged from reactors contains appreciable quantities of fissile (U-235 and Pu-239), fertile (U-238), and other radioactive materials, including reaction poisons, which is why the fuel had to be removed. These fissile and fertile materials can be chemically separated and recovered from the spent fuel. The recovered uranium and plutonium can, if economic and institutional conditions permit, be recycled for use as nuclear fuel. This is currently not done for civilian spent nuclear fuel in the United States.

Mixed oxide, or MOX fuel, is a blend of reprocessed uranium and plutonium and depleted uranium which behaves similarly, although not identically, to the enriched uranium feed for which most nuclear reactors were designed. MOX fuel is an alternative to low-enriched uranium (LEU) fuel used in the light water reactors which predominate nuclear power generation.

Currently, plants in Europe are reprocessing spent fuel from utilities in Europe and Japan. Reprocessing of spent commercial-reactor nuclear fuel is currently not permitted in the United States due to the perceived danger of nuclear proliferation. However the recently announced Global Nuclear Energy Partnership would see the U.S. form an international partnership to see spent nuclear fuel reprocessed in a way that renders the plutonium in it usable for nuclear fuel but not for nuclear weapons.

Partitioning and Transmutation

As an alternative to the disposal of the PUREX raffinate in glass or Synroc, the most radiotoxic elements can be removed through advanced reprocessing. After separation the minor actinides and some long lived fission products can be converted to short-lived isotopes by either neutron or photon irradiation. This is called transmutation.

Waste Disposal

A current concern in the nuclear power field is the safe disposal and isolation of either spent fuel from reactors or, if the reprocessing option is used, wastes from reprocessing plants. These materials must be isolated from the biosphere until the radioactivity contained in them has diminished to a safe level. In the U.S., under the Nuclear Waste Policy Act of 1982 as amended, the Department of Energy has responsibility for the development of the waste disposal system for spent nuclear fuel and high-level radioactive waste. Current plans call for the ultimate disposal of the wastes in solid form in a licensed deep, stable geologic structure called a deep geological repository. The Department of Energy chose Yucca Mountain as the location for the repository. However, its opening has been repeatedly delayed.

It is worth noting that some non-PLWR reactor designs, and in particular the ones using liquid thorium fuel in molten salt reactors, would produce virtually no long-lasting nuclear waste. It is also possible to burn rather than bury nuclear waste, for instance in Integral Fast Reactors or in variations of molten salt reactors.

A proposed type of nuclear reactor called a travelling wave reactor is claimed, if it were to be built, to be able to be fuelled by nuclear waste, and to be able to operate for 200 years without needing any refuelling.

Fuel Cycles—Once-through Nuclear Fuel Cycle

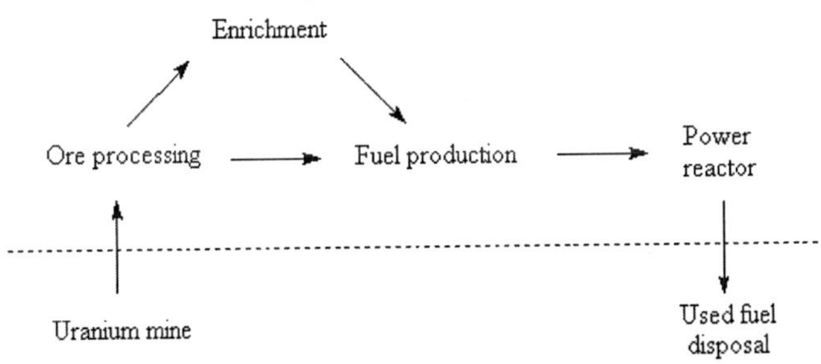

Figure 30: A once through (or open) fuel cycle

Not a cycle *per se*, fuel is used once and then sent to storage without further processing save additional packaging to provide for better isolation from the biosphere. This method is favoured by six countries: the United States, Canada, Sweden, Finland, Spain and South Africa. Some countries, notably Sweden and Canada, have designed repositories to permit future recovery of the material should the need arise, while others plan for permanent sequestration in a geological repository like the Yucca Mountain nuclear waste repository in the United States.

Plutonium Cycle

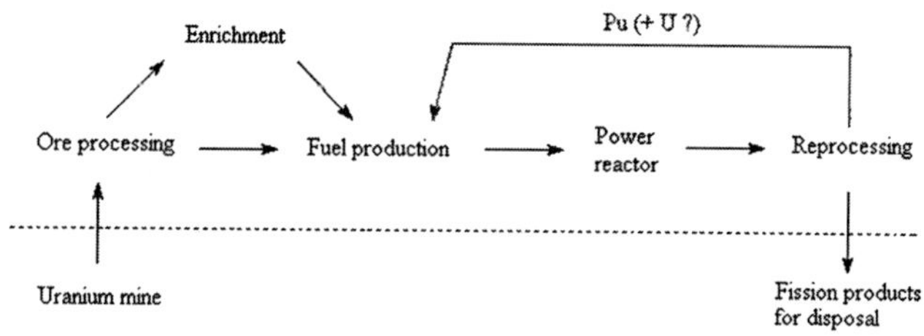

Figure 31: A fuel cycle in which plutonium is used for fuel

Several countries, including Japan, Switzerland, and previously Spain and Germany, are using or have used the reprocessing services offered by BNFL and COGEMA. Here, the fission products, minor actinides, activation products, and reprocessed uranium are separated from the reactor-grade plutonium, which can then be fabricated into MOX fuel. Because the proportion of the non-fissile even-mass isotopes of plutonium rises with each pass through the cycle, there are currently no plans to reuse plutonium from used MOX fuel for a third pass in a thermal reactor. However, if fast reactors become available, they may be able to burn these, or almost any other actinide isotopes.

Minor Actinides Recycling

It has been proposed that in addition to the use of plutonium, the minor actinides could be used in a critical power reactor. Tests are already being conducted in which americium is being used as a fuel.

A number of reactor designs, like the Integral Fast Reactor, have been designed for this rather different fuel cycle. In principle, it should be possible to derive energy from the fission of any actinide nucleus.

With a careful reactor design, all the actinides in the fuel can be consumed, leaving only lighter elements with short half-lives. Whereas this has been done in prototype plants, no such reactor has ever been operated on a large scale, and the first plants with full actinide recovery are expected to be ready for commercial deployment in 2015 at the earliest.

However, such schemes would most likely require advanced remote reprocessing methods due to the neutron emitting compounds formed.

For instance if curium is irradiated with neutrons it will form the very heavy actinides californium and fermium which undergo spontaneous fission. As a result, the neutron emission from a used fuel element which had included curium will be much higher, potentially posing a risk to workers at the back end of the cycle unless all reprocessing is done remotely.

This could be seen as a disadvantage, but on the other hand it also makes the nuclear material difficult to steal or divert, making it more resistant to nuclear proliferation

It so happens that the neutron cross-section of many actinides decreases with increasing neutron energy, but the ratio of fission to simple activation (neutron capture) changes in favour of fission as the neutron energy increases.

Thus with a sufficiently high neutron energy, it should be possible to destroy even curium without the generation of the transcurium metals. This could be very desirable as it would make it significantly easier to reprocess and handle the actinide fuel.

One promising alternative from this perspective is an accelerator driven subcritical reactor. Here a beam of either protons (United States and European designs) or electrons (Japanese design) is directed into a target.

In the case of protons, very fast neutrons will spall off the target, while in the case of the electrons, very high energy photons will be generated. These high-energy neutrons and photons will then be able to cause the fission of the heavy actinides.

Such reactors compare very well to other neutron sources in terms of neutron energy:

- Thermal 0 to 100 eV
- Epithermal 100 eV to 100 KeV
- Fast (from nuclear fission) 100 KeV to 3 MeV
- DD fusion 2.5 MeV
- DT fusion 14 MeV
- Accelerator driven core 200 MeV (lead driven by 1.6 GeV protons)
- Muon-catalysed fusion 7 GeV.

As an alternative, the curium-244, with a half-life of 18 years, could be left to decay into plutonium-240 before being used in fuel in a fast reactor.

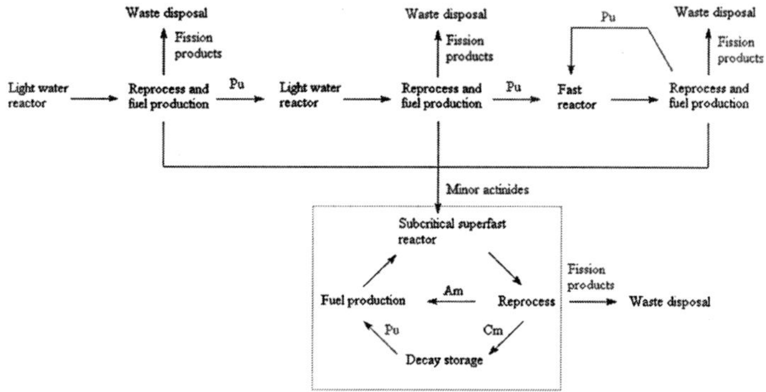

Figure 32: A pair of fuel cycles in which uranium and plutonium are kept separate from the minor actinides. The minor actinide cycle is kept within the green box.

Fuel or Targets for This Actinide Transmutation

To Date the Nature of the Fuel (Targets) for Actinide Transformation has not been Chosen. If actinides are transmuted in a Subcritical reactor it is likely that the fuel will have to be able to tolerate more thermal cycles than conventional fuel. An accelerator driven sub critical reactor is unlikely to be able to maintain a constant operation period for equally long times as a critical reactor, and each time the accelerator stops then the fuel will cool down.

On the other hand, if actinides are destroyed using a fast reactor, such as an Integral Fast Reactor, then the fuel will most likely not be exposed to many more thermal cycles than in a normal power station.

Depending on the matrix the process can generate more transuranics from the matrix. This could either be viewed as good (generate more fuel) or can be viewed as bad (generation of more *radiotoxic* transuranic elements). A series of different matrices exists which can control this production of heavy actinides.

Fissile nuclei, like Uranium-235, Plutonium-239 and Uranium-233 respond well to delayed neutrons and are thus important to keep a critical reactor stable, and this limits the amount of minor actinides that can be destroyed in a critical reactor. As a consequence it is important that the chosen matrix allows the reactor to keep the ratio of fissile to non-fissile nuclei high, as this enables it to destroy the long lived actinides safely. In contrast, the power output of a sub-critical reactor is limited by the intensity of the driving particle accelerator, and thus it need not contain any uranium or plutonium at all. In such a system it may be preferable to have an inert matrix that doesn't produce additional long-lived isotopes.

Actinides in an Inert Matrix

The actinides will be mixed with a metal which will not form more actinides, for instance an alloy of actinides in a solid such as zirconia could be used.

Actinides in a Thorium Matrix

Thorium will on neutron bombardment form uranium-233. U-233 is fissile, and has a larger fission cross section than both U-235 and U-238, and thus it is likely to produce very little additional actinides through neutron capture.

Actinides in a Uranium Matrix

If the actinides are incorporated into a uranium-metal or uranium-oxide matrix, then the neutron capture of U-238 is likely to generate new plutonium-239. An advantage of mixing the actinides with uranium and plutonium is that the large fission cross sections of U-235 and Pu-239 for the less energetic delayed-neutrons could make the reaction stable enough to be carried out in a critical fast reactor, which is likely to be both cheaper and simpler than an accelerator driven system.

Mixed Matrix

It is also possible to create a matrix made from a mix of the above mentioned materials. This is most commonly done in fast reactors where one may wish to keep the breeding ratio of new fuel high enough to keep powering the reactor, but still low enough that the generated actinides can be safely destroyed without transporting them to another site. One way to do this is to use fuel where actinides and uranium is mixed with inert zirconium, producing fuel elements with the desired properties.

Thorium Cycle

In the thorium fuel cycle thorium-232 absorbs a neutron in either a fast or thermal reactor. The thorium-233 beta decays to protactinium-233 and then to uranium-233, which in turn is used as fuel. Hence, like uranium-238, thorium-232 is a fertile material.

After starting the reactor with existing U-233 or some other fissile material such as U-235 or Pu-239, a breeding cycle similar to but more efficient than that with U-238 and plutonium can be created. The Th-232 absorbs a neutron to become Th-233 which quickly decays to protactinium-233. Protactinium-233 in turn decays with a half-life of 27 days to U-233. In some molten salt reactor designs, the Pa-233 is extracted and protected from neutrons (which could transform it to Pa-234 and then to U-234), until it has decayed to U-233. This is done in order to improve the breeding ratio which is low compared to fast reactors.

Thorium is at least 4-5 times more abundant in nature than all of uranium isotopes combined; thorium is fairly evenly spread around Earth with a lot of countries having huge supplies of it; preparation of thorium fuel does not require difficult and expensive enrichment processes; the thorium fuel cycle creates mainly Uranium-233 contaminated with Uranium-232 which makes it harder to use in a normal, pre-assembled nuclear weapon which is stable over long periods of time (unfortunately drawbacks are much lower for immediate use weapons or where final assembly occurs just prior to usage time); elimination of at least the transuranic portion of the nuclear waste problem is possible in MSR and other breeder reactor designs.

One of the earliest efforts to use a thorium fuel cycle took place at Oak Ridge National Laboratory in the 1960s. An experimental reactor was built based on molten salt reactor technology to study the feasibility of such an approach, using thorium fluoride salt kept hot enough to be liquid, thus eliminating the need for fabricating fuel elements. This effort culminated in the Molten-Salt Reactor Experiment that used ^{232}Th as the fertile material and ^{233}U as the fissile fuel. Due to a lack of funding, the MSR program was discontinued in 1976.

Current Industrial Activity

Currently the only isotopes used as nuclear fuel are uranium-235 (U-235), uranium-238 (U-238) and plutonium-239, although the proposed thorium fuel cycle has advantages. Some modern reactors, with minor modifications, can use thorium. Thorium is approximately three times more abundant in the Earth's crust than uranium (and 550 times more abundant than uranium-235). However, there has been little exploration for thorium resources, and thus the proved resource is small. Thorium is more plentiful than uranium in some countries, notably India.

Heavy water reactors and graphite-moderated reactors can use natural uranium, but the vast majority of the world's reactors require enriched uranium, in which the ratio of U-235 to U-238 is increased. In civilian reactors the enrichment is increased to as much as 5% U-235 and 95% U-238, but in naval reactors there is as much as 93% U-235.

The term *nuclear fuel* is not normally used in respect to fusion power, which fuses isotopes of hydrogen into helium to release energy.

Passive Fire Protection

Figure 33: Fire-resistance rated wall assembly with fire door, cable tray penetration and intumescent cable coating.

Passive fire protection (PFP) is an integral component of the three components of structural fire protection and fire safety in a building. PFP attempts to contain fires or slow the spread, through use of fire-resistant walls, floors, and doors (amongst other examples). PFP systems must comply with the associated Listing and approval use and compliance in order to provide the effectiveness expected by building codes.

Structural Fire Protection

Fire protection in a building, offshore facility or a ship is a system that includes:

- Active fire protection, which can include manual or automatic fire detection and fire suppression.
- Passive fire protection, which includes compartmentalisation of the overall building through the use of fire-resistance rated walls and floors. Organisation into smaller fire compartments, consisting of one or more rooms or floors, prevents or slows the spread of fire from the room of fire origin to other building spaces, limiting building damage and providing more time to the building occupants for emergency evacuation or to reach an area of refuge.
- Fire prevention includes minimising ignition sources, as well as educating the occupants and operators of the facility, ship or structure concerning operation and maintenance of fire-

related systems for correct function, and emergency procedures including notification for fire service response and emergency evacuation..

Main Characteristics

The aim for passive fire protection systems is typically demonstrated in fire testing the ability to maintain the item or the side to be protected at or below either 140 °C (for walls, floors and electrical circuits required to have a fire-resistance rating) or ca. 550 °C, which is considered the critical temperature for structural steel, above which it is in jeopardy of losing its strength, leading to collapse. This is based, in most countries, on the basic test standards for walls and floors, such as ASTM E119. Smaller components, such as fire dampers, fire doors, etc., follow suit in the main intentions of the basic standard for walls and floors. Fire testing involves live fire exposures upwards of 1100 °C, depending on the fire-resistance rating and duration one is after. More items than just fire exposures are typically required to be tested to ensure the survivability of the system under realistic conditions.

To accomplish these aims, many different types of materials are employed in the design and construction of systems. For instance, common endothermic building materials include concrete and gypsum wallboard. During fire testing of concrete floor slabs, water can be seen to boil out of a slab. Gypsum wall board typically loses all its strength during a fire. The use of endothermic materials is established and proven to be sound engineering practice. The chemically bound water inside these materials sublimes. During this process, the unexposed side cannot exceed the boiling point of water. Once the hydrates are spent, the temperature on the unexposed side of an endothermic fire barrier tends to rise rapidly. Too much water can be a problem, however. Concrete slabs that are too wet, will literally explode in a fire, which is why test laboratories insist on measuring water content of concrete and mortar in fire test specimens, before running any fire tests. PFP measures can also include intumescents and ablative materials. The point is, however, that whatever the nature of the materials, they on their own bear no rating. They must be organised into systems, which bear a rating when installed in accordance with certification listings or established catalogues, such as DIN 4102 Part 4 or the Canadian National Building Code.

Passive Fire Protection measures are intended to contain a fire in the fire compartment of origin, thus limiting the spread of fire and smoke for a limited period of time, as determined the local building code and fire code. Passive fire protection measures, such as firestops, fire walls, and fire doors, are tested to determine the fire resistance rating of the final assembly, usually expressed in terms of hours of fire resistance (e.g., ⅓, ¾, 1, 1½, 2, 3, 4 hour). A certification listing provides the limitations of the rating.

Contrary to active fire protection measures, *passive* fire protection means do not typically require electric or electronic activation or a degree of motion. Exceptions to that particular rule of thumb are fire dampers (fire-resistive closures within air ducts, excluding grease ducts) and fire door closers, which *must* move, open and shut in order to work, as well as all intumescent products, which swell, thus move, in order to function.

Passive fire protection (PFP), as the name suggest remains silent in your coating system till the eventuality of a fire. There are mainly two types of PFP. Intumescent fire protection and vermiculite fire protection. In vermiculite fire protection, the structural steel members are covered with vermiculite materials, mostly a very thick layer. This is a cheaper option as compared to an intumescent one, but is very crude and aesthetically unpleasant.

Moreover if the environment is corrosive in nature, then the vermiculite option is not advisable, as there is the possibility of water seeping into it (because of the porous nature of vermiculite), and there it is difficult to monitor for corrosion. Intumescent fireproofing is a layer of paint which is applied along with the coating system on the structural steel members. The thickness of this intumescent coating is dependent on the steel section used. For calculation of DFT (dry film thickness) a factor called *Hp/A* (heated perimetre divided by cross sectional area) is used. Intumescent coatings are applied as an intermediate coat in a coating system (primer, intermediate, and top/ finish coat). Because of the relatively low thickness of this intumescent coating (350-700 micrometres), nice finish, and anti-corrosive nature, intumescent coatings are preferred aesthetically and performance-wise.

It should be noted that in the eventuality of a fire, the steel structure will eventually collapse once the steel attains the critical core temperature (around 450 degrees Celsius or 850 degrees Fahrenheit). The PFP system will only delay this by creating a layer of char in between the steel and fire. Depending upon the requirement, PFP systems can provide fire ratings in excess of 120 minutes. PFP systems are highly recommended in infrastructure projects as they can save lives and property.

PFP in a building can be described as a group of systems within systems. An installed firestop, for instance, is a system that is based upon a product certification listing. It forms part of a fire-resistance rated wall or floor, and this wall or floor forms part of a fire compartment which forms an integral part of the overall fire safety plan of the building. The building itself, as a whole, can also be seen as a system.

Examples

Figure 34: This I beam has a fireproofing material sprayed onto it as a form of passive fire protection.

- fire-resistance rated walls
- Firewalls not only have a rating, they are also designed to subdivide buildings such that if collapse occurs on one side, this will not affect the other side. They can also be used to eliminate the need for sprinklers, as a trade-off.
- Fire-resistant glass glass using multi-layer intumescent technology or wire mesh embedded within the glass may be used in the fabrication of fire-resistance rated windows in walls or fire doors.
- fire-resistance rated floors
- occupancy separations (barriers designated as occupancy separations are intended to segregate parts of buildings, where different uses are on each side; for instance, apartments on one side and stores on the other side of the occupancy separation).

- closures (fire dampers) Sometimes firestops are treated in building codes identically to closures. Canada de-rates closures, where, for instance a 2 hour closure is acceptable for use in a 3 hour fire separation, so long as the fire separation is not an occupancy separation or firewall. The lowered rating is then referred to as a *fire protection rating*, both for firestops, unless they contain plastic pipes and regular closures.
- firestops
- grease ducts (These refer to ducts that lead from commercial cooking equipment such as ranges, deep fryers and double-decker and conveyor-equipped pizza ovens to grease duct fans.) In North America, grease ducts are made of minimum 16 gauge (1.6 mm) sheet metal, all welded, and certified openings for cleaning, whereby the ducting is either inherently manufactured to have a specific fire-resistance rating, OR it is ordinary 16 gauge ductwork with an exterior layer of purpose-made and certified fireproofing. Either way, North American grease ducts must comply with NFPA96 requirements.
- cable coating (application of fire-retardants, which are either endothermic or intumescent, to reduce flamespread and smoke development of combustible cable-jacketing)
- spray fireproofing (application of intumescent or endothermic paints, or fibrous or cementitious plasters to keep substrates such as structural steel, electrical or mechanical services, valves, liquefied petroleum gas (LPG) vessels, vessel skirts, bulkheads or decks below either 140 °C for electrical items or ca. 500 °C for structural steel elements to maintain operability of the item to be protected)
- fireproofing cladding (boards used for the same purpose and in the same applications as spray fireproofing) Materials for such cladding include perlite, vermiculite, calcium silicate, gypsum, intumescent epoxy, Durasteel (cellulose-fibre reinforced concrete and punched sheet-metal bonded composite panels), MicroTherm
- enclosures (boxes or wraps made of fireproofing materials, including fire-resistive wraps and tapes to protect speciality valves and other items deemed to require protection against fire and heat—an analogy for this would be a safe) or the

provision of circuit integrity measures to keep electrical cables operational during an accidental fire.

Regulations

The most important goal of PFP is identical to that of all fire protection: life safety. This is mainly accomplished by maintaining structural integrity for a time during the fire, and limiting the spread of fire and the effects thereof (e.g., heat and smoke). Property protection and continuity of operations are usually secondary objectives in codes. Exceptions include nuclear facilities and marine applications, as evacuation may be more complex or impossible. Nuclear facilities, both buildings and ships, must also ensure the nuclear reactor does not experience a nuclear meltdown. In this case, fixing the reactor may be more important than evacuation for key safety personnel.

Examples of testing that underlies certification listing:

- *Netherlands:* NEN 6068
- *Germany:* DIN 4102
- *United Kingdom:* BS 476
- *Canada:* ULC-S101
- *United States:* ASTM E119

Each of these test procedures have very similar fire endurance regimes and heat transfer limitations. Differences include the hose-stream tests, which are unique to Canada and the United States, whereas Germany includes a very rigorous impact test during the fire for firewalls. Germany is unique in including heat induced expansion and collapse of ferrous cable trays into account for firestops, resulting in the favouring of firestop mortars, which tend to hold the penetrating cable tray in place, whereas "softseals", typically made of rockwool and elastomeric toppings, have been demonstrated in testing by Otto-Graf_institut to be torn open and rendered inoperable when the cable tray expands, pushes in and then collapses. Spin-offs from these basic tests cover closures, firestops and more. Furnace operations, thermocoupling and reporting requirements remain uniform within each country.

In exterior applications for the offshore and the petroleum sectors, the fire endurance testing uses a higher temperature and faster heat rise, whereas in interior applications, such as office buildings, factories and residential, the fire endurance is based upon experiences gained from burning wood. The interior fire time/temperature curve is referred to as "ETK" (Einheitstemperaturkurve = standard time/temperature curve) or the "building elements" curve, whereas the high temperature variety is called the hydrocarbon curve as it is based on burning oil and gas products, which burn hotter and faster. The most severe, and most rarely used, of all fire exposure tests is the British "jetfire" test, which has been used to some extent in the UK and Norway but is not typically found in common regulations.

Typically, during the construction of buildings, fire protective systems must conform to the requirements of building code that was in effect on the day that the building permit was applied for. Enforcement for compliance with building codes is typically the responsibility of municipal building departments. Once construction is complete, the building must maintain its design basis by remaining in compliance with the current fire code, which is enforced by the fire prevention officers of the municipal fire department. An up to date fire protection plan, containing a complete inventory and maintenance details of all fire protection components, including firestops, fireproofing, fire sprinklers, fire detectors, fire alarm systems, fire extinguishers, etc. are typical requirements for demonstration of compliance with applicable laws and regulations. In order to know whether or not one's building is in compliance with fire safety regulations, it is helpful to know what systems one has in place and what their installation and maintenance are based upon.

Changes to fire protection systems or items affecting the structural or fire-integrity or use (occupancy) of a building is subject to regulatory scrutiny. A contemplated change to a facility requires a building permit, or, if the change is very minor, a review by the local fire prevention officer. Such reviews by the Authority Having Jurisdiction (AHJ) also help to prevent potential problems that may not be apparent to a building owner or contractors. Large and very common deficiencies in existing buildings include the disabling of fire door closers through propping the doors open and running rugs through them and perforating fire-resistance rated walls and floors without proper firestopping. Example pictures of code violations can be seen here.

Common, Accredited Product Certification and Testing Organisations

- Europe:
    - *Testing:* Efectis
- UK:
    - *Testing:* Chiltern International Fire
    - Certification: BM TRADA Certification

- Netherlands:
    - o *Testing:* Efectis Nederland
- Germany:
    - o *Testing:* iBMB/ TU Braunschweig
    - o *Testing:* BAM Berlin
    - o *Testing:* MPA Dortmund
    - o *Certification:* Deutsches Institut für Bautechnik (DIBt)
- *Canada:* Underwriters' Laboratories of Canada (ULC)
- *United States:* Underwriters Laboratories, FM Global
- Norway:
    - o Testing: SINTEF NBL

"Old" versus "New"

Generally, one differentiates between "old" and "new" barrier systems. "Old" systems have been tested and verified by governmental authorities including DIBt, the British Standards Institute (BSI) and the National Research Council's Institute for Research in Construction. These organisations each publish in codes and standards, wall and floor assembly details that can be used with generic, standardised components, to achieve quantified fire-resistance ratings. Architects routinely refer to these details in drawings to enable contractors to build passive fire protection barriers of certain ratings.

The "old" systems are sometimes added to, through testing performed in governmental laboratories such as those maintained by Canada's Institute for Research in Construction, which then publishes the results in Canada's National Building Code (NBC). Germany and the UK, by comparison, publish their "old" systems in respective standards, DIN4102 Part 4 (Germany) and BS476 (United Kingdom). "New" systems are typically based on certification listings, whereby the installed configuration must comply with the tolerances set out in the certification listing. The United Kingdom is an exception to this, whereby certification, although not testing, is optional.

Countries where Certification is Optional

Fire tests in the UK are reported in the form of test results, but contrary to North America and Germany, building authorities do not require written proof that the materials that have been installed on site are actually identical to the materials and products that were used in the test. The test report is also often interpreted by engineers, as the test results are not communicated in the form of uniformly structured listings. In the UK, and other countries which do not require certification, the proof that the manufacturer has not substituted other materials apart from those used in the original testing is based on trust in the ethics or the culpability of the manufacturer.

While in North America and in Germany, product certification is the key to the success and legal defensibility of passive fire protection barriers, alternate quality control certifications of specific installation companies and their work is *available*, though not a legislative or regulatory requirement. Still, the question of how one can be sure, apart from faith in the vendor, that what was tested is identical to that which has been bought and installed is a matter of personal judgment.

The most highly publicised example of PFP systems which were not subject of certification and were declared *inoperable* by the Authority Having Jurisdiction is the Thermo-Lag scandal, which was brought to light by whistleblower Gerald W. Brown, who notified the Nuclear Regulatory Commission of the inadequacy of fire testing for circuit integrity measures in use in licensed nuclear power plants. This led to a congressional enquiry, significant press coverage and a large amount of remedial work on the part of the industry to mitigate the problem. There is no known case a similar instance for PFP systems which were under the follow-up regime of organisations holding national accreditation for product certification, such as DIBt or Underwriters Laboratories.

Chapter 3

Safety Engineering

Safety engineering is an applied science strongly related to systems engineering / industrial engineering and the subset System Safety Engineering. Safety engineering assures that a life-critical system behaves as needed even when components fail.

Overview

Ideally, safety-engineers take an early design of a system, analyse it to find what faults can occur, and then propose safety requirements in design specifications up front and changes to existing systems to make the system safer. In an early design stage, often a fail-safe system can be made acceptably safe with a few sensors and some software to read them. Probabilistic fault-tolerant systems can often be made by using more, but smaller and less-expensive pieces of equipment. Far too often, rather than actually influencing the design, safety engineers are assigned to prove that an existing, completed design is safe. If a safety engineer then discovers significant safety problems late in the design process, correcting them can be very expensive. This type of error has the potential to waste large sums of money.

The exception to this conventional approach is the way some large government agencies approach safety engineering from a more proactive and proven process perspective, known as "system safety". The system safety philosophy is to be applied to complex and critical systems, such as commercial airliners, complex weapon systems, spacecraft, rail and transportation systems, air traffic control system and other complex and safety-critical industrial systems. The proven system safety methods and techniques are to prevent, eliminate and control hazards and risks through designed influences by a collaboration of key engineering disciplines and product teams. Software safety is a fast growing field since modern systems functionality are increasingly being put under control of software. The whole concept of system safety and software safety, as a subset of systems engineering, is to influence safety-critical systems designs by conducting several types of hazard analyses to identify risks and to specify design safety features and procedures to strategically mitigate risk to acceptable levels before the system is certified.

Additionally, failure mitigation can go beyond design recommendations, particularly in the area of maintenance. There is an entire realm of safety and reliability engineering known as Reliability Centred Maintenance (RCM), which is a discipline that is a direct result of analysing potential failures within a system and determining maintenance actions that can mitigate the risk of failure. This methodology is used extensively on aircraft and involves understanding the failure modes of the serviceable replaceable assemblies in addition to the means to detect or predict an impending failure. Every automobile owner is familiar with this concept when they take in their car to have the oil changed or brakes checked. Even filling up one's car with fuel is a simple example of a failure mode (failure due to fuel exhaustion), a means of detection (fuel gauge), and a maintenance action (filling the car's fuel tank).

For large scale complex systems, hundreds if not thousands of maintenance actions can result from the failure analysis. These maintenance actions are based on conditions (e.g., gauge reading or leaky valve), hard conditions (e.g., a component is known to fail after 100 hrs of operation with 95% certainty), or require inspection to determine the maintenance action (e.g., metal fatigue). The RCM concept then analyses each individual maintenance item for its risk contribution to safety, mission, operational readiness, or cost to repair if a failure does occur.

Then the sum total of all the maintenance actions are bundled into maintenance intervals so that maintenance is not occurring around the clock, but rather, at regular intervals. This bundling process introduces further complexity, as it might stretch some maintenance cycles, thereby increasing risk, but reduce others, thereby potentially reducing risk, with the end result being a comprehensive maintenance schedule, purpose built to reduce operational risk and ensure acceptable levels of operational readiness and availability.

Analysis Techniques

Analysis techniques can be split into two categories: qualitative and quantitative methods. The both approaches share the goal of finding causal dependencies between a hazard on system level and failures of individual components. Qualitative approaches focus on the question "What must go wrong, such that a system hazard may occur?", while quantitative methods aim at providing estimations about probabilites, rates and/or severity of consequences.

Traditionally, safety analysis techniques rely solely on skill and expertise of the safety engineer. In the last decade model-based approaches have become prominent. In contrast to traditional methods, model-based techniques try to derive relationships between causes and consequences from some sort of model of the system.

Traditional Methods for Safety Analysis

The two most common fault modelling techniques are called failure mode and effects analysis and fault tree analysis. These techniques are just ways of finding problems and of making plans to cope with failures, as in probabilistic risk assessment. One of the earliest complete studies using this technique on a commercial nuclear plant was the WASH-1400 study, also known as the Reactor Safety Study or the Rasmussen Report.

Failure Modes and Effects Analysis

Failure Mode and Effects Analysis (FMEA) is a bottom-up, inductive analytical method which may be performed at either the functional or piece-part level. For functional FMEA, failure modes are identified for each function in a system or equipment item, usually with the help of a functional block diagram. For piece-part FMEA, failure modes are identified for each piece-part component (such as a valve, connector, resistor, or diode). The effects of the failure mode are described, and assigned a probability based on the failure rate and failure mode ratio of the function or component.

Failure modes with identical effects can be combined and summarised in a Failure Mode Effects Summary. When combined with criticality analysis, FMEA is known as Failure Mode, Effects, and Criticality Analysis or FMECA, pronounced "fuh-MEE-kuh".

Fault Tree Analysis

Fault tree analysis (FTA) is a top-down, deductive analytical method. In FTA, initiating primary events such as component failures, human errors, and external events are traced through Boolean logic gates to an undesired top event such as an aircraft crash or nuclear reactor core melt. The intent is to identify ways to make top events less probable, and verify that safety goals have been achieved.

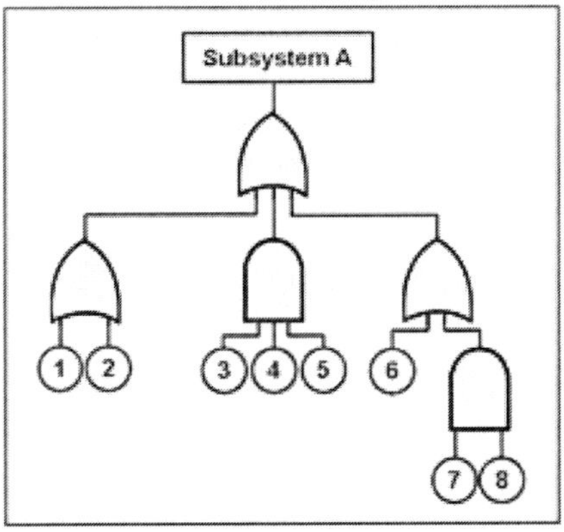

Figure 1: A fault tree diagram

Fault trees are a logical inverse of success trees, and may be obtained by applying de Morgan's theorem to success trees (which are directly related to reliability block diagrams).

FTA may be qualitative or quantative. When failure and event probabilites are unknown, qualitative fault trees may be analysed for minimal cut sets.

For example, if any minimal cut set contains a single base event, then the top event may be caused by a single failure. Quantitative FTA is used to compute top event probability, and usually requires computer software such as CAFTA from the Electric Power Research Institute or SAPHIRE from the Idaho National Laboratory.

Some industries use both fault trees and event trees. An event tree starts from an undesired initiator (loss of critical supply, component failure etc.) and follows possible further system events through to a series of final consequences. As each new event is considered, a new node on the tree is added with a split of probabilities of taking either branch. The probabilities of a range of "top events" arising from the initial event can then be seen.

Safety Certification

Usually a failure in safety-certified systems is acceptable if, on average, less than one life per 10^9 hours of continuous operation is lost to failure.

Most Western nuclear reactors, medical equipment, and commercial aircraft are certified to this level. The cost versus loss of lives has been considered appropriate at this level (by FAA for aircraft systems under Federal Aviation Regulations).

Preventing Failure

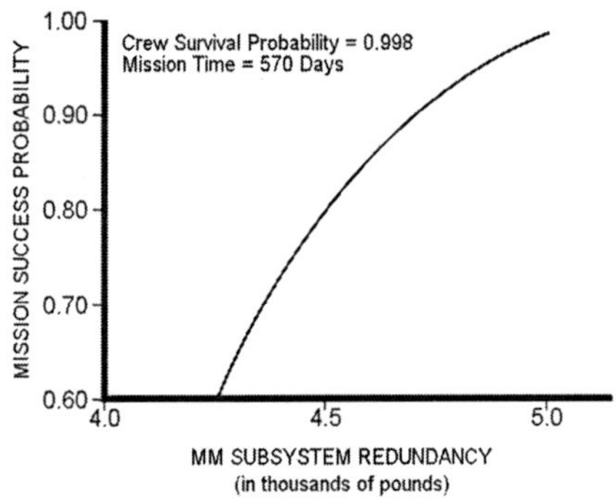

Figure 2: A NASA graph shows the relationship between the survival of a crew of astronauts and the amount of redundant equipment in their spacecraft (the "MM", Mission Module).

Once a failure mode is identified, it can usually be mitigated by adding extra or redundant equipment to the system. For example, nuclear reactors contain dangerous radiation, and nuclear reactions can cause so much heat that no substance might contain them. Therefore reactors have emergency core cooling systems to keep the temperature down, shielding to contain the radiation, and engineered barriers (usually several, nested, surmounted by a containment building) to prevent accidental leakage. Safety-critical systems are commonly required to permit no single event or component failure to result in a catastrophic failure mode.

Most biological organisms have a certain amount of redundancy: multiple organs, multiple limbs, etc.

For any given failure, a fail-over or redundancy can almost always be designed and incorporated into a system.

Safety and Reliability

Probabilistic risk assessment has created a close relationship between safety and reliability. Component reliability, generally defined in terms of component failure rate, and external event probability are both used in quantitative safety assessment methods such as FTA. Related probabilistic methods are used to determine system Mean Time Between Failure (MTBF), system availability, or probability of mission success or failure. Reliability analysis has a broader scope than safety analysis, in that non-critical failures are considered. On the other hand, higher failure rates are considered acceptable for non-critical systems.

Safety generally cannot be achieved through component reliability alone. Catastrophic failure probabilities of 10^{-9} per hour correspond to the failure rates of very simple components such as resistors or capacitors. A complex system containing hundreds or thousands of components might be able to achieve a MTBF of 10,000 to 100,000 hours, meaning it would fail at 10^{-4} or 10^{-5} per hour. If a system failure is catastrophic, usually the only practical way to achieve 10^{-9} per hour failure rate is through redundancy. Two redundant systems with independent failure modes, each having an MTBF of 100,000 hours, could achieve a failure rate on the order of 10^{-10} per hour because of the multiplication rule for independent events.

When adding equipment is impractical (usually because of expense), then the least expensive form of design is often "inherently fail-safe". That is, change the system design so its failure modes are not catastrophic. Inherent fail-safes are common in medical equipment, traffic and railway signals, communications equipment, and safety equipment.

The typical approach is to arrange the system so that ordinary single failures cause the mechanism to shut down in a safe way (for nuclear power plants, this is termed a passively safe design, although more than ordinary failures are covered). Alternately, if the system contains a hazard source such as a battery or rotor, then it may be possible to remove the hazard from the system so that its failure modes cannot be catastrophic. The U.S. Department of Defence Standard Practice for System Safety (MIL–STD–882) places the highest priority on elimination of hazards through design selection.

One of the most common fail-safe systems is the overflow tube in baths and kitchen sinks. If the valve sticks open, rather than causing an overflow and damage, the tank spills into an overflow. Another common example is that in an elevator the cable supporting the car keeps spring-loaded brakes open. If the cable breaks, the brakes grab rails, and the elevator cabin does not fall.

Some systems can never be made fail safe, as continuous availability is needed. For example, loss of engine trust in flight is dangerous. Redundancy, fault tolerance, or recovery procedures are used for these situations (e.g. multiple independent controlled and fuel fed engines). This also makes the system less sensitive for the reliability prediction errors or quality induced uncertainty for the separate items. On the other hand, failure detection & correction and avoidance of common cause failures becomes here increasingly important to ensure system level reliability.

Containing Failure

It is common practice to plan for the failure of safety systems through containment and isolation methods. The use of isolating valves, also known as the block and bleed manifold, is very common in isolating pumps, tanks, and control valves that may fail or need routine maintenance. In addition, nearly all tanks containing oil or other hazardous chemicals are required to have containment barriers set up around them to contain 100% of the volume of the tank in the event of a catastrophic tank failure. Similarly, in a long pipeline, there are remote-closing valves at regular intervals so that a leak can be isolated. Fault isolation boundaries are similarly designed into critical electronic systems or computer software. The goal of all containment systems is to provide means of mitigating the consequences of failure. Fault isolation might also refer to the extend to which detected failures might be isolated for sucessfull recovery. The isolation level shows the system identure level at which the failure cause can be recovered (often by replacement of a line replaceable unit).

Nuclear Safety

Nuclear safety covers the actions taken to prevent nuclear and radiation accidents or to limit their consequences. This covers nuclear power plants as well as all other nuclear facilities, the transportation of nuclear materials, and the use and storage of nuclear materials for medical, power, industry, and military uses.

The nuclear power industry has improved the safety and performance of reactors, and has proposed new safer (but generally untested) reactor designs but there is no guarantee that the reactors will be designed, built and operated correctly. Mistakes do occur and the designers of reactors at Fukushima in Japan did not anticipate that a tsunami generated by an earthquake would disable the backup systems that were supposed to stabilise the reactor after the earthquake. According to UBS AG, the Fukushima I nuclear accidents have cast doubt on whether even an advanced economy like Japan can master nuclear safety. Catastrophic scenarios involving terrorist attacks are also conceivable.

An interdisciplinary team from MIT have estimated that given the expected growth of nuclear power from 2005 – 2055, at least four serious nuclear accidents would be expected in that period. To date, there have been five serious accidents (core damage) in the world since 1970 (one at Three Mile Island in 1979; one at Chernobyl in 1986; and three at Fukushima-Daiichi in 2011), corresponding to the beginning of the operation of generation II reactors. This leads to on average one serious accident happening every eight years worldwide.

Nuclear weapon safety, as well as the safety of military research involving nuclear materials, is generally handled by agencies different from those that oversee civilian safety, for various reasons, including secrecy.

Agencies

Figure 3: IAEA headquarters in Vienna, Austria

Internationally the International Atomic Energy Agency "works with its Member States and multiple partners worldwide to promote safe, secure and peaceful nuclear technologies." Some scientists say that the 2011 Japanese nuclear accidents have revealed that the nuclear industry lacks sufficient oversight, leading to renewed calls to redefine the mandate of the IAEA so that it can better police nuclear power plants worldwide. There are several problems with the IAEA says Najmedin Meshkati of University of Southern California:

It recommends safety standards, but member states are not required to comply; it promotes nuclear energy, but it also monitors nuclear use; it is the sole global organisation overseeing the nuclear energy industry, yet it is also weighed down by checking compliance with the Nuclear Non-Proliferation Treaty (NPT).

Many nations utilising nuclear power have special institutions overseeing and regulating nuclear safety. Civilian nuclear safety in the U.S. is regulated by the Nuclear Regulatory Commission (NRC). The safety of nuclear plants and materials controlled by the U.S. government for research, weapons production, and those powering naval vessels is not governed by the NRC. In the UK nuclear safety is regulated by the Office for Nuclear Regulation (ONR) and the Defence Nuclear Safety Regulator (DNSR). The Australian Radiation Protection and Nuclear Safety Agency (ARPANSA) is the Federal

Government body that monitors and identifies solar radiation and nuclear radiation risks in Australia. It is the main body dealing with ionising and non-ionising radiation and publishes material regarding radiation protection.

Other agencies include:

- Autorite de sûrete nucleaire
- Canadian Nuclear Safety Commission
- Radiological Protection Institute of Ireland
- Federal Atomic Energy Agency in Russia
- Kernfysische dienst, (NL)
- Pakistan Nuclear Regulatory Authority
- Bundesamt für Strahlenschutz, (DE)
- Atomic Energy Regulatory Board (India)

Nuclear Power Plant

Complexity: Nuclear power plants are some of the most sophisticated and complex energy systems ever designed. Any complex system, no matter how well it is designed and engineered, cannot be deemed failure-proof. Stephanie Cooke has said that:

The reactors themselves were enormously complex machines with an incalculable number of things that could go wrong. When that happened at Three Mile Island in 1979, another fault line in the nuclear world was exposed. One malfunction led to another, and then to a series of others, until the core of the reactor itself began to melt, and even the world's most highly trained nuclear engineers did not know how to respond. The accident revealed serious deficiencies in a system that was meant to protect public health and safety. A fundamental issue related to complexity is that nuclear power systems have exceedingly long lifetimes. The timeframe involved from the start of construction of a commercial nuclear power station, through to the safe disposal of its last radioactive waste, may be 100 to 150 years.

Failure Modes of Nuclear Power Plants

There are concerns that a combination of human and mechanical error at a nuclear facility could result in significant harm to people and the environment: Operating nuclear reactors contain large amounts of radioactive fission products which, if dispersed, can pose a direct radiation hazard, contaminate soil and vegetation, and be ingested by humans and animals. Human exposure at high enough levels can cause both short-term illness and death and longer-term death by cancer and other diseases. It is impossible for a commercial nuclear reactor to explode like a nuclear bomb since the fuel is never sufficiently enriched for this to occur.

Nuclear reactors can fail in a variety of ways. Should the instability of the nuclear material generate unexpected behaviour, it may result in an uncontrolled power excursion. Normally, the cooling system in a reactor is designed to be able to handle the excess heat this causes; however, should the reactor also experience a loss-of-coolant accident, then the fuel may melt or cause the vessel it is contained in to overheat and melt. This event is called a nuclear meltdown. After shutting down, for some time the reactor still needs external energy to power its cooling systems. Normally this energy is provided by the power grid to that the plant is connected, or by emergency diesel generators. Failure to provide power for the cooling systems, as happened in Fukushima I, can cause serious accidents.

Nuclear safety rules in the United States "do not adequately weigh the risk of a single event that would knock out electricity from the grid and from emergency generators, as a quake and tsunami recently did in Japan", Nuclear Regulatory Commission officials said in June 2011. Intentional cause of such failures may be the result of nuclear terrorism.

Vulnerability of Nuclear Plants to Attack

Nuclear reactors become preferred targets during military conflict and, over the past three decades, have been repeatedly attacked during military air strikes, occupations, invasions and campaigns:

- In September 1980, Iran bombed the Al Tuwaitha nuclear complex in Iraq.
- In June 1981, an Israeli air strike completely destroyed Iraq's Osirak nuclear research facility.
- Between 1984 and 1987, Iraq bombed Iran's Bushehr nuclear plant six times.
- In Iraq in 1991, the U.S. bombed three nuclear reactors and an enrichment pilot facility.
- In 1991, Iraq launched Scud missiles at Israel's Dimona nuclear power plant.
- In September 2007, Israel bombed a Syrian reactor under construction.

In the U.S., plants are surrounded by a double row of tall fences which are electronically monitored. The plant grounds are patrolled by a sizeable force of armed guards. The NRC's "Design Basis Threat" criteria for plants is a secret, and so what size of attacking force the plants are able to protect against is unknown. However, to scram (make an emergency shutdown) a plant takes fewer than 5 seconds while unimpeded restart takes hours, severely hampering a terrorist force in a goal to release radioactivity.

Attack from the air is an issue that has been highlighted since the September 11 attacks in the U.S. However, it was in 1972 when three hijackers took control of a domestic passenger flight along the east coast of the U.S. and threatened to crash the plane into a U.S. nuclear weapons plant in Oak Ridge, Tennessee. The plane got as close as 8,000 feet above the site before the hijackers' demands were met. The most important barrier against the release of radioactivity in the event of an aircraft strike on a nuclear power plant is the containment building and its missile shield. Current NRC Chairman Dale Klein has said "Nuclear power plants are inherently robust structures that our studies show provide adequate protection in a hypothetical attack by an aeroplane. The NRC has also taken actions that require nuclear power plant operators to be able to manage large fires or explosions—no matter what has caused them."

In addition, supporters point to large studies carried out by the U.S. Electric Power Research Institute that tested the robustness of both reactor and waste fuel storage and found that they should be able to sustain a terrorist attack comparable to the September 11 terrorist attacks in the U.S. Spent fuel is usually housed inside the plant's "protected zone" or a spent nuclear fuel shipping cask; stealing it for use in a "dirty bomb" is extremely difficult. Exposure to the intense radiation would almost certainly quickly incapacitate or kill anyone who attempts to do so.

In September 2010, analysis of the Stuxnet computer worm suggested that it was designed to sabotage a nuclear power plant. Such a cyber attack would bypass the physical safeguards in place and so the exploit demonstrates an important new vulnerability.

Plant Location

In many countries, plants are often located on the coast, in order to provide a ready source of cooling water for the essential service water system. As a consequence the design needs to take the risk of flooding and tsunamis into account. The World Energy Council (WEC) argues disaster risks are changing and increasing the likelihood of disasters such as earthquakes, cyclones, hurricanes, typhoons, ûooding. Climate change and increased temperatures, lower precipitation levels and an increase in the frequency and severity of droughts may lead to fresh water shortages. Seawater is corrosive and so nuclear energy supply is likely to be negatively affected by the fresh water shortage. This generic problem may become increasingly significant over time. Failure to calculate the risk of flooding correctly lead to a Level 2 event on the International Nuclear Event Scale during the 1999 Blayais Nuclear Power Plant flood, while flooding caused by the 2011 Tôhoku earthquake and tsunami lead to the Fukushima I nuclear accidents.

The design of plants located in seismically active zones also requires the risk of earthquakes and tsunamis to be taken into account. Japan, India, China and the USA are among the countries to have plants in earthquake-prone regions. Damage caused to Japan's Kashiwazaki-Kariwa Nuclear Power Plant during the 2007 Chûetsu offshore earthquake underlined concerns expressed by experts in Japan prior to the Fukushima accidents, who have warned of a *genpatsu-shinsai* (domino-effect nuclear power plant earthquake disaster).

Nuclear Safety Systems

The three primary objectives of nuclear safety systems as defined by the Nuclear Regulatory Commission are to shut down the reactor, maintain it in a shutdown condition, and prevent the release of radioactive material during events and accidents. These objectives are accomplished using a variety of equipment, which is part of different systems, of which each performs specific functions.

Hazards of Nuclear Material

Figure 4: Spent nuclear fuel stored underwater and uncapped at the Hanford site in Washington, USA.

The world's nuclear fleet creates about 10,000 metric tons of high-level spent nuclear fuel each year. High-level radioactive waste management concerns management and disposal of highly radioactive materials created during production of nuclear power. The technical issues in accomplishing this are daunting, due to the extremely long periods radioactive wastes remain deadly to living organisms. Of particular concern are two long-lived fission products, Technetium-99 (half-life 220,000 years) and Iodine-129 (half-life 15.7 million years), which dominate spent nuclear fuel radioactivity after a few thousand years.

The most troublesome transuranic elements in spent fuel are Neptunium-237 (half-life two million years) and Plutonium-239 (half-life 24,000 years). Consequently, high-level radioactive waste requires sophisticated treatment and management to successfully isolate it from the biosphere. This usually necessitates treatment, followed by a long-term management strategy involving permanent storage, disposal or transformation of the waste into a non-toxic form.

Governments around the world are considering a range of waste management and disposal options, usually involving deep-geologic placement, although there has been limited progress toward implementing long-term waste management solutions. This is partly because the timeframes in question when dealing with radioactive waste range from 10,000 to millions of years, according to studies based on the effect of estimated radiation doses.

New Nuclear Technologies

Nuclear proponents have tried bolster public support by offering newer, safer, reactor designs. These designs include those that incorporate passive safety and Small Modular Reactors. While these reactor designs "are intended to inspire trust, they may have an unintended effect: creating distrust of older reactors that lack the touted safety features".

The next nuclear plants to be built will likely be Generation III or III+ designs, and a few such are already in operation in Japan. Generation IV reactors would have even greater improvements in safety. These new designs are expected to be passively safe or nearly so, and perhaps even inherently safe (as in the PBMR designs).

Some improvements made (not all in all designs) are having three sets of emergency diesel generators and associated emergency core cooling systems rather than just one pair, having quench tanks (large coolant-filled tanks) above the core that open into it automatically, having a double containment (one containment building inside another), etc.

However, safety risks may be the greatest when nuclear systems are the newest, and operators have less experience with them. Nuclear engineer David Lochbaum explained that almost all serious nuclear accidents occurred with what was at the time the most recent technology. He argues that "the problem with new reactors and accidents is twofold: scenarios arise that are impossible to plan for in simulations; and humans make mistakes". As one director of a U.S. research laboratory put it, "fabrication, construction, operation, and maintenance of new reactors will face a steep learning curve: advanced technologies will have a heightened risk of accidents and mistakes. The technology may be proven, but people are not".

Safety Culture and Human Errors

One relatively prevalent notion in discussions of nuclear safety is that of safety culture. The International Nuclear Safety Advisory Group, defines the term as "the personal dedication and accountability of all individuals engaged in any activity which has a bearing on the safety of nuclear power plants". The goal is "to design systems that use human capabilities in appropriate ways, that protect systems from human frailties, and that protect humans from hazards associated with the system".

At the same time, there is some evidence that operational practices are not easy to change. Operators almost never follow instructions and written procedures exactly, and "the violation of rules appears to be quite rational, given the actual workload and timing constraints under which the operators must do their job". Many attempts to improve nuclear safety culture "were compensated by people adapting to the change in an unpredicted way".

According to Areva's Southeast Asia and Oceania director, Selena Ng, Japan's Fukushima nuclear disaster is "a huge wake-up call for a nuclear industry that hasn't always been sufficiently transparent about safety issues". She said "There was a sort of complacency before Fukushima and I don't think we can afford to have that complacency now".

An assessment conducted by the *Commissariat a l'Energie Atomique* (CEA) in France concluded that no amount of technical innovation can eliminate the risk of human-induced errors associated with the operation of nuclear power plants. Two types of mistakes were deemed most serious: errors committed during field operations, such as maintenance and testing, that can cause an accident; and human errors made during small accidents that cascade to complete failure.

According to Mycle Schneider, reactor safety depends above all on a 'culture of security', including the quality of maintenance and training, the competence of the operator and the workforce, and the rigour of regulatory oversight. So a better-designed, newer reactor is not always a safer one, and older reactors are not necessarily more dangerous than newer ones. The 1978 Three Mile Island accident in the United States occurred in a reactor that had started operation only three months earlier, and the Chernobyl disaster occurred after only two years of operation. A serious loss of coolant occurred at the French Civaux-1 reactor in 1998, less than five months after start-up.

However safe a plant is designed to be, it is operated by humans who are prone to errors. Laurent Stricker, a nuclear engineer and chairman of the World Association of Nuclear Operators says that operators must guard against complacency and avoid overconfidence. Experts say that the "largest single internal factor determining the safety of a plant is the culture of security among regulators, operators and the workforce — and creating such a culture is not easy".

Risks

The routine health risks and greenhouse gas emissions from nuclear fission power are small relative to those associated with coal, but there are several "catastrophic risks":

The extreme danger of the radioactive material in power plants and of nuclear technology in and of itself is so well-known that the US government was prompted (at the industry's urging) to enact provisions that protect the nuclear industry from bearing the full burden of such inherently risky nuclear operations. The Price-Anderson Act limits industry's liability in the case of accidents, and the 1982 Nuclear Waste Policy Act charges the federal government with responsibility for permanently storing nuclear waste.

Population density is one critical lens through which other risks have to be assessed, says Laurent Stricker, a nuclear engineer and chairman of the World Association of Nuclear Operators:

The KANUPP plant in Karachi, Pakistan, has the most people — 8.2 million — living within 30 kilometres of a nuclear plant, although it has just one relatively small reactor with an output of 125 megawatts.

Next in the league, however, are much larger plants — Taiwan's 1,933-megawatt Kuosheng plant with 5.5 million people within a 30-kilometre radius and the 1,208-megawatt Chin Shan plant with 4.7 million; both zones include the capital city of Taipei.

- International Nuclear Events Scale
- Comparative Risk Assessment
- Statistical Risk Assessment
- Probabilistic risk assessment
    - o *Severe Accident Risks: An Assessment for Five U.S. Nuclear Power Plants* NUREG-1150 1991
    - o *Calculation of Reactor Accident Consequences* CRAC-II 1982
    - o *Rasmussen Report: Reactor Safety Study* WASH-1400 1975
    - o *The Brookhaven Report: Theoretical Possibilities and Consequences of Major Accidents in Large Nuclear Power Plants* WASH-740 1957

The AP1000 has a maximum core damage frequency of 5.09×10^{-7} per plant per year. The Evolutionary Power Reactor (EPR) has a maximum core damage frequency of 4×10^{-7} per plant per year. General Electric has recalculated maximum core damage frequencies per year per plant for its nuclear power plant designs:

BWR/4 — 1×10^{-5}

BWR/6 — 1×10^{-6}

ABWR — 2×10^{-7}

ESBWR — 3×10^{-8}

Beyond Design Basis Events

As Fukushima showed, external threats — such as earthquakes, tsunamis, fires, flooding, tornadoes and terrorist attacks — are some of the greatest risk factors for a serious nuclear accident. Yet, nuclear plant operators have normally considered these accident sequences (called 'beyond design basis' events) so unlikely that they have not built in complete safeguards.

Forecasting the location of the next earthquake or the size of the next tsunami is an imperfect art. Nuclear plants situated outside known geological danger zones "could pose greater accident threats in the event of an earthquake than those inside, as the former could have weaker protection built in". The Fukushima I plant, for example, was "located in an area designated, on Japan's seismic risk map, as having a relatively low chance of a large earthquake and tsunami; when the 2011 tsunami arrived, it was in excess of anything its engineers had planned for".

Transparency and Ethics

According to Stephanie Cooke, it is difficult to know what really goes on inside nuclear power plants because the industry is shrouded in secrecy. Corporations and governments control what information is made available to the public. Cooke says "when information is made available, it is often couched in jargon and incomprehensible prose".

Kennette Benedict has said that nuclear technology and plant operations continue to lack transparency and to be relatively closed to public view:

> *Despite victories like the creation of the Atomic Energy Commission, and later the Nuclear Regular Commission, the secrecy that began with the Manhattan Project has tended to permeate the civilian nuclear program, as well as the military and defence programs.*

In 1986, Soviet officials held off reporting the Chernobyl disaster for several days. The operators of the Fukushima plant, Tokyo Electric Power Co, were also criticised for not quickly disclosing information on radiation leaks from the plant. Russian President Dmitry Medvedev said there must be greater transparency in nuclear emergencies.

Historically many scientists and engineers have made decisions on behalf of potentially affected populations about whether a particular level of risk and uncertainty is acceptable for them. Many nuclear engineers and scientists that have made such decisions, even for good reasons relating to long term energy availability, now consider that doing so without informed consent is wrong, and that nuclear power safety and nuclear technologies should be based fundamentally on morality, rather than purely on technical, economic and business considerations.

Non-Nuclear Futures: The Case for an Ethical Energy Strategy is a 1975 book by Amory B. Lovins and John H. Price. The main theme of the book is that the most important parts of the nuclear power debate are not technical disputes but relate to personal values, and are the legitimate province of every citizen, whether technically trained or not.

Nuclear and Radiation Accidents—2011 Fukushima I Accidents

The 40-year-old Fukushima I Nuclear Power Plant, built in the 1970s, endured Japan's worst earthquake on record in March 2011 but had its power and back-up generators knocked out by a 7-metre tsunami that followed. The designers of the reactors at Fukushima did not anticipate that a tsunami generated by an earthquake would disable the backup systems that were supposed to stabilise the reactor after the earthquake. Nuclear reactors are such "inherently complex, tightly coupled systems that, in rare, emergency situations, cascading interactions will unfold very rapidly in such a way that human operators will be unable to predict and master them".

Lacking electricity to pump water needed to cool the atomic core, engineers vented radioactive steam into the atmosphere to release pressure, leading to a series of explosions that blew out concrete walls around the reactors. Radiation readings spiked around Fukushima as the disaster widened, forcing the evacuation of 200,000 people and causing radiation levels to rise on the outskirts of Tokyo, 135 miles (210 kilometres) to the south, with a population of 30 million.

Back-up diesel generators that might have averted the disaster were positioned in a basement, where they were overwhelmed by waves. The cascade of events at Fukushima had been foretold in a report published in the U.S. several decades ago:

> *The 1990 report by the U.S. Nuclear Regulatory Commission, an independent agency responsible for safety at the country's power plants, identified earthquake-induced diesel generator failure and power outage leading to failure of cooling systems as one of the "most likely causes" of nuclear accidents from an external event.*

While the report was cited in a 2004 statement by Japan's Nuclear and Industrial Safety Agency, it seems adequate measures to address the risk were not taken by Tokyo Electric. Katsuhiko Ishibashi, a seismology professor at Kobe University, has said that Japan's history of nuclear accidents stems from an overconfidence in plant engineering. In 2006, he resigned from a government panel on nuclear reactor safety, because the review process was rigged and "unscientific".

According to the International Atomic Energy Agency, Japan "underestimated the danger of tsunamis and failed to prepare adequate backup systems at the Fukushima Daiichi nuclear plant". This repeated a widely held criticism in Japan that "collusive ties between regulators and industry led to weak oversight and a failure to ensure adequate safety levels at the plant". The IAEA also said that the Fukushima disaster exposed the lack of adequate backup systems at the plant. Once power was completely lost, critical functions like the cooling system shut down. Three of the reactors "quickly overheated, causing meltdowns that eventually led to explosions, which hurled large amounts of radioactive material into the air".

Louise Frechette and Trevor Findlay have said that more effort is needed to ensure nuclear safety and improve responses to accidents:

The multiple reactor crises at Japan's Fukushima nuclear power plant reinforce the need for strengthening global instruments to ensure nuclear safety worldwide. The fact that a country that has been operating nuclear power reactors for decades should prove so alarmingly improvisational in its response and so unwilling to reveal the facts even to its own people, much less the International Atomic Energy Agency, is a reminder that nuclear safety is a constant work-in-progress.

David Lochbaum, chief nuclear safety officer with the Union of Concerned Scientists, has repeatedly questioned the safety of the Fukushima I Plant's General Electric Mark 1 reactor design, which is used in almost a quarter of the United States' nuclear fleet.

A report from the Japanese Government to the IAEA says the "nuclear fuel in three reactors probably melted through the inner containment vessels, not just the core". The report says the "inadequate" basic reactor design — the Mark-1 model developed by General Electric — included "the venting system for the containment vessels and the location of spent fuel cooling pools high in the buildings, which resulted in leaks of radioactive water that hampered repair work".

Following the Fukushima emergency, the European Union decided that reactors across all 27 member nations should undergo safety tests.

According to UBS AG, the Fukushima I nuclear accidents are likely to hurt the nuclear power industry's credibility more than the Chernobyl disaster in 1986:

The accident in the former Soviet Union 25 years ago 'affected one reactor in a totalitarian state with no safety culture,' UBS analysts including Per Lekander and Stephen Oldfield wrote in a report today. 'At Fukushima, four reactors have been out of control for weeks — casting doubt on whether even an advanced economy can master nuclear safety.'

The Fukushima Accident Exposed Some Troubling Nuclear Safety Issues:

Despite the resources poured into analysing crustal movements and having expert committees determine earthquake risk, for instance, researchers never considered the possibility of a magnitude-9 earthquake followed by a massive tsunami. The failure of multiple safety features on nuclear power plants has raised questions about the nation's engineering prowess. Government flip-flopping on acceptable levels of radiation exposure confused the public, and health professionals provided little guidance. Facing a dearth of reliable information on radiation levels, citizens armed themselves with dosimetres, pooled data, and together produced radiological contamination maps far more detailed than anything the government or official scientific sources ever provided.

1986 Chernobyl Disaster

Figure 5: Map showing Caesium-137 contamination in Belarus, Russia, and Ukraine as of 1996.

The Chernobyl disaster was a nuclear accident that occurred on 26 April 1986 at the Chernobyl Nuclear Power Plant in Ukraine. An explosion and fire released large quantities of radioactive contamination into the atmosphere, which spread over much of Western USSR and Europe. It is considered the worst nuclear power plant accident in history, and is one of only two classified as a level 7 event on the International Nuclear Event Scale (the other being the Fukushima Daiichi nuclear disaster). The battle to contain the contamination and avert a greater catastrophe ultimately involved over 500,000 workers and cost an estimated 18 billion rubles, crippling the Soviet economy. The accident raised concerns about the safety of the nuclear power industry, slowing its expansion for a number of years.

UNSCEAR has conducted 20 years of detailed scientific and epidemiological research on the effects of the Chernobyl accident. Apart from the 57 direct deaths in the accident itself, UNSCEAR predicted in 2005 that up to 4,000 additional cancer deaths related to the accident would appear "among the 600 000 persons receiving more significant exposures (liquidators working in 1986–87, evacuees, and residents of the most contaminated areas)". Russia, Ukraine, and Belarus have been burdened with the continuing and substantial decontamination and health care costs of the Chernobyl disaster.

Other Accidents

Serious nuclear and radiation accidents include the Chalk River accidents (1952, 1958 & 2008), Mayak disaster (1957), Windscale fire (1957), SL-1 accident (1961), Soviet submarine K-19 accident (1961), Three Mile Island accident (1979), Church Rock uranium mill spill (1979), Soviet submarine K-431 accident (1985), Goiânia accident (1987), Zaragoza radiotherapy accident (1990), Costa Rica radiotherapy accident (1996), Tokaimura nuclear accident (1999), Sellafield THORP leak (2005), and the Flerus IRE Cobalt-60 spill (2006).

Health Impacts

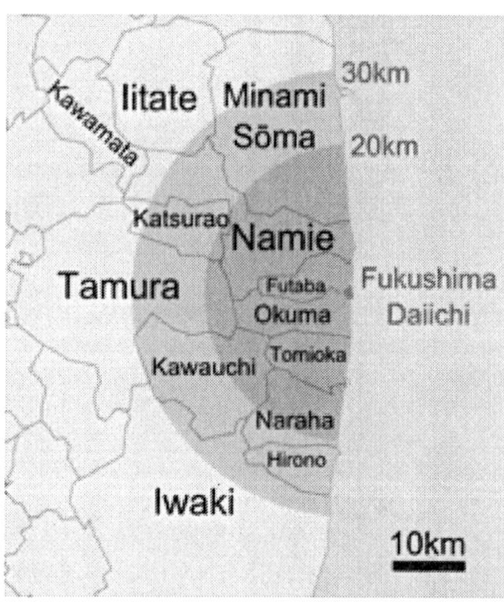

Figure 6: Japan towns, villages, and cities around the Fukushima Daiichi nuclear plant. The 20km and 30km areas had evacuation and sheltering orders, and additional administrative districts that had an evacuation order are highlighted.

In spite of accidents like Chernobyl, studies have shown that nuclear deaths are mostly in uranium mining and that nuclear energy has generated far fewer deaths than the high pollution levels that result from the use of conventional fossil fuels.

Stephanie Cooke says that it is not useful to make comparisons just in terms of number of deaths, as the way people live afterwards is also relevant, as in the case of the 2011 Japanese nuclear accidents:

You have people in Japan right now that are facing either not returning to their homes forever, or if they do return to their homes, living in a contaminated area for basically ever. And knowing that whatever food they eat, it might be contaminated and always living with this sort of shadow of fear over them that they will die early because of cancer and induced by Caesium or Strontium or some other radionuclide that's laced their vegetables. It affects millions of people, it affects our land, it affects our atmosphere, we know now the radio nuclides from Fukushima are going into the sea. It doesn't just kill now, it kills later, and it could kill centuries later. Because the stuff that that's depositing, doesn't just end, it has a long, long life. It's affecting future generations, it's not just affecting this generation. So I'm not a great fan of coal-burning. I don't think any of these great big massive plants that spew pollution into the air are good. But I don't think it's really helpful to make these comparisons just in terms of number of deaths. The Fukushima accident forced more than 80,000 residents to evacuate from neighbourhoods around the plant.

Developing Countries

There are concerns about developing countries "rushing to join the so-called nuclear renaissance without the necessary infrastructure, personnel, regulatory frameworks and safety culture". Some countries with nuclear aspirations, like Nigeria, Kenya, Bangladesh and Venezuela, have no significant industrial experience and will require at least a decade of preparation even before breaking ground at a reactor site.

The speed of the nuclear construction program in China has raised safety concerns. The challenge for the government and nuclear companies is to "keep an eye on a growing army of contractors and subcontractors who may be tempted to cut corners". China is advised to maintain nuclear safeguards in a business culture where quality and safety are sometimes sacrificed in favour of cost-cutting, profits, and corruption. China has asked for international assistance in training more nuclear power plant inspectors.

Design Basis Accident

A design basis accident (DBA) or *maximum credible accident (MCA)* is a postulated accident that a nuclear facility must be designed and built to withstand without loss to the systems, structures, and components necessary to assure public health and safety.

Subtypes of DBAs are:

- *design-basis criticality:* "A criticality accident that is the most severe design basis accident of that type applicable to the area under consideration."
- *design-basis earthquake (DBE):* "That earthquake for which the safety systems are designed to remain functional both during and after the event, thus assuring the ability to shut down and maintain a safe configuration."
- *design-basis event (DBE):* "A postulated event used in the design to establish the acceptable performance requirements of the structures, systems, and components."
- *design-basis explosion:* "An explosion that is the most severe design basis accident of that type applicable to the area under consideration."
- *design-basis fire:* "A fire that is the most severe design basis accident of this type. In postulating such a fire, failure of automatic and manual fire suppression provisions shall be assumed except for those safety class items or systems that are specifically designed to remain available (structurally or functionally) through the event."
- *design-basis flood:* "A flood that is the most severe design basis accident of that type applicable to the area under consideration."
- *design-basis tornado (DBT):* "A tornado that is the most severe design basis accident of that type applicable to the area under consideration."

Circumstances like the 2011 Tôhoku earthquake and tsunami were considered unlikely or impossible, and so the resulting Fukushima I nuclear accidents are described using this terminology as "beyond design basis accidents" or "non-design-basis accidents".

Accidents caused by poor design, failure to follow listed safety procedures or other forms of human error are not considered to be beyond design basis accidents. The terminology can be unclear, however, because a poorly handled design basis accident can result in conditions beyond what was considered likely, causing a beyond design basis accident. For this reason, some industry experts have criticized the use of design basis terminology. The Three Mile Island accident and the Chernobyl disaster are examples of design basis accidents becoming non design basis accidents largely due to failures to follow the written operating procedures.

Beyond Design Basis Events

As the Fukushima I nuclear accidents showed, external threats — such as earthquakes, tsunamis, fires, flooding, tornadoes and terrorist attacks — are some of the greatest risk factors for a serious nuclear accident. Some accident sequences are considered so unlikely that complete safeguards against the accident sequence have not been built in.

Forecasting the location of the next earthquake or the size of the next tsunami is an imperfect art. Nuclear plants situated outside known geological danger zones "could pose greater accident threats in the event of an earthquake than those inside, as the former could have weaker protection built in". The Fukushima I plant, for example, was "located in an area designated, on Japan's seismic risk map, as having a relatively low chance of a large earthquake and tsunami; when the 2011 tsunami arrived, it was in excess of anything its engineers had planned for".

Combined Cycle

In electric power generation a combined cycle is an assembly of heat engines that work in tandem off the same source of heat, converting it into mechanical energy, which in turn usually drives electrical generators. The principle is that the exhaust of one heat engine is used as the heat source for another, thus extracting more useful energy from the heat, increasing the system's overall efficiency. This works because heat engines are only able to use a portion of the energy their fuel generates (usually less than 50%).

The remaining heat (e.g., hot exhaust fumes) from combustion is generally wasted. Combining two or more thermodynamic cycles results in improved overall efficiency, reducing fuel costs. In stationary power plants, a successful, common combination is the Brayton cycle (in the form of a turbine burning natural gas or synthesis gas from coal) and the Rankine cycle (in the form of a steam power plant). Multiple stage turbine or steam cylinders are also common.

Historically successful combined cycles have used hot cycles with mercury vapour turbines, magnetohydrodynamic generators or molten carbonate fuel cells, with steam plants for the low temperature bottoming cycle. Bottoming cycles operating from a steam condenser's heat are theoretically possible, but uneconomical because of the very large, expensive equipment needed to extract energy from the small temperature differences between condensing steam and outside air or water. However, it is common in cold climates (such as Finland) to drive community heating systems from a power plant's condenser heat. Such cogeneration systems can yield theoretical efficiencies above 95%.

In automotive and aeronautical engines, turbines have been driven from the exhausts of Otto, Diesel, and Crower cycles. These are called turbo-compound engines. Aside from turbochargers, they have failed commercially because their mechanical complexity and weight are less economical than multistage turbines. Stirling engines are also a good theoretical fit for this application.

In a combined cycle power plant (CCPP), or combined cycle gas turbine (CCGT) plant, a gas turbine generator generates electricity and heat in the exhaust is used to make steam, which in turn drives a steam turbine to generate additional electricity. This last step enhances the efficiency of electricity generation. Many new gas power plants in North America and Europe are of this type. Such an arrangement used for marine propulsion is called *combined gas (turbine) and steam (turbine) (COGAS).*

Design Principle

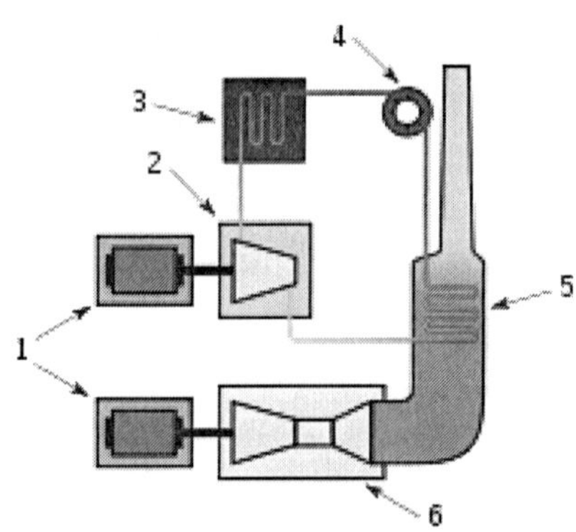

Figure 7: Working principle of a combined cycle power plant (Legend: 1-Electric generators, 2-Steam turbine, 3-Condenser, 4-Pump, 5-Boiler/heat exchanger, 6-Gas turbine)

In a thermal power station water is the working medium. High pressure steam requires strong, bulky components. High temperatures require expensive alloys made from nickel or cobalt, rather than inexpensive steel. These alloys limit practical steam temperatures to 655 °C while the lower temperature of a steam plant is fixed by the boiling point of water. With these limits, a steam plant has a fixed upper efficiency of 35 to 42%.

An open circuit gas turbine cycle has a compressor, a combustor and a turbine. For gas turbines the amount of metal that must withstand the high temperatures and pressures is small, and lower quantities of expensive materials can be used. In this type of cycle, the input temperature to the turbine (the firing temperature), is relatively high (900 to 1,400 °C). The output temperature of the flue gas is also high (450 to 650 °C). This is therefore high enough to provide heat for a second cycle which uses steam as the working fluid; (a Rankine cycle).

In a combined cycle power plant, the heat of the gas turbine's exhaust is used to generate steam by passing it through a heat recovery steam generator (HRSG) with a live steam temperature between 420 and 580 °C. The condenser of the Rankine cycle is usually cooled by water from a lake, river, sea or cooling towers. This temperature can be as low as 15 °C

In an automotive powerplant, an Otto, Diesel, Atkinson or similar engine would provide one part of the cycle and the waste heat would power a Rankine cycle steam or Stirling engine, which could either power ancillaries (such as the alternator) or be connected to the crankshaft by a turbo compounding system.

Typical Size of CCGT Plants

For large scale power generation, a typical set would be a 270 MW gas turbine coupled to a 130 MW steam turbine giving 400 MW. A typical power station might consist of between 1 and 6 such sets.

Plant size is important in the cost of the plant. The larger plant sizes benefit from economies of scale (lower initial cost per kilowatt) and improved efficiency.

A single shaft combined cycle plant comprises a gas turbine and a steam turbine driving a common generator. In a multi-shaft combined cycle plant, each gas turbine and each steam turbine has its own generator. The single shaft design provides slightly less initial cost and slightly better efficiency than if the gas and steam turbines had their own generators. The multi-shaft design enables 2 or more gas turbines to operate in conjunction with a single steam turbine, which can be more economical than a number of single shaft units.

The primary disadvantage of multiple stage combined cycle power plants is that the number of steam turbines, condensers and condensate systems - and perhaps the number of cooling towers and circulating water systems - increases to match the number of gas turbines. For a multiple shaft combined cycle power plant there is only one steam turbine, condenser and the rest of the heat sink for up to three gas turbines; only their size increases. Having only one large steam turbine and heat sink results in low cost because of economies of scale. A larger steam turbine also allows the use of higher pressures and results in a more efficient steam cycle. Thus the overall plant size and the associated number of gas turbines required have a major impact on whether a single shaft combined cycle power plant or a multiple shaft combined cycle power plant is more economical.

Gas turbines of about 150 MW size are already in operation manufactured by at least four separate groups - General Electric and its licensees, Alstom, Siemens, and Westinghouse/Mitsubishi. These groups are also developing, testing and/or marketing gas turbine sizes of about 200 MW. Combined cycle units are made up of one or more such gas turbines, each with a waste heat steam generator arranged to supply steam to a single steam tubine, thus forming a combined cycle block or unit. Typical Combined cycle block sizes offered by three major manufacturers (Alstom, General Electric and Siemens) are roughly in the range of 50 MW to 500 MW and costs are about $600/kW.

Efficiency of CCGT Plants

To avoid confusion, the efficiency of heat engines and power stations should be stated HHV (aka Gross Heating Value) or LCV (aka Net Heating value), and whether Gross output at the generator terminals or Net Output at the power station fence are being considered.

In general in service Combined Cycle efficiencies are over 50 percent on a lower heating value and Gross Output basis. Most combined cycle units, especially the larger units, have peak, steady state efficiencies of 55 to 59%. Research aimed at 1370°C (2500°F) turbine inlet temperature has led to even more efficient combined cycles and 60 percent efficiency has been reached in the combined cycle unit of Baglan Bay, a GE H-technology gas turbine with a NEM 3 pressure reheat boiler, utilising steam from the HRSG to cool the turbine blades. Siemens AG announced in may 2011 to have achieved a 60.75% net efficiency with a 578 megawatts SGT5-8000H gas turbine at the Irsching Power Station.

By combining both gas and steam cycles, high input temperatures and low output temperatures can be achieved. The efficiency of the cycles add, because they are powered by the same fuel source. So, a combined cycle plant has a thermodynamic cycle that operates between the gas-turbine's high firing temperature and the waste heat temperature from the condensers of the steam cycle. This large range means that the Carnot efficiency of the cycle is high. The actual efficiency, while lower than this, is still higher than that of either plant on its own. The actual efficiency achievable is a complex area.

The electric efficiency of a combined cycle power station, calculated as electric energy produced as a percent of the lower heating value of the fuel consumed, may be as high as 58 percent when operating new, ie unaged, and at continuous output which are ideal conditions. As with single cycle thermal units, combined cycle units may also deliver low temperature heat energy for industrial processes, district heating and other uses. This is called cogeneration and such power plants are often referred to as a Combined Heat and Power (CHP) plant.

Boosting Efficiency

The efficiency of CCGT and GT can be boosted by pre-cooling combustion air. This is practised in hot climates and also has the effect of increasing power output. This is achieved by evapourative cooling of water using a moist matrix placed in front of the turbine, or by using Ice storage air conditioning. The latter has the advantage of greater improvements due to the lower temperatures available. Furthermore, ice storage can be used as a means of load control or load shifting since ice can be made during periods of low power demand and, potentially in the future the anticipated high availability of other resources such as renewables during certain periods.

Supplementary Firing and Blade Cooling

Supplementary firing may be used in combined cycles (in the HRSG) raising exhaust temperatures from 600°C (GT exhaust) to 800 or even 1000°C. Using supplemental firing will however not raise the combined cycle efficiency for most combined cycles. For single boilers it may raise the efficiency if fired to 700- 750°C - for multiple boilers however, supplemental firing is often used to improve peak power production of the unit, or to enable higher steam production to compensate for failure of a second unit.

Maximum supplementary firing refers to the maximum fuel that can be fired with the oxygen available in the gas turbine exhaust. The steam cycle is conventional with reheat and regeneration. Hot gas turbine exhaust is used as the combustion air. Regenerative air preheater is not required. A fresh air fan which makes it possible to operate the steam plant even when the gas turbine is not in operation, increases the availability of the unit.

The use of large supplementary firing in Combined Cycle Systems with high gas turbine inlet temperatures causes the efficiency to drop. For this reason the Combined Cycle Plants with maximum supplementary firing are only of minimal importance today, in comparison to simple Combined Cycle installations. However, they have two advantages that is a) coal can be burned in the steam generator as the supplementary fuel, b) has very good part load efficiency.

The HRSG can be designed with supplementary firing of fuel after the gas turbine in order to increase the quantity or temperature of the steam generated. Without supplementary firing, the efficiency of the combined cycle power plant is higher, but supplementary firing lets the plant respond to fluctuations of electrical load. Supplementary burners are also called *duct burners*.

More fuel is sometimes added to the turbine's exhaust. This is possible because the turbine exhaust gas (flue gas) still contains some oxygen. Temperature limits at the gas turbine inlet force the turbine to use excess air, above the optimal stoichiometric ratio to burn the fuel. Often in gas turbine designs part of the compressed air flow bypasses the burner and is used to cool the turbine blades.

Supplementary firing raises the temperature of the exhaust gas from 800 to 900 degree Celsius. Relatively high flue gas temperature raises the condition of steam (84 bar, 525 degree Celsius) thereby improving the efficiency of steam cycle.

Fuel for Combined Cycle Power Plants

The turbines used in Combined Cycle Plants are commonly fuelled with natural gas. However, global natural gas reserves are expected to be fully consumed by 2070. Despite this fact, it is becoming the fuel of choice for an increasing amount of private investors and consumers because it is more versatile than coal or oil and can be used in 90% of energy applications. Chile which once depended on hydropower for

70% of its electricity supply, is now boosting its gas supplies to reduce reliance on its drought afflicted hydro dams. Similarly China is tapping its gas reserves to reduce reliance on coal, which is currently burned to generate 80% of the country's electric supply.

Where the extension of a gas pipeline is impractical or cannot be economically justified, electricity needs in remote areas can be met with small scale Combined Cycle Plants, using renewable fuels. Instead of natural gas, Combined Cycle Plants can be filled with biogas derived from agricultural and forestry waste, which is often readily available in rural areas.

Combined cycle plants are usually powered by natural gas, although fuel oil, synthesis gas or other fuels can be used. The supplementary fuel may be natural gas, fuel oil, or coal. Biofuels can also be used. Integrated solar combined cycle power stations combine the energy harvested from solar radiation with another fuel to cut fuel costs and environmental impact. The first such system to come online is Yazd power plant, Iran and more are under construction at Hassi R'mel, Algeria and Ain Beni Mathar, Morocco. Next generation nuclear power plants are also on the drawing board which will take advantage of the higher temperature range made available by the Brayton top cycle, as well as the increase in thermal efficiency offered by a Rankine bottoming cycle.

Low-Grade Fuel for Turbines: Gas turbines burn mainly natural gas and light oil. Crude oil, residual, and some distillates contain corrosive components and as such require fuel treatment equipment. In addition, ash deposits from these fuels result in gas turbine debating's of up to 15 percent they may still be economically attractive fuels however, particularly in combined-cycle plants.

Sodium and potassium are removed from residual, crude and heavy distillates by a water washing procedure. A simpler and less expensive purification system will do the same job for light crude and light distillates. A magnesium additive system may also be needed to reduce the corrosive effects if vanadium is present. Fuels requiring such treatment must have a separate fuel-treatment plant and a system of accurate fuel monitoring to assure reliable, low-maintenance operation of gas turbines.

Configuration of CCGT Plants

The combined-cycle system includes single-shaft and multi-shaft configurations. The single-shaft system consists of one gas turbine, one steam turbine, one generator and one Heat Recovery Steam

Generator (HRSG), with the gas turbine and steam turbine coupled to the single generator in a tandem arrangement on a single shaft. Key advantages of the single-shaft arrangement are operating simplicity, smaller footprint, and lower startup cost. Single-shaft arrangements, however, will tend to have less flexibility and equivalent reliability than multi-shaft blocks. Additional operational flexibility is provided with a steam turbine which can be disconnected, using an synchro-self-shifting (SSS) Clutch, for start up or for simple cycle operation of the gas turbine.

Multi-shaft systems have one or more gas turbine-generators and HRSGs that supply steam through a common header to a separate single steam turbine-generator. In terms of overall investment a multi-shaft system is about 5% higher in costs.

Single- and multiple-pressure non-reheat steam cycles are applied to combined-cycle systems equipped with gas turbines having rating point exhaust gas temperatures of approximately 540 °C or less. Selection of a single- or multiple-pressure steam cycle for a specific application is determined by economic evaluation which considers plant installed cost, fuel cost and quality, plant duty cycle, and operating and maintenance cost. Multiple-pressure reheat steam cycles are applied to combined-cycle systems with gas turbines having rating point exhaust gas temperatures of approximately 600 °C.

The most efficient power generation cycles are those with unfired HRSGs with modular pre-engineered components. These unfired steam cycles are also the lowest in cost. Supplementary-fired combined-cycle systems are provided for specific application. The primary regions of interest for cogeneration combined-cycle systems are those with unfired and supplementary fired steam cycles. These systems provide a wide range of thermal energy to electric power ratio and represent the range of thermal energy capability and power generation covered by the product line for thermal energy and power systems by Engr. Bilal Pervez

Integrated Gasification Combined Cycle (IGCC)

An integrated gasification combined cycle, or IGCC, is a power plant using synthesis gas (syngas). Syngas can be produced from a number of sources, including coal and biomass.

Integrated Solar Combined Cycle (ISCC)

An integrated solar combined cycle, or ISCC, is a power plant using solar thermal collectors. This is typically in the form of parabolic troughs.

Automotive Use

Any turbocharged engine is effectively a combined cycle with the turbo charger extracting extra energy from the exhaust gases. Theoretically, this extracted energy could be used to drive the wheels, but it is more practical to use it to force air into the engine which reduces the suction loss and thereby improves the efficiency overall. On large marine diesels turbo-compounding has been employed where the turbocharger physically pushes the engine around via some sort of gearing arrangement.

Combined cycles have traditionally only been used in large power plants. BMW, however, has proposed that automobiles use exhaust heat to drive steam turbines. This can even be connected to the car or truck's cooling system to save space and weight, but also to provide a condenser in the same location as the radiator and preheating of the water using heat from the engine block. However, stirling engines can also be used if light weight is a priority (as in a sports car or racing application), because they use a gas such as air rather than water as the working fluid.

It may be possible to use the pistons in a reciprocating engine for both combustion and steam expansion like in the Crower six stroke.

Aeromotive Use

Some versions of the Wright R-3350 were produced as turbo-compound engines. Three turbines driven by exhaust gases, known as *power recovery turbines*, provided nearly 600 hp at takeoff. These turbines added power to the engine crankshaft through bevel gears and fluid couplings.

Chapter 4

Heat Recovery Steam Generator

A heat recovery steam generator or HRSG is an energy recovery heat exchanger that recovers heat from a hot gas stream. It produces steam that can be used in a process or used to drive a steam turbine.

General Usage

A common application for an HRSG is in a combined-cycle power station, where hot exhaust from a gas turbine is fed to an HRSG to generate steam which in turn drives a steam turbine. This combination produces electricity more efficiently than either the gas turbine or steam turbine alone. Another application for an HRSG is in diesel engine combined cycle power plants, where hot exhaust from a diesel engine, as primary source of energy, is fed to an HRSG to generate steam which in turn drives a steam turbine. The HRSG is also an important component in cogeneration plants. Cogeneration plants typically have a higher overall efficiency in comparison to a combined cycle plant. This is due to the loss of energy associated with the steam turbine.

HRSGs

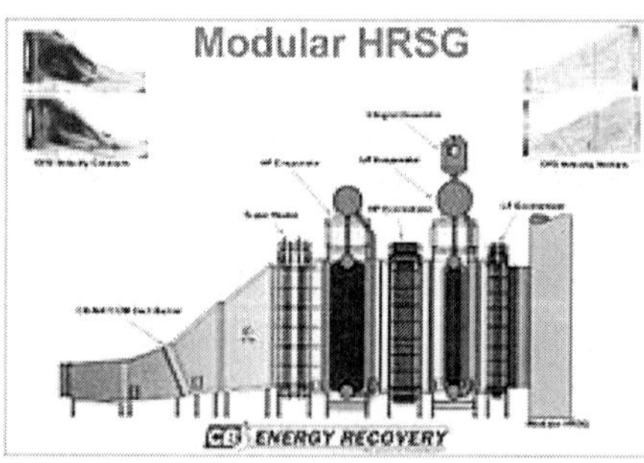

Figure 1: Modular HRSG GA

HRSGs consist of Four major components: the Evapourator, Superheater, Economizer and Water preheater. The different components are put together to meet the operating requirements of the unit.

Modular HRSGs can be categorised by a number of ways such as direction of exhaust gases flow or number of pressure levels. Based on the flow of exhaust gases, HRSGs are categorised into vertical and horizontal types. In horizontal type HRSGs, exhaust gas flows horizontally over vertical tubes whereas in vertical type HRSGs, exhaust gas flow vertically over horizontal tubes. Based on pressure levels, HRSGs can be categorised into single pressure and multi pressure. Single pressure HRSGs have only one steam drum and steam is generated at single pressure level whereas multi pressure HRSGs employ two (double pressure) or three (triple pressure) steam drums.

As such triple pressure HRSGs consist of three sections: an LP (low pressure) section, a reheat/IP (intermediate pressure) section, and an HP (high pressure) section. Each section has a steam drum and an evapourator section where water is converted to steam. This steam then passes through superheaters to raise the temperature and pressure past the saturation point.

Packaged HRSGs

Packaged HRSGs are designed to be shipped as a fully assembled unit from the factory. They can be used in waste heat or turbine (usually under 20 MW) applications. The packaged HRSG can have a water-cooled furnace, which allows for higher supplemental firing and better overall efficiency.

Variations

Some HRSGs include supplemental, or duct firing. These additional burners provide additional energy to the HRSG, which produces more steam and hence increases the output of the steam turbine. Generally, duct firing provides electrical output at lower capital cost. It is therefore often utilised for peaking operations.

HRSGs can also have diverter valves to regulate in the inlet flow into the HRSG. This allows the gas turbine to continue to operate when there is no steam demand or if the HRSG needs to be taken offline.

Emissions controls may also be located in the HRSG. Some may contain a Selective Catalytic Reduction system to reduce nitrogen oxides (a large contributor to the formation of smog and acid rain) and/or a catalyst to remove carbon monoxide. The inclusion of an SCR dramatically effects the layout of the HRSG. NOx catalyst performs best in temperatures between 650 °F (340 °C) and 750 °F (400 °C). This usually means that the evapourator section of the HRSG will have to be split and the SCR placed in between the two sections. Some low temperature NOx catalysts have recently come to market that allows for the SCR to be placed between the Evapourator and Economizer sections (350 °F - 500 °F (175 °C - 260 °C)).

Figure 2: Packaged HRSG

Once-through Steam Generators (OTSG)

A specialised type of HRSG without boiler drums is the once-through steam generator. In this design, the inlet feedwater follows a continuous path without segmented sections for economizers, evapourators and superheaters. This provides a high degree of flexibility as the sections are allowed to grow or contract based on the heat load being received from the gas turbine. The absence of drums allows for quick changes in steam production and fewer variables to control, and is ideal for cycling and base load operation.

Applications

- Heat recovery can be used extensively in energy projects.
- In the energy-rich Persian Gulf region, the steam from the HRSG is used for desalination plants.
- Universities are ideal candidates for HRSG applications. They can use a gas turbine to produce high reliability electricity for campus use. The HRSG can recover the heat from the gas turbine to produce steam/hot water for district heating or cooling.
- Large ocean vessels (e.g. Emma Maersk) make use of heat recovery

Manufacturers

- Vogt Power/Babcock Power
- IHI Corporation
- CMI Energy
- Victory Energy
- Nooter/Eriksen
- NEM
- Doosan Heavy Industries & Construction
- BHI Co.
- BHEL
- 3DCAD (I) PVT LTD
- Thermax Ltd
- Aalborg Engineering A/S
- DKME
- Innovative Steam Technologies
- Larsen & Toubro Ltd
- Deltak USA a world leader
- Greshams Ltd
- Rentech Boiler Inc.
- AZARAB Industery Co.
- DESCON Engineering Limited
- GREENS POWER EQUIPMENT INDIA PVT.LTD.

Hydrogen-cooled Turbogenerator

Figure 3: A power plant turbogenerator

Figure 4: A power plant turbogenerator; the yellow cylinder is the electrical generator, the blue device is the steam turbine, the yellow rectangular object in front is the excitation generator

A hydrogen-cooled turbo generator is a turbo generator with gaseous hydrogen as a coolant. Hydrogen-cooled turbo generators are designed to provide a low-drag atmosphere and cooling for single-shaft and combined-cycle applications in combination with steam turbines. Because of the high thermal conductivity and other favourable properties of hydrogen gas this is the most common type in its field today.

History

Based on the air-cooled turbo generator, gaseous hydrogen went into service as a coolant in the rotor and the stator in 1937 at Dayton, Ohio, in October by the Dayton Power & Light Co allowing an increase in specific utilisation and a 99.0 % efficiency.

Design

The use of gaseous hydrogen as a coolant is based on its properties, namely low density, high specific heat, and highest thermal conductivity at 0.168 W/(m·K) of all gases; it is 7-10 times better coolant than air. Another advantage of hydrogen is its easy detection by hydrogen sensors. A hydrogen-cooled generator can be significantly smaller, and therefore less expensive, than an air-cooled one. For stator cooling, water can be used.

Helium with a thermal-conductivity of 0.142 W/(m·K) was considered as coolant as well, however its high cost hinders its adoption despite its non-flammability.

Generally, three cooling approaches are used. For generators up to 300 MW, air cooling can be used. Between 250-450 MW hydrogen cooling is employed. For the highest power generators, up to 1800 MW, hydrogen and water cooling is used; the rotor is hydrogen-cooled, the stator windings are made of hollow copper tubes cooled with water circulating through them.

The generators produce high voltage; the choice of voltage depends on the tradeoff between demands to electrical insulation and demands to handling high electric current. For generators up to 40 MVA, the voltage is 6.3 kV; large generators with power above 1000 MW generate voltages up to 27 kV; voltages between 2.3-30 kV are used depending on the size of the generator. The generated power is left to a nearby station transformer, where it is converted to the electric power transmission line voltage (typically between 115 and 1200 kV).

To control the centrifugal forces at high rotational speeds, the rotor is mounted horizontally and its diametre typically does not exceed 1.25 metre; the required large size of the coils is achieved by their length. The generators operate typically at 3000 rpm for 50 Hz and 3600 rpm for 60 Hz systems for two-pole machines, half of that for four-pole machines.

The turbogenerator contains also a smaller generator producing direct current excitation power for the rotor coil. Older generators used dynamos and slip rings for DC injection to the rotor, but the moving mechanical contacts were subject to wear. Modern generators have the excitation generator on the same shaft as the turbine and main generator; the diodes needed are located directly on the rotor. The excitation current on larger generators can reach 10 kA. The amount of excitation power ranges between 0.5-3% of the generator output power.

The rotor usually contains caps or cage made of nonmagnetic material; its role is to provide a low impedance path for eddy currents which occur when the three phases of the generator are unevenly loaded. In such cases, eddy currents are generated in the rotor, and the resulting Joule heating could in extreme cases destroy the generator. Hydrogen gas is circulated in a closed loop to remove heat from the active parts then it is cooled by gas-to-water heat exchangers on the stator frame. The working pressure is up to 6 bar.

An on-line thermal conductivity detector (TCD) analyser is used with three measuring ranges. The first range (80-100% H_2) to monitor the hydrogen purity during normal operation. The second (0-100% H_2) and third (0-100% CO_2) measuring ranges allow safe opening of the turbines for maintenance.

Hydrogen has very low viscosity, a favourable property for reducing drag losses in the rotor; these losses can be significant, as the rotors have large diametre and high rotational speed. Every reduction in the purity of the hydrogen coolant increases windage losses in the turbine; as air is 14 times more dense than hydrogen, each 1% of air corresponds to about 14% increase of density of the coolant and the associated increase of viscosity and drag. A purity drop from 97 to 95% in a large generator can increase windage losses by 32%; this equals to 685 kW for a 907 MW generator. The windage losses also increase heat losses of the generator and the associated cooling problems.

The absence of oxygen in the atmosphere within significantly reduces the damage of the windings insulation by eventual corona discharges; these can be problematic as the generators typically operate at high voltage, often 20 kV.

The bearings have to be leak-tight. A hermetic seal, usually a liquid seal, is employed; a turbine oil at pressure higher than the hydrogen inside is typically used. A metal, e.g. brass, ring is pressed by springs onto the generator shaft, the oil is forced under pressure between the ring and the shaft; part of the oil flows into the hydrogen side of the generator, another part to the air side. The oil entrains a small amount of air; as the oil is recirculated, some of the air is carried over into the generator. This causes a gradual air contamination buildup and requires maintaining hydrogen purity. Scavenging systems are used for this purpose; gas (mixture of entrained air and hydrogen, released from the oil) is collected in the holding tank for the sealing oil, and released into the atmosphere; the hydrogen losses have to be replenished, either from gas cylinders or from on-site hydrogen generators. Degradation of bearings leads to higher oil leaks, which increases the amount of air transferred into the generator; increased oil consumption can be detected by a flow metre associated to each bearing.

Presence of water in hydrogen has to be avoided, as it causes deterioration to hydrogen cooling properties, corrosion of the generator parts, arcing in the high voltage windings, and reduces the lifetime of the generator. A desiccant-based dryer is usually included in the gas circulation loop, typically with a moisture probe in the dryer's outlet, sometimes also in its inlet. Presence of moisture is also an indirect evidence for air leaking into the generator compartment. Another option is optimising the hydrogen scavenging, so the dew point is kept within the generator manufacturer specifications. The water is usually introduced into the generator atmosphere as an impurity in the turbine oil; another route is via leaks in water cooling systems.

The flammability limits (4-75% of hydrogen in air at normal temperature, wider at high temperatures), its autoignition temperature at 571°C, its very low minimum ignition energy, and its tendency to form explosive mixtures with air, require provisions to be made for maintaining the hydrogen content within the generator above the upper or below the flammability limit at all times, and other hydrogen safety measures. When filled with hydrogen, overpressure has to be maintained as inlet of air into the generator could cause a dangerous explosion in confined space. The generator enclosure is purged before opening it for maintenance, and before refilling the generator with hydrogen. During shutdown, hydrogen is purged by an inert gas, then the inert gas is replaced by air; the opposite sequence is used before startup. Carbon dioxide or nitrogen can be used for this purpose, as they do not form combustible mixtures with hydrogen and are inexpensive. Gas purity sensors are used to indicate the end of the purging cycle, which shortens the startup and shutdown times and reduces consumption of the purging gas. Carbon dioxide is favoured as due to very high density difference it is easily displaced by hydrogen.

Hydrogen is often produced on-site in electrolyzers, as this reduces the need for stored amount of compressed hydrogen and allows storage in lower pressure tanks, with associated safety benefits and lower costs. Some gaseous hydrogen has to be kept for refilling the generator but it can be also generated on-site.

As technology evolves no materials susceptible to hydrogen embrittlement are used in the generator design. Not adhering to this can lead to equipment failure.

Precooled Jet Engine

A precooled jet engine is a concept for high speed jet engines that features a cryogenic fuel-cooled heat exchanger immediately after the air intake, to precool the air entering the engine. After gaining heat and vapourising in the heat exchanger system, the fuel (e.g. H_2) is burnt in the combustor. Precooled jet engines have never flown, but are predicted to have much higher thrust and efficiency at speeds up to Mach 5.5. Precooled jet engines were described by Robert P. Carmichael in 1955.

Unlike Liquid Air Cycle Engines (LACE), precooled engines simply cool, but do not liquify the air.

A potential application for a precooled engine is as part of the powerplant for a space launcher vehicle, or for a very long range, very high speed aircraft.

Advantages of Precoolers

One main advantage of precooling is that, for a given overall pressure ratio, there is a significant reduction in compressor delivery temperature (T3), which delays the onset of the T3 limit as flight speed increases. Consequently sea-level conditions (corrected flow) can be maintained after the precooler over a very wide range of flight speeds, thus maximising net thrust even at high speeds.

Another advantage is that the compressor and ducting after the inlet is subject to much lower and more consistent temperatures, and hence may be made of light alloys. This greatly reduces the weight of the engine, which further improves the thrust/weight ratio.

The fuel typically proposed for precooled jet engines is usually hydrogen, since hydrogen is liquid at deeply cryogenic temperatures, and over its useful range has a very high total specific heat capacity, including the latent heat of vapourisation, higher than water.

However, the low density of liquid hydrogen has negative effects on the rest of the vehicle, and the vehicle physically becomes very large, although the weight on the undercarriage and wing loading may remain low.

Another issue is that the amount of hydrogen needed to cool the air is high, (and using enough to liquifying it which would avoid the need for a turbocompressor greatly reduces the Isp), more than can be burnt in the engine. This means that the engine has a lot of excess hydrogen that must be disposed of, and precooled jet engines may need to use ramjet-like burners around the central precooled engine core.

History and Status of Precoolers

Precoolers have Never Flown: The state of the art is that a lab scale precooler has been developed in the UK, and tested successfully under representative conditions. Rejection of condensed water has been achieved, which is an important achievement in that it prevents ice from blocking the system.

Similar Systems

Precoolers were first proposed as part of the research in America on Project Suntan- a liquid hydrogen fuelled aircraft. Robert P.

Carmichael in 1955 devised several engine cycles that could be used with hydrogen fuel, and this was one.

Interest in precooled engines saw a brief emergence in the UK in 1982, when Alan Bond (formerly of the Blue Streak missile project) created a LACE-like design he called SATAN. The primary difference of these systems is that the air is only cooled, rather than liquified, and thus the gases are not actually separated, apparently giving significantly greater overall performance, due to a reduction in the amount of hydrogen used for cooling.

At the same time, John Scott and Bob Parkinson at British Aerospace had started some preliminary work on reusable launch systems. The two teams met and created HOTOL, which would use the BAe designed airframe with a Rolls Royce version of Bond's engine, known as the RB545. In 1986 the project was given an official go-ahead to the tune of 2 million pounds for research, but the program was later killed in 1989 when the project encountered problems and the government ended funding.

The principal designers continued development on their own, but the RB545 had been classified top secret and could not be used. Instead Bond developed another version that is more advanced, known as SABRE (ostensibly for *Synergic Air BReathing Engine*) which is meant for their Skylon design. Funding has not been terribly forthcoming (surprising, considering the design's potential to power an SSTO craft) and development continues at a relatively low level; papers and laboratory work are ongoing. A study of an aircraft powered by a SABRE class engine is underway under LAPCAT partially funded by the EU looking towards hypersonic intercontinental travel (Brussels to Sydney in 2–4 hours non stop).

Liquid Air Cycle Engine

A Liquid Air Cycle Engine (LACE) is a type of spacecraft propulsion engine that attempts to increase its efficiency by gathering part of its oxidiser from the atmosphere. A liquid air cycle engine uses cryogenic hydrogen fuel to liquify the air.

In a LOX/LH2 bipropellant rocket the liquid oxygen needed for combustion is the majority of the weight of the spacecraft on lift-off, so if some of this can be collected from the air on the way, it might dramatically lower the take-off weight of the spacecraft.

LACE was studied to some extent in the USA during the late 1950s and early 1960s, and by late 1960 Marquardt had a testbed system running. However, as NASA moved to ballistic capsules during Project Mercury, funding for research into winged vehicles slowly disappeared, and LACE work along with it.

Principle of Operation

Conceptually LACE works by compressing and then quickly liquefying the air. Compression is achieved through the ram-air effect in an intake similar to that found on a high-speed aircraft like Concorde, where intake ramps create shock waves that compress the air. The LACE design then blows the compressed air over a heat exchanger, in which the liquid hydrogen fuel is flowing. This rapidly cools the air, and the various constituents quickly liquefy. By careful mechanical arrangement the liquid oxygen can be removed from the other parts of the air, notably water, nitrogen and carbon dioxide, at which point it can be fed into the engine as usual. The hydrogen is so much lighter than oxygen that the now-warm hydrogen is often dumped overboard instead of being re-used as fuel, at a net gain.

The use of a winged launch vehicle allows using lift rather than thrust to overcome gravity, which greatly reduces gravity losses.

One issue with the LACE system is that in order to appreciably reduce the mass of the oxygen carried at launch, a LACE vehicle needs to spend more time in the lower atmosphere to collect enough oxygen to supply the engines. This leads to greatly increased vehicle heating and drag losses, which therefore increases fuel consumption to offset the drag losses and the additional mass of the thermal protection system. This increased fuel consumption offsets somewhat the savings in oxidiser mass; these losses are in turn offset by the higher Isp (Specific impulse) of the air-breathing engine. Thus, the engineering trade-offs involved are quite complex, and highly sensitive to the design assumptions made.

Other issues are introduced by the relative material and logistical properties of LOX versus LH2. LOX is quite cheap; LH2 is nearly two orders of magnitude more expensive. LOX is dense (1.141 kg/L), whereas LH2 has a very low density (0.0678 kg/L) and is therefore very bulky. (The extreme bulkiness of the LH2 tankage tends to increase vehicle drag by increasing the vehicle's frontal area.) Finally, LOX tanks are relatively lightweight and fairly cheap, while the deep cryogenic nature and extreme physical properties of LH2 mandate that LH2 tanks and plumbing must be large and use heavy, expensive, exotic materials and insulation. Hence, much as the costs of using LH2 rather than a hydrocarbon fuel may well outweigh the Isp benefit of using LH2 in a Single-Stage-To-Orbit rocket, the costs of using more LH2 as a propellant and air-liquefaction coolant in LACE may well outweigh the benefits gained by not needing to carry as much LOX on-board.

Most significantly, the LACE system is far heavier than a pure rocket engine having the same thrust (air-breathing engines of almost all types have relatively poor thrust-to-weight ratios compared to rockets), and the performance of launch vehicles of all types is particularly affected by increases in vehicle dry mass (such as engines) that must be carried all the way to orbit, as opposed to oxidiser mass that would be burnt off over the course of the flight.

Moreover, the lower thrust-to-weight ratio of an air-breathing engine as compared to a rocket significantly decreases the launch vehicle's maximum possible acceleration, and increases gravity losses since more time must be spent to accelerate to orbital velocity. Also, the higher inlet and airframe drag losses of a lifting, air-breathing vehicle launch trajectory as compared to a pure rocket on a ballistic launch trajectory introduces an additional penalty term $1 / (1 + 1 / ((L / D)(a / g)))$ into the rocket equation known as the "Air-Breather's Burden."

The "air-breather's burden" term implies that unless the lift-to-drag ratio (L / D) and the acceleration of the vehicle as compared to gravity (a / g) are both implausibly large for a hypersonic air-breathing vehicle, the advantages of the higher Isp of the air-breathing engine and the savings in LOX mass are largely lost.

Thus, the advantages, or disadvantages, of the LACE design continue to be a matter of some debate.

History

LACE was studied to some extent in the United States of America during the late 1950s and early 1960s, where it was seen as a "natural" fit for a winged spacecraft project known as the Aerospaceplane. At the time the concept was known as LACES, for *Liquid Air Collection Engine System*, or ACES for *Air Collection and Enrichment System*. Both Marquardt and General Dynamics were involved in the research, and by late 1960 Marquardt had a testbed system running that was capable of running a 275 lbf (1.2 kN) thrust engine for minutes at a time. However, as NASA moved to ballistic capsules during Project Mercury, funding for research into winged vehicles slowly disappeared, and LACE along with it.

Scramjet

A scramjet (*supersonic combustion ramjet*) is a variant of a ramjet airbreathing jet engine in which combustion takes place in supersonic airflow. As in ramjets, a scramjet relies on high vehicle speed to forcefully compress and decelerate the incoming air before combustion (hence *ram*jet), but whereas a ramjet decelerates the air to subsonic velocities before combustion, airflow in a scramjet is supersonic throughout the entire engine. This allows the scramjet to efficiently operate at extremely high speeds: theoretical projections place the top speed of a scramjet between Mach 12 and Mach 24, which is near orbital velocity. The fastest air-breathing aircraft is a SCRAM jet design, the NASA X-43A which reached Mach 9.8. For comparison, the second fastest air-breathing aircraft, the manned SR-71 Blackbird, has a cruising speed of Mach 3.2.

The scramjet is composed of three basic components: a converging inlet, where incoming air is compressed and decelerated; a combustor, where gaseous fuel is burned with atmospheric oxygen to produce heat; and a diverging nozzle, where the heated air is accelerated to produce thrust. Unlike a typical jet engine, such as a turbojet or turbofan engine, a scramjet does not use rotating, fan-like components to compress the air; rather, the achievable speed of the aircraft moving through the atmosphere causes the air to compress within the inlet. As such, no moving parts are needed in a scramjet. In comparison, typical turbojet engines require inlet fans, multiple stages of rotating compressor fans, and multiple rotating turbine stages, all of which add weight, complexity, and a greater number of failure points to the engine.

Due to the nature of their design, scramjet operation is limited to near-hypersonic velocities. As they lack mechanical compressors, scramjets require the high kinetic energy of a hypersonic flow to compress the incoming air to operational conditions. Thus, a scramjet-powered vehicle must be accelerated to the required velocity by some other means of propulsion, such as turbojet, railgun, or rocket engines. In the flight of the experimental scramjet-powered Boeing X-51A, the test craft was lifted to flight altitude by a Boeing B-52 Stratofortress before being released and accelerated by a detachable rocket to near Mach 4.5.

While scramjets are conceptually simple, actual implementation is limited by extreme technical challenges. Hypersonic flight within the atmosphere generates immense drag, and temperatures found on the aircraft and within the engine can be much greater than that of the surrounding air. Maintaining combustion in the supersonic flow presents additional challenges, as the fuel must be injected, mixed, ignited, and burned within milliseconds. While scramjet technology has been under development since the 1950s, only very recently have scramjets successfully achieved powered flight.

History

During and after World War II, tremendous amounts of time and effort were put into researching high-speed jet- and rocket-powered aircraft. The Bell X-1 attained supersonic flight in 1947 and, by the early 1960s, rapid progress towards faster aircraft suggested that operational aircraft would be flying at "hypersonic" speeds within a few years. Except for specialised rocket research vehicles like the North American X-15 and other rocket-powered spacecraft, aircraft top speeds have remained level, generally in the range of Mach 1 to Mach 3.

In the 1950s and 1960s a variety of experimental scramjets engines were built and ground tested in US and the UK. In 1964, Dr. Frederick S. Billig and Dr. Gordon L. Dugger submitted a patent application for a supersonic combustion ramjet based on Billig's Ph.D. thesis. This patent was issued in 1981 following the removal of an order of secrecy.

In 1981 tests were made in Australia under the guidance of Professor Ray Stalker in the T3 ground test facility at ANU.

First successful flight test of Scramjet was performed by Russia in 1991. It was axisymmetric hydrogen-fuelled dual-mode scramjet developed by Central Institute of Aviation Motors (CIAM), Moscow from late 1970s. The scramjet flight was ûown captive-carry atop the SA-5 surface-to-air missile that included an experiment ûight support unit known as the "Hypersonic Flying Laboratory" (HFL), "Kholod". Then during the period since 1992 up to 1998 an additional 6 flight tests of the axisymmetric high-speed scramjet-demonstrator were conducted by CIAM together with France and then with NASA, USA. Maximum flight velocity greater than Mach 6.4 was achieved and Scramjet operation during 77 seconds was demonstrated. These flight test series also provided insight into autonomous hypersonic flight controls.

Investigations of hypersonic flight have continued over the past few decades. At present, the US military and NASA have formulated a "National Hypersonics Strategy" to investigate a range of options for hypersonic flight.

Different U.S. organisations have accepted hypersonic flight as a common goal. The U.S. Army desires hypersonic missiles that can attack mobile missile launchers quickly. NASA believes hypersonics could help develop economical, reusable launch vehicles. The Air Force is interested in a range of hypersonic systems, from air-launched cruise missiles to orbital spaceplanes, that the service believes could bring about a true "aerospace force."

The problem is complicated by the release of previously classified material and by partial publication, where claims are made, but specific parts of an experiment are kept secret. Additionally experimental difficulties in verifying that supersonic combustion actually occurred, or that actual net thrust was produced mean that at least four consortia have legitimate claims to "firsts", with several nations and institutions involved in each consortium. On June 15, 2007, the US Defence Advanced Research Project Agency (DARPA) and the Australian Defence Science and Technology Organisation (DSTO), announced a successful scramjet flight at Mach 10 using rocket engines to boost the test vehicle to hypersonic speeds, at the Woomera Rocket Range in Central Australia.

No scramjet powered vehicle has yet been produced outside an experimental program.

Design Principles

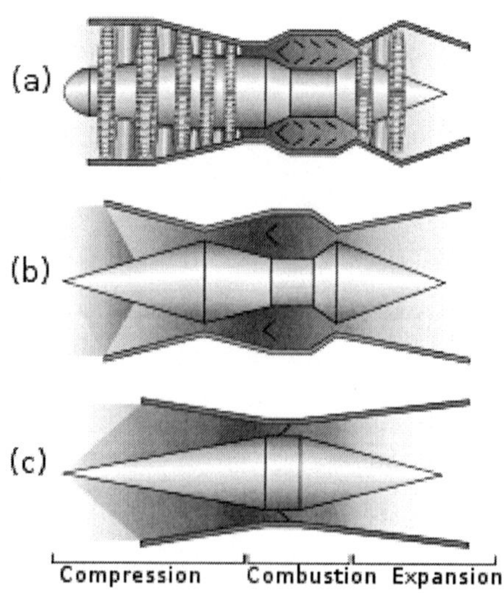

Figure 5: The compression, combustion, and expansion regions of: (a) turbojet, (b) ramjet, and (c) scramjet engines.

Scramjet engines are a type of jet engine, and rely on the combustion of fuel and an oxidiser to produce thrust. Similar to conventional jet engines, scramjet-powered aircraft carry the fuel on board, and obtain the oxidiser by the ingestion of atmospheric oxygen (as compared to rockets, which carry both fuel and an oxidising agent). This requirement limits scramjets to suborbital atmospheric flight, where the oxygen content of the air is sufficient to maintain combustion.

Basic Principles

Scramjets are designed to operate in the hypersonic flight regime, beyond the reach of turbojet engines, and, along with ramjets, fill the gap between the high efficiency of turbojets and the high speed of rocket engines. Turbomachinery-based engines, while highly efficient at subsonic speeds, become increasingly inefficient at transonic speeds, as the compressor fans found in turbojet engines require subsonic speeds to operate. While the flow from transonic to low supersonic speeds can be decelerated to these conditions, doing so at supersonic speeds results in a tremendous increase in temperature and a loss in the total enthalpy of the flow. Around Mach 3–4, turbomachinery is no longer useful, and ram-style compression becomes the preferred method.

Ramjets utilise high-speed characteristics of air to literally 'ram' air through an inlet nozzle into the combustor. At transonic and supersonic flight speeds, the air upstream of the nozzle is not able to move out of the way quickly enough, and is compressed within the nozzle before being diffused into the combustor. Combustion in a ramjet takes place at subsonic velocities, similar to turbojets, but the combustion products are then accelerated through a convergent-divergent nozzle to supersonic speeds. As they have no mechanical means of compression, ramjets cannot start from a standstill, and generally do not achieve sufficient compression until supersonic flight. The lack of intricate turbomachinery allows ramjets to deal with the temperature rise associated with decelerating a supersonic flow to subsonic speeds, but this only goes so far: at near-hypersonic velocities, the temperature rise and inefficiencies discourage decelerating the flow to the magnitude found in ramjet engines.

Scramjet engines operate on the same principles as ramjets, but do not decelerate the flow to subsonic velocities. Rather, a scramjet combustor is supersonic: the inlet decelerates the flow to a lower Mach number for combustion, after which it is accelerated to an even higher Mach number through the nozzle. By limiting the amount of deceleration, temperatures within the engine are kept at a tolerable level, from both a material and combustive standpoint. Even so, current scramjet technology requires the use of high-energy fuels and active cooling schemes to maintain sustained operation, often using hydrogen and regenerative cooling techniques.

Theory

All scramjet engines have an intake, which compresses the incoming air, fuel injectors, a combustion chamber, and a divergent thrust nozzle. Sometimes engines also include a region which acts as a flame holder, although the high stagnation temperatures mean that an area of focused waves may be used, rather than a discrete engine part as seen in turbine engines. Other engines use pyrophoric fuel additives, such as silane, to avoid such issues. An isolator between the inlet and combustion chamber is often included to improve the homogeneity of the flow in the combustor and to extend the operating range of the engine.

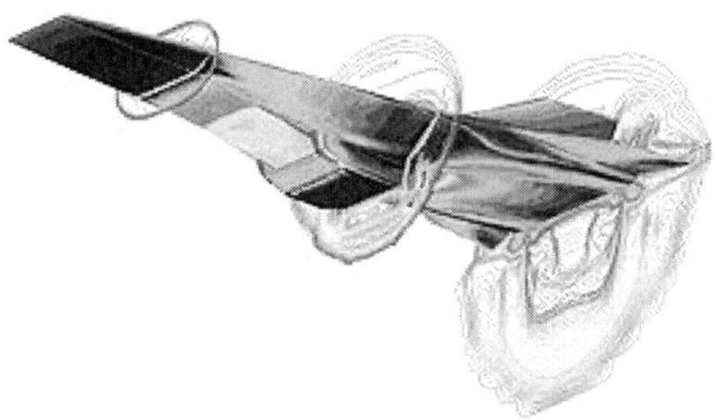

Figure 6: Computational fluid dynamics (CFD) image of the NASA X-43A with scramjet attached to the underside at Mach 7

A scramjet is reminiscent of a ramjet. In a typical ramjet, the supersonic inflow of the engine is decelerated at the inlet to subsonic speeds and then reaccelerated through a nozzle to supersonic speeds to produce thrust. This deceleration, which is produced by a normal shock, creates a total pressure loss which limits the upper operating point of a ramjet engine.

For a scramjet, the kinetic energy of the freestream air entering the scramjet engine is large comparable to the energy released by the reaction of the oxygen content of the air with a fuel (say hydrogen). Thus the heat released from combustion at Mach 25 is around 10% of the total enthalpy of the working fluid. Depending on the fuel, the kinetic energy of the air and the potential combustion heat release will be equal at around Mach 8. Thus the design of a scramjet engine is as much about minimising drag as maximising thrust.

This high speed makes the control of the flow within the combustion chamber more difficult. Since the flow is supersonic, no upstream influence propagates within the freestream of the combustion chamber. Thus throttling of the entrance to the thrust nozzle is not a usable control technique. In effect, a block of gas entering the combustion chamber must mix with fuel and have sufficient time for initiation and reaction, all the while travelling supersonically through the combustion chamber, before the burned gas is expanded through the thrust nozzle. This places stringent requirements on the pressure and temperature of the flow, and requires that the fuel injection and mixing be extremely efficient. Usable dynamic pressures lie in the range 20 to 200 kPa (0.2-2 bar), where

$$q = \frac{1}{2}\rho v^2$$

where

q is the dynamic pressure of the gas

ρ (rho) is the density of the gas

v is the velocity of the gas

To keep the combustion rate of the fuel constant, the pressure and temperature in the engine must also be constant. This is problematic because the airflow control systems that would facilitate this are not physically possible in a scramjet launch vehicle due to the large speed and altitude range involved, meaning that it must travel at an altitude specific to its speed. Because air density reduces at higher altitudes, a scramjet must climb at a specific rate as it accelerates to maintain a constant air pressure at the intake. This optimal climb/descent profile is called a "constant dynamic pressure path". It is thought that scramjets might be operable up to an altitude of 75 km.

Fuel injection and management is also potentially complex. One possibility would be that the fuel be pressurised to 100 bar by a turbo pump, heated by the fuselage, sent through the turbine and accelerated to higher speeds than the air by a nozzle. The air and fuel stream are crossed in a comb like structure, which generates a large interface. Turbulence due to the higher speed of the fuel leads to additional mixing. Complex fuels like kerosene need a long engine to complete combustion.

The minimum Mach number at which a scramjet can operate is limited by the fact that the compressed flow must be hot enough to burn the fuel, and have pressure high enough that the reaction be finished before the air moves out the back of the engine. Additionally, in order to be called a scramjet, the compressed flow must still be supersonic after combustion. Here two limits must be observed: Firstly, since when a supersonic flow is compressed it slows down, the level of compression must be low enough (or the initial speed high enough) not to slow the gas below Mach 1. If the gas within a scramjet goes below Mach 1 the engine will "choke", transitioning to subsonic flow in the combustion chamber. This effect is well known amongst experimenters on scramjets since the waves caused by choking are easily observable. Additionally, the sudden increase in pressure and temperature in the engine can lead to an acceleration of the combustion, leading to the combustion chamber exploding.

Secondly, the heating of the gas by combustion causes the speed of sound in the gas to increase (and the Mach number to decrease) even though the gas is still travelling at the same speed. Forcing the speed of air flow in the combustion chamber under Mach 1 in this way is called "thermal choking". It is clear that a pure scramjet can operate at Mach numbers of 6-8, but in the lower limit, it depends on the definition of a scramjet. Certainly there are designs where a ramjet transforms into a scramjet over the Mach 3-6 range (Dual-mode scramjets). In this range however, the engine is still receiving significant thrust from subsonic combustion of "ramjet" type.

The high cost of flight testing and the unavailability of ground facilities have hindered scramjet development. A large amount of the experimental work on scramjets has been undertaken in cryogenic facilities, direct-connect tests, or burners, each of which simulates one aspect of the engine operation. Further, vitiated facilities, storage heated facilities, arc facilities and the various types of shock tunnels each have limitations which have prevented perfect simulation of scramjet operation. The HyShot flight test showed the relevance of the 1:1 simulation of conditions in the T4 and HEG shock tunnels, despite having cold models and a short test time. The NASA-CIAM tests provided similar verification for CIAM's C-16 V/K facility and the Hyper-X project is expected to provide similar verification for the Langley AHSTF, CHSTF and 8 ft (2.4 m) HTT.

Computational fluid dynamics has only recently reached a position to make reasonable computations in solving scramjet operation problems. Boundary layer modelling, turbulent mixing, two-phase flow, flow separation, and real-gas aerothermodynamics continue to be problems on the cutting edge of CFD. Additionally, the modelling of kinetic-limited combustion with very fast-reacting species such as hydrogen makes severe demands on computing resources. Reaction schemes are numerically stiff requiring reduced reaction schemes.

Much of scramjet experimentation remains classified. Several groups including the US Navy with the SCRAM engine between 1968–1974, and the Hyper-X program with the X-43A have claimed successful demonstrations of scramjet technology. Since these results have not been published openly, they remain unverified and a final design method of scramjet engines still does not exist.

The final application of a scramjet engine is likely to be in conjunction with engines which can operate outside the scramjet's operating range. Dual-mode scramjets combine subsonic combustion with supersonic combustion for operation at lower speeds, and rocket-based combined cycle (RBCC) engines supplement a traditional rocket's propulsion with a scramjet, allowing for additional oxidiser to be added to the scramjet flow. RBCCs offer a possibility to extend a scramjet's operating range to higher speeds or lower intake dynamic pressures than would otherwise be possible.

Advantages and Disadvantages of Scramjets

1. Does not have to carry oxygen
2. No rotating parts makes it easier to manufacture
3. Has a higher specific impulse (change in momentum per unit of propellant) than a conventional engine; could provide between 1000 and 4000 seconds, while a rocket only provides 600 seconds or less
4. Higher speed could mean cheaper access to outer space in the future

Special Cooling and Materials

Unlike a rocket that quickly passes mostly vertically through the atmosphere or a turbojet or ramjet that flies at much lower speeds, a hypersonic airbreathing vehicle optimally flies a "depressed trajectory", staying within the atmosphere at hypersonic speeds. Because scramjets have only mediocre thrust-to-weight ratios, acceleration would be limited. Therefore time in the atmosphere at hypersonic speed would be considerable, possibly 15–30 minutes.

Similar to a reentering space vehicle, heat insulation would be a formidable task. The time in the atmosphere would be greater than that for a typical space capsule, but less than that of the space shuttle.

New materials offer good insulation at high temperature, but they often sacrifice themselves in the process. Therefore studies often plan on "active cooling", where coolant circulating throughout the vehicle skin prevents it from disintegrating. Often the coolant is the fuel itself, in much the same way that modern rockets use their own fuel and oxidiser as coolant for their engines. All cooling systems add weight and complexity to a launch system. The cooling of scramjets in this way may result in greater efficiency, as heat is added to the fuel prior to entry into the engine, but result in increased complexity and weight which ultimately could outweigh any performance gains.

Vehicle Performance

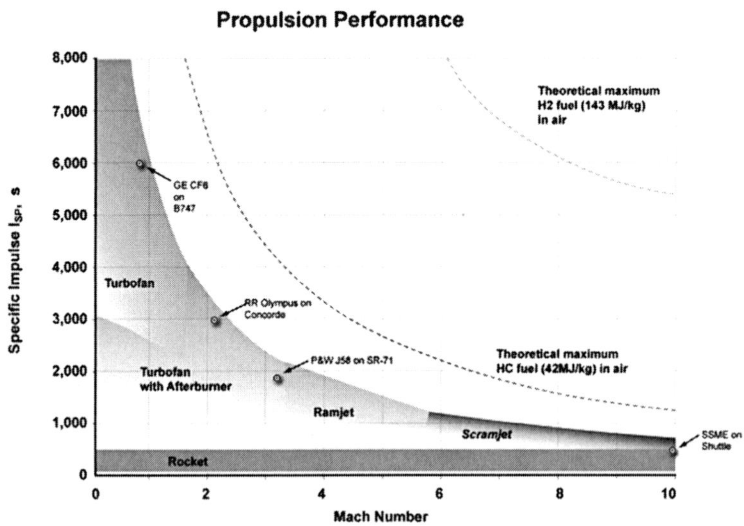

Figure 7: The specific impulse of various engines

The performance of a launch system is complex and depends greatly on its weight. Normally craft are designed to maximise range (R), orbital radius (R) or payload mass fraction (Γ) for a given engine and fuel. This results in tradeoffs between the efficiency of the engine (takeoff fuel weight) and the complexity of the engine (takeoff dry weight), which can be expressed by the following:

$$\Pi_e + \Pi_f + \frac{1}{\Gamma} = 1$$

Where :

- $\Pi_e = \frac{m_{\text{empty}}}{m_{\text{initial}}}$ is the empty mass fraction, and represents the weight of the superstructure, tankage and engine.
- $\Pi_f = \frac{m_{\text{fuel}}}{m_{\text{initial}}}$ is the fuel mass fraction, and represents the weight of fuel, oxidiser and any other materials which are consumed during the launch.
- $\Gamma = \frac{m_{\text{initial}}}{m_{\text{payload}}}$ is initial mass ratio, and is the inverse of the payload mass fraction. This represents how much payload the vehicle can deliver to a destination.

A scramjet increases the mass of the engine Π_e over a rocket, and decreases the mass of the fuel Π_f. It can be difficult to decide whether this will result in an increased Γ (which would be an increased payload delivered to a destination for a constant vehicle takeoff weight). The logic behind efforts driving a scramjet is (for example) that the reduction in fuel decreases the total mass by 30%, while the increased engine weight adds 10% to the vehicle total mass. Unfortunately the uncertainty in the calculation of any mass or efficiency changes in a vehicle is so great that slightly different assumptions for engine efficiency or mass can provide equally good arguments for or against scramjet powered vehicles.

Additionally, the drag of the new configuration must be considered. The drag of the total configuration can be considered as the sum of the vehicle drag (D) and the engine installation drag (D_e). The installation drag traditionally results from the pylons and the coupled flow due to the engine jet, and is a function of the throttle setting. Thus it is often written as:

$D_e = \phi_e F$ Where:

- ϕ_e is the loss coefficient
- F is the thrust of the engine

For an engine strongly integrated into the aerodynamic body, it may be more convenient to think of (D_e) as the difference in drag from a known base configuration.

The overall engine efficiency can be represented as a value between 0 and 1 (η_0), in terms of the specific impulse of the engine:

$$\eta_0 = \frac{g_0 V_0}{h_{PR}} \cdot I_{sp} = \frac{\text{Thrust Power}}{\text{Chemical energy rate}}$$

Where:

- g_0 is the acceleration due to gravity at ground level
- V_0 is the vehicle speed
- I_{sp} is the specific impulse
- h_{PR} is fuel heat of reaction

Specific impulse is often used as the unit of efficiency for rockets, since in the case of the rocket, there is a direct relation between specific impulse, specific fuel consumption and exhaust velocity. This direct relation is not generally present for airbreathing engines, and so specific impulse is less used in the literature. Note that for an airbreathing engine, both η_0 and I_{sp} are a function of velocity.

The specific impulse of a rocket engine is independent of velocity, and common values are between 200 and 600 seconds (450s for the space shuttle main engines). The specific impulse of a scramjet varies with velocity, reducing at higher speeds, starting at about 1200s, although values in the literature vary.

For the simple case of a single stage vehicle, the fuel mass fraction can be expressed as:

$$\Pi_f = 1 - \exp\left[-\frac{\left(\frac{V_{initial}^2}{2} - \frac{V_i^2}{2} \right) + \int g\,dr}{\eta_0 h_{PR} \left(1 - \frac{D + D_e}{F} \right)} \right]$$

Where this can be expressed for single stage transfer to orbit as:

$$\Pi_f = 1 - \exp\left[-\frac{g_0 r_0 \left(1 - \frac{1}{2} \frac{r_0}{r} \right)}{\eta_0 h_{PR} \left(1 - \frac{D + D_e}{F} \right)} \right]$$

or for level atmospheric flight from air launch (missile flight):

$$\Pi_f = 1 - \exp\left[-\frac{g_0 R}{\eta_0 h_{PR} (1 - \phi_e) \frac{C_L}{C_D}} \right]$$

Where R is the range, and the calculation can be expressed in the form of the Breguet range formula:

$$\Pi_f = 1 - e^{-BR}$$

$$B = \frac{g_0}{\eta_0 h_{PR} \left(1 - \phi_e\right) \frac{C_L}{C_D}}$$

Where:

- C_L is the lift coefficient
- C_D is the drag coefficient

This extremely simple formulation, used for the purposes of discussion assumes:

- Single stage vehicle
- No aerodynamic lift for the transatmospheric lifter

However they are true generally for all engines.

Initial Propulsion Requirements

A scramjet cannot produce efficient thrust unless boosted to high speed, around Mach 5, although depending on the design it could act as a ramjet at low speeds. A horizontal take-off aircraft would need conventional turbofan or rocket engines to take off, sufficiently large to move a heavy craft. Also needed would be fuel for those engines, plus all engine-associated mounting structure and control systems. Turbofan engines are heavy and cannot easily exceed about Mach 2-3, so another propulsion method would be needed to reach scramjet operating speed. That could be ramjets or rockets. Those would also need their own separate fuel supply, structure, and systems. Many proposals instead call for a first stage of droppable solid rocket boosters, which greatly simplifies the design.

Testing Difficulties

Figure 8: Test of Pratt & Whitney Rocketdyne SJY61 scramjet engine for the Boeing X-51

Unlike jet or rocket propulsion systems facilities which can be tested on the ground, testing scramjet designs use extremely expensive hypersonic test chambers or expensive launch vehicles, both of which lead to high instrumentation costs. Tests using launched test vehicles very typically end with destruction of the test item and instrumentation.

Advantages and Disadvantages for Orbital Vehicles

Propellant: An advantage of a hypersonic airbreathing (typically scramjet) vehicle like the X-30 is avoiding or at least reducing the need for carrying oxidiser. For example the space shuttle external tank holds 616,432 kg of liquid oxygen (LOX) and 103,000 kg of liquid hydrogen (LH2) while having an empty weight of 30,000 kg. The orbiter gross weight is 109,000 kg with a maximum payload of about 25,000 kg and to get the assembly off the launch pad the shuttle uses two very powerful solid rocket boosters with a weight of 590,000 kg each. If the oxygen could be eliminated, the vehicle could be lighter at liftoff and maybe carry more payload. That would be an advantage, but the central motivation in pursuing hypersonic airbreathing vehicles would be to reduce cost.

On the other hand, scramjets spend more time in the atmosphere and require more hydrogen fuel to deal with aerodynamic drag. Whereas liquid oxygen is quite a dense fluid, liquid hydrogen has much lower density and takes up much more volume. This means that the vehicle using this fuel becomes much bigger and gives even more drag.

Thrust-to-weight Ratio

One issue is that scramjet engines are predicted to have exceptionally poor thrust-to-weight ratio, when installed in a launch vehicle, of around 2. A rocket has the advantage that its engines have *very* high thrust-weight ratios (~100:1), while the tank to hold the liquid oxygen approaches a tankage ratio of ~100:1 also.

Thus a rocket can achieve a very high mass fraction, which improves performance. By way of contrast the projected thrust/weight ratio of scramjet engines of about 2 mean a very much larger percentage of the takeoff mass is engine (ignoring that this fraction increases anyway by a factor of about four due to the lack of onboard oxidiser). In addition the vehicle's lower thrust does not necessarily avoid the need for the expensive, bulky, and failure prone high performance turbopumps found in conventional liquid-fuelled rocket engines, since most scramjet designs seem to be incapable of orbital speeds in airbreathing mode, and hence extra rocket engines are needed.

Need for Additional Propulsion to Reach Orbit

Scramjets might be able to accelerate from approximately Mach 5-7 to around somewhere between half of orbital velocity and orbital velocity (X-30 research suggested that Mach 17 might be the limit compared to an orbital speed of Mach 25, and other studies put the upper speed limit for a pure scramjet engine between Mach 10 and 25, depending on the assumptions made). Generally, another propulsion system (very typically rocket is proposed) is expected to be needed for the final acceleration into orbit. Since the delta-V is moderate and the payload fraction of scramjets high, lower performance rockets such as solids, hypergolics, or simple liquid fuelled boosters might be acceptable.

Reentry

The scramjet's heat-resistant underside potentially doubles as its reentry system if a single-stage-to-orbit vehicle using non-ablative, non-active cooling is visualised. If an ablative shielding is used on the engine it will probably not be usable after ascent to orbit. If active cooling is used the loss of all fuel during the burn to orbit will also mean the loss of all cooling for the thermal protection system.

Costs

Reducing the amount of fuel and oxidiser does not necessarily improve costs as rocket propellants are comparatively very cheap. Indeed, the unit cost of the vehicle can be expected to end up far higher, since aerospace hardware cost is about two orders of magnitude higher than liquid oxygen, fuel and tankage, and scramjet hardware seems to be much heavier than rockets for any given payload. Still, if scramjets enable reusable vehicles, this could theoretically be a cost benefit. Whether equipment subject to the extreme conditions of a scramjet can be reused sufficiently many times is unclear; all flown scramjet tests only survive for short periods and have never been designed to survive a flight to date.

The eventual cost of such a vehicle is the subject of intense debate since even the best estimates disagree whether a scramjet vehicle would be advantageous. It is likely that a scramjet vehicle would need to lift more load than a rocket of equal takeoff weight in order to be equally as cost efficient (if the scramjet is a non-reusable vehicle).

Issues

Space launch vehicles may or may not benefit from having a scramjet stage. A scramjet stage of a launch vehicle theoretically provides a specific impulse of 1000 to 4000 s whereas a rocket provides less than 450 s while in the atmosphere, potentially permitting much cheaper access to space. A scramjet's specific impulse decreases rapidly with speed, however, and the vehicle would suffer from a relatively low lift to drag ratio.

The installed thrust to weight ratio of scramjets compares very unfavourably with the 50-100 of a typical rocket engine. This is compensated for in scramjets partly because the weight of the vehicle would be carried by aerodynamic lift rather than pure rocket power (giving reduced 'gravity losses'), but scramjets would take much longer to get to orbit due to lower thrust which greatly offsets the advantage. The takeoff weight of a scramjet vehicle is significantly reduced over that of a rocket, due to the lack of onboard oxidiser, but increased by the structural requirements of the larger and heavier engines.

Whether this vehicle would be reusable or not is still a subject of debate and research.

Applications

An aircraft using this type of jet engine could dramatically reduce the time it takes to travel from one place to another, potentially putting any place on Earth within a 90 minute flight. However, there are questions about whether such a vehicle could carry enough fuel to make useful length trips, and there are issues with sonic booms.

Scramjet vehicle has been proposed for a single stage to tether vehicle, where a Mach 12 spinning orbital tether would pick up a payload from a vehicle at around 100 km and carry it to orbit.

Progress in the 2000s

Figure 9: Artist's conception of the NASA X-43 with scramjet attached to the underside

In the 2000s, significant progress was made in the development of hypersonic technology, particularly in the field of scramjet engines.

US efforts are probably the best funded, and the Hyper-X team claimed the first flight of a thrust-producing scramjet powered vehicle with full aerodynamic maneuvering surfaces in 2004 with the X-43A. However, the first group to demonstrate a scramjet working in an atmospheric test was a project by a joint British and Australian team from UK defence company QinetiQ and the University of Queensland.

The HyShot project demonstrated scramjet combustion on July 30, 2002. The scramjet engine worked effectively and demonstrated supersonic combustion in action. However, the engine was not designed to provide thrust to propel a craft. It was designed more or less as a technology demonstrator.

On Friday, June 15, 2007, the US Defence Advanced Research Project Agency (DARPA), in cooperation with the Australian Defence Science and Technology Organisation (DSTO), announced a successful scramjet flight at Mach 10 using rocket engines to boost the test vehicle to hypersonic speeds.

A series of scramjet ground test were completed at NASA Langley Arc-Heated Scramjet Test Facility (AHSTF) at simulated Mach 8 flight conditions. These experiments were used to support HiFIRE flight 2.

On May 22, 2009, Woomera hosted the first successful test flight of a hypersonic aircraft in HIFire. The launch was one of 10 planned test flights. The series of up to 10 planned hypersonic flight experiments are part of a joint research program between the Defence Science and Technology Organisation and the US Air Force, designated as the Hypersonic International Flight Research Experimentation (HIFiRE). HIFiRE is investigating hypersonics technology (the study of flight exceeding five times the speed of sound) and its application to advanced scramjet-powered space launch vehicles — the objective is to support the new Boeing X-51 scramjet demonstrator while also building a strong base of flight test data for quick-reaction space launch development and hypersonic "quick-strike" weapons.

On 22 and 23 March 2010, Australian and American defence scientists successfully tested a (HIFiRE) hypersonic rocket. It reached an atmospheric velocity of "more than 5,000 kilometres per hour" after taking off from the Woomera Test Range in outback South Australia.

On May 27, 2010, NASA and the United States Air Force successfully flew the X-51A Waverider for approximately 200 seconds at Mach 5, setting a new world record hypersonic airspeed. The Waverider flew autonomously before losing acceleration for an unknown reason and destroying itself as planned. The test was declared a success. The X-51A was carried aboard a B-52, accelerated to Mach 4.5 via a solid rocket booster, and then ignited the Pratt & Whitney Rocketdyne scramjet engine to reach Mach 5 at 70,000 feet. However, a second flight June 13, 2011 was ended prematurely when the engine lit briefly on ethylene but failed to transition to its primary JP7 fuel, failing to reach full power.

On 16 November 2010, Australian scientists successfully demonstrated that the high-speed flow in a naturally non-burning scramjet engine can be ignited using a pulsed laser source.

The US Army successfully tested the scramjet powered "Advanced Hypersonic Weapon" (AHW) on November 18, 2011. The weapon reached speeds over Mach 5 and hit a target around 2500 miles distant.

Pulse Detonation Engine

Figure 10: A pulsed detonation engine ground demonstrator operating at a frequency of 35 Hz. Ground demonstrator engines are subject to intense heating, creating the need for active water cooling of this engine. Fuel and oxidiser are supplied to the engine using a valving system that matches with the operating frequency.

A pulse detonation engine, or "PDE", is a type of propulsion system that uses detonation waves to combust the fuel and oxidiser mixture. The engine is pulsed because the mixture must be renewed in the combustion chamber between each detonation wave initiated by an ignition source. Theoretically, a PDE can operate from subsonic up to a hypersonic flight speed of roughly Mach 5. An ideal PDE design can have a thermodynamic efficiency higher than other designs like turbojets and turbofans because a detonation wave rapidly compresses the mixture and adds heat at constant volume. Consequently, moving parts like compressor spools are not necessarily required in the engine, which could significantly reduce overall weight and cost. PDEs have been considered for propulsion for over 70 years. Key issues for further development include fast and efficient mixing of the fuel and oxidiser, the prevention of autoignition, and integration with an inlet and nozzle.

To date, no practical PDE has been put into production, but several testbed engines have been built and one was successfully integrated into a low-speed demonstration aircraft that flew in sustained PDE powered flight in 2008. In June 2008, the Defence Advanced Research Projects Agency (DARPA) unveiled Blackswift which was intended to use this technology to reach speeds of up to Mach 6. However the project was cancelled soon afterward, in October 2008.

Concept

All regular jet engines and most rocket engines operate on the *deflagration* of fuel, that is, the rapid but subsonic combustion of fuel. The pulse detonation engine is a concept currently in active development to create a jet engine that operates on the supersonic *detonation* of fuel.

The basic operation of the PDE is similar to that of the pulse jet engine; air is mixed with fuel to create a flammable mixture that is then ignited. The resulting combustion greatly increases the pressure of the mixture to approximately 100 atmospheres (10 MPa), which then expands through a nozzle for thrust. To ensure that the mixture exits to the rear, thereby pushing the aircraft forward, a series of shutters are used to close off the front of the engine. Careful tuning of the inlet ensures the shutters close at the right time to force the air to travel in one direction only through the engine.

The main difference between a PDE and a traditional pulse jet engine is that the mixture does not undergo subsonic combustion but instead, supersonic detonation. In the PDE, the oxygen and fuel combination process is supersonic, effectively an explosion instead of burning. The other difference is that the shutters are replaced by more sophisticated valves. In some PDE designs from General Electric, the shutters are eliminated through careful timing, using the pressure differences between the different areas of the engine to ensure the "shot" is ejected rearward. The main side effect of the change in cycle is that the PDE is considerably more efficient. In the pulse jet engine the combustion pushes a considerable amount of the fuel/air mix (the *charge*) out the rear of the engine before it has had a chance to burn (thus the trail of flame seen on the V-1 flying bomb). Even while inside the engine the mixture's volume is continually changing, which is an inefficient way to burn fuel. In contrast the PDE deliberately uses a high-speed combustion process that burns all of the charge while it is still inside the engine at a constant volume.

This is said to increase the amount of heat produced per unit of fuel above any other engines, although conversion of that energy into thrust would remain inefficient. A combustion process able to produce more heat per unit of fuel would, of course, be incredibly valuable in countless applications. Another side effect, not yet demonstrated in practical use, is the cycle time.

A traditional pulsejet tops out at about 250 pulses per second due to the cycle time of the mechanical shutters, but the aim of the PDE is thousands of pulses per second, so fast that it is basically continuous from an engineering perspective. This should help smooth out the otherwise highly vibrational pulsejet engine — many small pulses will create less volume than a smaller number of larger pulses for the same net thrust. Unfortunately, detonations are many times louder than deflagrations.

The major difficulty with a pulse-detonation engine is starting the detonation. While it is possible to start a detonation directly with a large spark, the amount of energy input is very large and is not practical for an engine. The typical solution is to use a deflagration-to-detonation transition (DDT)—that is, start a high-energy deflagration, and have it accelerate down a tube to the point where it becomes fast enough to become a detonation. Alternatively the detonation can be sent around a circle and valves ensure that only the highest peak power can leak into exhaust.

This process is far more complicated than it sounds, due to the resistance the advancing wavefront encounters (similar to wave drag). DDTs occur far more readily if there are obstacles in the tube. The most widely used is the "Shchelkin spiral", which is designed to create the most useful eddies with the least resistance to the moving fuel/air/exhaust mixture. The eddies lead to the flame separating into multiple fronts, some of which go backwards and collide with other fronts, and then accelerate into fronts ahead of them.

The behaviour is difficult to model and to predict, and research is ongoing. As with conventional pulsejets, there are two main types of designs: valved and valveless. Designs with valves encounter the same difficult-to-resolve wear issues encountered with their pulsejet equivalents. Valveless designs typically rely on abnormalities in the air flow to ensure a one-way flow, and are very hard to achieve in a regular DDT.

NASA maintains a research program on the PDE, which is aimed at high-speed, about Mach 5, civilian transport systems. However most PDE research is military in nature, as the engine could be used to develop a new generation of high-speed, long-range reconnaissance aircraft that would fly high enough to be out of range of any current anti-aircraft defences, while offering range considerably greater than the SR-71, which required a massive tanker support fleet to use in operation.

While most research is on the high speed regime, newer designs with much higher pulse rates in the hundreds of thousands appear to work well even at subsonic speeds. Whereas traditional engine designs always include tradeoffs that limit them to a "best speed" range, the PDE appears to outperform them at all speeds. Both Pratt & Whitney and General Electric now have active PDE research programs in an attempt to commercialise the designs.

Key difficulties in pulse detonation engines are achieving DDT without requiring a tube long enough to make it impractical and drag-imposing on the aircraft (adding a U-bend into the tube extinguishes the detonation wave); reducing the noise (often described as sounding like a jackhammer); and damping the severe vibration caused by the operation of the engine.

First PDE Powered Flight

Figure 11: In-flight picture of the pulsed detonation powered, and heavily modified, Rutan Long-EZ on January 31, 2008.

The first flight of an aircraft powered by a pulse detonation engine took place at the Mojave Air & Space Port on January 31, 2008. The project was developed by the Air Force Research Laboratory and Innovative Scientific Solutions, Inc. The aircraft selected for the flight was a heavily modified Scaled Composites Long-EZ, named *Borealis*. The engine consisted of four tubes producing pulse detonations at a frequency of 80 Hz, creating up to 200 pounds of thrust (890 newtons). Many fuels were considered and tested by the engine developers in recent years, but a refined octane was used for this flight. A small rocket system was used to facilitate the liftoff of the Long-EZ, but the PDE operated under its own power for 10 seconds at an altitude of approximately 100 feet (30 m). Obviously, this flight took place at a low speed whereas the appeal of the PDE engine concept lies more at high speeds, but the demonstration showed that a PDE can be integrated into an aircraft frame without experiencing structural problems from the 195-200 dB detonation waves. No more flights are planned for the modified Long-EZ, but the success is likely to fuel more funding for PDE research. The aircraft itself has been moved to the National Museum of the United States Air Force for display.

Popular Culture

- In the sci-fi novel *Aelita* (1923), two Russians travel to Mars in a pulse detonation rocket utilising "a fine powder of unusual explosive force" (p. 19).
- In the drama television series *JAG*, the Season Nine episode "The One That Got Away" (original air date October 17, 2003) features the Aurora — which, in the show is a super-secret hypersonic aircraft under development by the CIA, that uses a pulse-detonation engine.
- In the movie *Stealth* (2005), the advanced fighters use pulse-detonation engines with scramjet boosters.
- The PDE has been used as a story point in a number of modern novels such as Dan Brown's thriller, *Deception Point* (the second page of the book states that all technologies in the story are non-fictional and exist, albeit without referencing any sources), and Victor Koman's science fiction polemic, *Kings of the High Frontier*.
- In X-COM, the Interceptor is equipped with a PDE, however even this huge propulsion force is inferior compared to the alien Gravity Wave Engines.

- In the SyFy television series, Eureka, season 4 episode 10 features a flying sleigh described as being powered by a PDE.

Mercury Vapour Turbine

A Mercury vapour turbine is a form of heat engine that uses mercury to drive the thermal cycle. A mercury vapour turbine has been used in conjunction with a steam turbine for generating electricity. This example of combined cycle generation does not seem to have been widely adopted, probably because of high capital cost and the obvious toxic hazard if the mercury leaked into the environment.

The mercury cycle offers an efficiency increase compared to a steam-only cycle because energy can be injected into the Rankine Cycle at higher temperature. Metallurgical developments have allowed steam-only plants to increase in efficiency over time, making the mercury vapour turbine obsolete.

Historical Example

The Electrical Year Book, 1937, contained the following description of a mercury vapour turbine operating in commercial use:

The advantage of operating a mercury-vapour turbine in conjunction with a steam power plant lies in the fact that the complete cycle can be worked over a very wide range of temperature without employing any abnormal pressure. The exhaust from the mercury turbine is used to raise steam for the steam turbine. The Hartford Electric Light Co. (U.S.A.) has a 10,000kW turbo-generator driven by mercury vapour, which reaches the turbine at 70 lb. per sq. in. (gauge), 880°F.

The mercury vapour is condensed at 445°F and raises 129,000 lb. steam per hr. at 280 lb. per sq. in. pressure. The latter is superheated to 735°F and passed to the steam turbines. During 4 months continuous operation, this plant averaged about 0.715 lb. of coal per kWh of net output, about 43% of the output being from the mercury turbine generator and 57% from the steam plant. On maintained full-load the heat output averages 9800 B.Th.U. per net kWh. It is believed that maintenance costs will be lower than in ordinary steam plant. The back-pressure on the mercury turbine is fixed by the steam boiler pressure; only a small vacuum pump is needed, as there is no air or other gas in the mercury system.

Chapter 5

Combined Gas and Steam

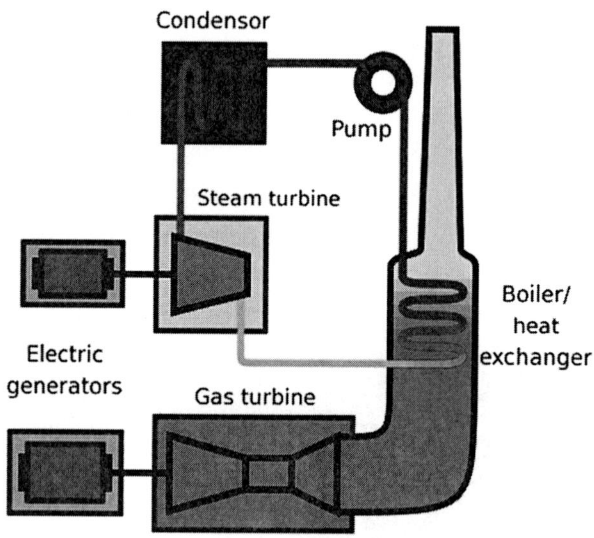

Figure 1: A turbo-electric COGAS power-plant

Combined gas and steam (COGAS) is the name given to marine compound powerplants comprising gas and steam turbines, the latter being driven by steam generated using the heat from the exhaust of the gas turbines. In this way, some of the otherwise lost energy can be reclaimed and the specific fuel consumption of the plant can be decreased. Large (land-based) electric powerplants built using this combined cycle can reach conversion efficiencies of over 58%.

If the turbines do not drive a propeller shafts directly and instead a turbo-electric transmission is used, the system is also known as COGES.

COGAS differs from many other combined marine propulsion systems in that it is not intended to operate on one system alone. While this is possible, it will not operate efficiently this way, as with

Combined diesel and gas systems when run solely on diesel engines. Especially COGAS should not be confused with Combined steam and gas (COSAG) power plants, which employ traditional, oil-fired boilers for steam turbine propulsion for normal cruising, and supplement this with gas turbines for faster reaction times and higher dash speed.

COGAS has been proposed as upgrade for ships that use gas turbines as their main (or only) engines, e.g. in COGOG or COGAG mode, like the Arleigh Burke class destroyers, but currently no naval ship uses this concept. However some modern cruise ships are equipped with COGES. E.g. Celebrity Cruises' *Millennium* and other ships of her class use turbo-electric plants with two General Electric LM2500+ gas turbines and one steam-turbine.

BMW is currently researching combined gas and steam for automotive use, using their turbosteamer system. This uses the waste heat of combustion from the exhaust and turns it into steam to produce torque which is input into the crankshaft.

Turbo-compound Engine

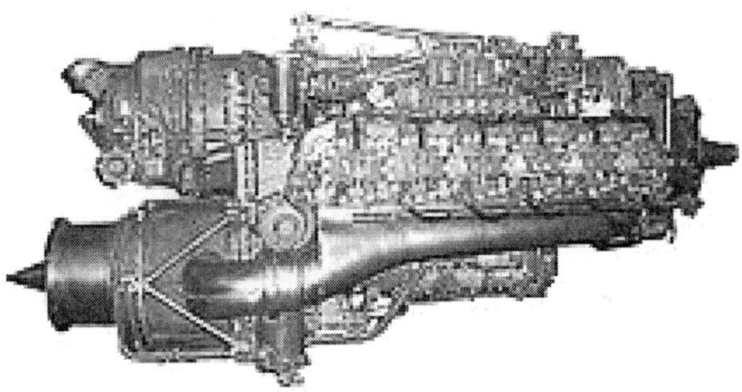

Figure 2: The Napier Nomad engine. The power-recovery turbine sits underneath a two-stroke diesel engine.

A turbo-compound engine is a reciprocating engine that employs a blowdown turbine to recover energy from the exhaust gases. The turbine is usually mechanically connected to the crankshaft but electric and hydraulic systems have been investigated as well. The turbine increases the output of the engine without increasing its fuel consumption, thus reducing the specific fuel consumption. The turbine is referred to as a blowdown turbine (or power-recovery turbine), as it recovers the energy developed in the exhaust manifold during blowdown, that is the first period of the exhaust process when the piston still is on its expansion stroke (this is possible since the exhaust valves open before bottom dead centre).

Figure 3: Wright R-3350 Duplex-Cyclone Turbo-Compound radial engine.

When a blowdown turbine is attached to an engine it will not reduce power due to exhaust gas flow restriction, since a blowdown turbine is a velocity turbine, not a pressure turbine as is a turbo supercharger. The exhaust restriction imparted by the three blowdown turbines used on the Wright R3350 is equal to a well-designed jet stack system used on a conventional radial engine. However, the blowdown turbines recover about 550 hp (410 kW) at METO (maximum continuous except for take-off) power.

Turbo-compounding was used on on several aeroplane engines after World War II, including the Napier Nomad and the Wright R3350 being examples. In the case of the R-3350, maintenance crews sometimes nicknamed the turbine the "Parts Recovery Turbine" due to its negative effect on engine reliability. Turbo-compound versions of the Napier Deltic, Rolls-Royce Crecy, and Allison V-1710 were constructed but none was developed beyond the prototype stage. It was realised in many cases the power produced by the simple turbine was approaching that of the enormously complex and maintenance-intensive piston engine to which it was attached. As a result, turbo-compound aero engines were soon supplanted by turboprop and turbojet engines.

Some modern heavy truck diesel manufacturers have incorporated turbo-compounding into their modern designs. Examples include the Detroit Diesel DD15 and Scania in production from 2001.

Turbo-compound Engines

Detroit Diesel

- DD15

Napier

- Napier Nomad

Wright Aeronautical

- Wright R-3350 - The turbo-compound version was the only turbo-compound aero-engine to see mass production and widespread usage.

Dobrynin

Dobrynin VD-4K

Gas Turbine

Figure 4: A typical axial-flow gas turbine turbojet, the J85, sectioned for display. Flow is left to right, multistage compressor on left, combustion chambers centre, two-stage turbine on right

A gas turbine, also called a combustion turbine, is a type of internal combustion engine. It has an upstream rotating compressor coupled to a downstream turbine, and a combustion chamber in-between.

Energy is added to the gas stream in the combustor, where fuel is mixed with air and ignited. In the high pressure environment of the combustor, combustion of the fuel increases the temperature. The products of the combustion are forced into the turbine section. There, the high velocity and volume of the gas flow is directed through a nozzle over the turbine's blades, spinning the turbine which powers the compressor and, for some turbines, drives their mechanical output. The energy given up to the turbine comes from the reduction in the temperature and pressure of the exhaust gas.

Energy can be extracted in the form of shaft power, compressed air or thrust or any combination of these and used to power aircraft, trains, ships, generators, or even tanks.

History

- *150:* Hero's Engine (*aeolipile*) — Apparently, Hero's steam engine was taken to be no more than a toy, and thus its full potential not realised for centuries.
- *1500:* The "Chimney Jack" was drawn by Leonardo da Vinci: Hot air from a fire rises through a single-stage axial turbine rotor mounted in the exhaust duct of the fireplace and turning the roasting spit by gear/ chain connection.
- *1551:* Taqi al-Din invented a rudimentary form of an impulse steam turbine, which he used to power a self-rotating spit.
- *1629:* Jets of steam rotated an impulse turbine that then drove a working stamping mill by means of a bevel gear, developed by Giovanni Branca.
- *1678:* Ferdinand Verbiest built a model carriage relying on a steam jet for power.
- *1791:* A patent was given to John Barber, an Englishman, for the first true gas turbine. His invention had most of the elements present in the modern day gas turbines. The turbine was designed to power a horseless carriage.
- *1872:* A gas turbine engine was designed by Dr. Franz Stolze, but the engine never ran under its own power.
- *1894:* Sir Charles Parsons patented the idea of propelling a ship with a steam turbine, and built a demonstration vessel, the *Turbinia*, easily the fastest vessel afloat at the time. This principle of propulsion is still of some use.
- *1895:* Three 4-ton 100 kW Parsons radial flow generators were installed in Cambridge Power Station, and used to power the first electric street lighting scheme in the city.
- *1899:* Charles Gordon Curtis patented the first gas turbine engine in the USA ("Apparatus for generating mechanical power", Patent No. US635,919).
- *1900:* Sanford Alexander Moss submitted a thesis on gas turbines. In 1903, Dr. Moss became an engineer for General Electric's Steam Turbine Department in Lynn, Massachusetts.

While there, he applied some of his concepts in the development of the turbosupercharger. His design used a small turbine wheel, driven by exhaust gases, to turn a supercharger.

- *1903:* A Norwegian, Ægidius Elling, was able to build the first gas turbine that was able to produce more power than needed to run its own components, which was considered an achievement in a time when knowledge about aerodynamics was limited. Using rotary compressors and turbines it produced 11 hp (massive for those days). His work was later used by Sir Frank Whittle.
- *1906:* The Armengaud-Lemale turbine engine in France with water-cooled combustion chamber.
- *1910:* Holzwarth impulse turbine (pulse combustion) achieved 150 kilowatts.
- *1913:* Nikola Tesla patents the Tesla turbine based on the boundary layer effect.
- *1918:* One of the leading gas turbine manufacturers of today, General Electric, started their gas turbine division.
- *1920:* The practical theory of gas flow through passages was developed into the more formal (and applicable to turbines) theory of gas flow past airfoils by Dr. A. A. Griffith.
- *1930:* Sir Frank Whittle patented the design for a gas turbine for jet propulsion. The first successful use of his engine was in April 1937.
- *1932:* BBC Brown, Boveri & Cie of Switzerland starts selling axial compressor and turbine turbosets as part of the turbocharged steam generating Velox boiler. Following the gas turbine principle, the steam evapouration tubes are arranged within the gas turbine combustion chamber; the first Velox plant was erected in Mondeville, France.
- *1934:* Raúl Pateras de Pescara patented the free-piston engine as a gas generator for gas turbines.
- *1936:* Hans von Ohain and Max Hahn in Germany were developing their own patented engine design.
- *1939:* First 4 MW utility power generation gas turbine from BBC Brown, Boveri & Cie. for an emergency power station in Neuchâtel, Switzerland.

Theory of Operation

Gases passing through an ideal gas turbine undergo three thermodynamic processes. These are isentropic compression, isobaric (constant pressure) combustion and isentropic expansion. Together these make up the Brayton cycle.

In a practical gas turbine, gases are first accelerated in either a centrifugal or radial compressor. These gases are then slowed using a diverging nozzle known as a diffuser; these processes increase the pressure and temperature of the flow. In an ideal system this is isentropic. However, in practice energy is lost to heat, due to friction and turbulence. Gases then pass from the diffuser to a combustion chamber, or similar device, where heat is added. In an ideal system this occurs at constant pressure (isobaric heat addition). As there is no change in pressure the specific volume of the gases increases. In practical situations this process is usually accompanied by a slight loss in pressure, due to friction. Finally, this larger volume of gases is expanded and accelerated by nozzle guide vanes before energy is extracted by a turbine. In an ideal system these are gases expanded isentropically and leave the turbine at their original pressure. In practice this process is not isentropic as energy is once again lost to friction and turbulence.

If the device has been designed to power a shaft as with an industrial generator or a turboprop, the exit pressure will be as close to the entry pressure as possible. In practice it is necessary that some pressure remains at the outlet in order to fully expel the exhaust gases. In the case of a jet engine only enough pressure and energy is extracted from the flow to drive the compressor and other components. The remaining high pressure gases are accelerated to provide a jet that can, for example, be used to propel an aircraft.

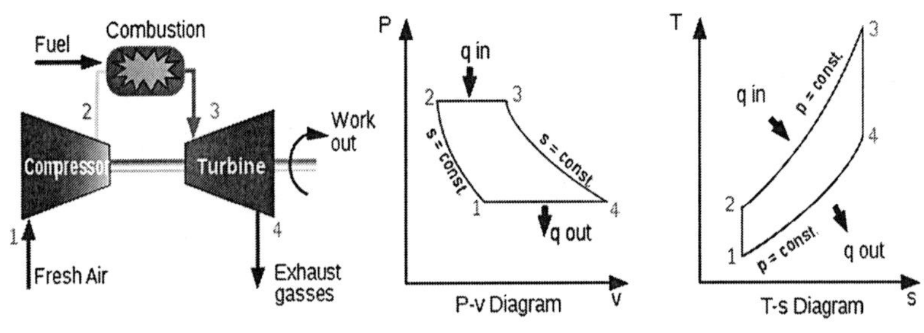

Figure 5: Brayton cycle

As with all cyclic heat engines, higher combustion temperatures can allow for greater efficiencies. However, temperatures are limited by ability of the steel, nickel, ceramic, or other materials that make up the engine to withstand high temperatures and stresses. To combat this many turbines feature complex blade cooling systems.

As a general rule, the smaller the engine the higher the rotation rate of the shaft(s) must be to maintain tip speed. Blade tip speed determines the maximum pressure ratios that can be obtained by the turbine and the compressor. This in turn limits the maximum power and efficiency that can be obtained by the engine. In order for tip speed to remain constant, if the diametre of a rotor is reduced by half, the rotational speed must double. For example large Jet engines operate around 10,000 rpm, while micro turbines spin as fast as 500,000 rpm.

Mechanically, gas turbines can be considerably less complex than internal combustion piston engines. Simple turbines might have one moving part: the shaft/compressor/turbine/alternative-rotor assembly, not counting the fuel system. However, the required precision manufacturing for components and temperature resistant alloys necessary for high efficiency often make the construction of a simple turbine more complicated than piston engines. More sophisticated turbines (such as those found in modern jet engines) may have multiple shafts (spools), hundreds of turbine blades, movable stator blades, and a vast system of complex piping, combustors and heat exchangers. Thrust bearings and journal bearings are a critical part of design. Traditionally, they have been hydrodynamic oil bearings, or oil-cooled ball bearings. These bearings are being surpassed by foil bearings, which have been successfully used in micro turbines and auxiliary power units.

Types of Gas Turbines—Jet Engines

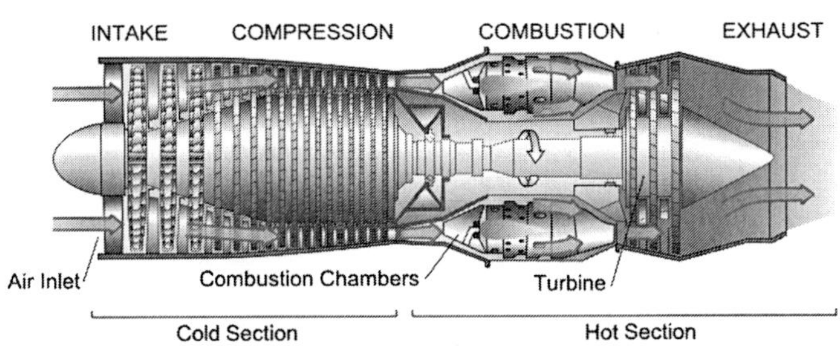

Figure 6: Diagram of a gas turbine jet engine

Airbreathing jet engines are gas turbines optimised to produce thrust from the exhaust gases, or from ducted fans connected to the gas turbines. Jet engines that produce thrust primarily from the direct impulse of exhaust gases are often called turbojets, whereas those that generate most of their thrust from the action of a ducted fan are often called turbofans or (rarely) fan-jets.

Gas turbines are also used in many liquid propellant rockets, the gas turbines are used to power a turbopump to permit the use of lightweight, low pressure tanks, which saves considerable dry mass.

Turboprop Engines

A turboprop engine is a type of turbine engine which drives an external aircraft propeller using a reduction gear. Turboprop engines are generally used on small subsonic aircraft, but some large military and civil aircraft, such as the Airbus A400M, Lockheed L-188 Electra and Tupolev Tu-95, have also used turboprop power.

Aeroderivative Gas Turbines

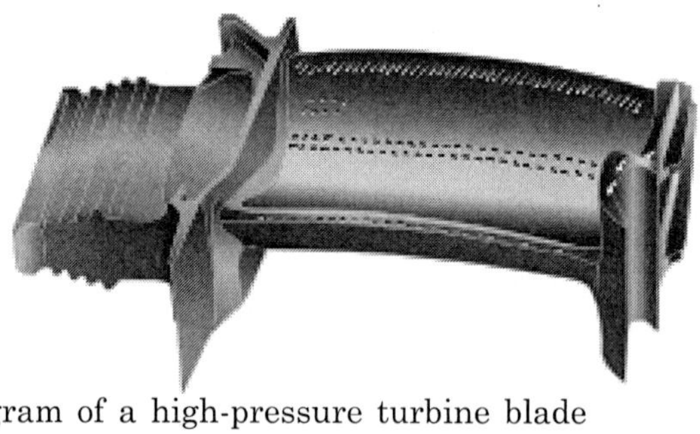

Figure 7: Diagram of a high-pressure turbine blade

Aeroderivatives are also used in electrical power generation due to their ability to be shut down, and handle load changes more quickly than industrial machines. They are also used in the marine industry to reduce weight. The General Electric LM2500, General Electric LM6000, Rolls-Royce RB211 and Rolls-Royce Avon are common models of this type of machine.

Amateur Gas Turbines

Increasing numbers of gas turbines are being used or even constructed by amateurs.

In its most straightforward form, these are commercial turbines acquired through military surplus or scrapyard sales, then operated for display as part of the hobby of engine collecting. In its most extreme form, amateurs have even rebuilt engines beyond professional repair and then used them to compete for the Land Speed Record. The simplest form of self-constructed gas turbine employs an automotive turbocharger as the core component. A combustion chamber is fabricated and plumbed between the compressor and turbine sections.

More sophisticated turbojets are also built, where their thrust and light weight are sufficient to power large model aircraft. The Schreckling design constructs the entire engine from raw materials, including the fabrication of a centrifugal compressor wheel from plywood, epoxy and wrapped carbon fibre strands.

Several small companies now manufacture small turbines and parts for the amateur. Most turbojet-powered model aircraft are now using these commercial and semi-commercial microturbines, rather than a Schreckling-like home-build.

Auxiliary Power Units

APUs are small gas turbines designed for auxiliary power of larger machines, such as those inside an aircraft. They supply compressed air for aircraft ventilation (with an appropriate compressor design), start-up power for larger jet engines, and electrical and hydraulic power.

Industrial Gas Turbines for Power Generation

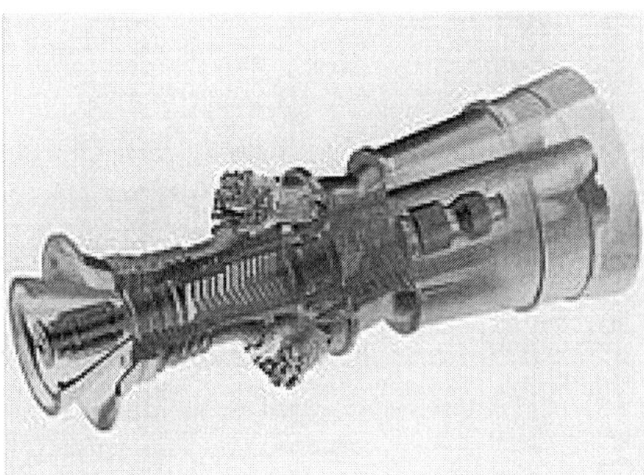

Figure 8: GE H series power generation gas turbine: in combined cycle configuration, this 480-megawatt unit has a rated thermal efficiency of 60%.

Industrial gas turbines differ from aeroderivative in that the frames, bearings, and blading are of heavier construction. Industrial gas turbines range in size from truck-mounted mobile plants to enormous, complex systems. They can be particularly efficient—up to 60%—when waste heat from the gas turbine is recovered by a heat recovery steam generator to power a conventional steam turbine in a combined cycle configuration. They can also be run in a cogeneration configuration: the exhaust is used for space or water heating, or drives an absorption chiller for cooling or refrigeration. Such engines require a dedicated enclosure, both to protect the engine from the elements and the operators from the noise.

The construction process for gas turbines can take as little as several weeks to a few months, compared to years for base load power plants. Their other main advantage is the ability to be turned on and off within minutes, supplying power during peak demand. Since single cycle (gas turbine only) power plants are less efficient than combined cycle plants, they are usually used as peaking power plants, which operate anywhere from several hours per day to a few dozen hours per year, depending on the electricity demand and the generating capacity of the region.

In areas with a shortage of base load and load following power plant capacity or low fuel costs, a gas turbine power plant may regularly operate during most hours of the day. A large single cycle gas turbine typically produces 100 to 400 megawatts of power and have 35–40% thermal efficiency.

Compressed Air Energy Storage

One modern development seeks to improve efficiency in another way, by separating the compressor and the turbine with a compressed air store. In a conventional turbine, up to half the generated power is used driving the compressor. In a compressed air energy storage configuration, power, perhaps from a wind farm or bought on the open market at a time of low demand and low price, is used to drive the compressor, and the compressed air released to operate the turbine when required.

Turboshaft Engines

Turboshaft engines are often used to drive compression trains (for example in gas pumping stations or natural gas liquefaction plants) and are used to power almost all modern helicopters. The first shaft bears the compressor and the high speed turbine (often referred to as "Gas Generator" or "Ng"), while the second shaft bears the low speed turbine (or "Power Turbine" or "Nf" - the 'f' stands for 'free wheeling turbine' on helicopters specifically due to the fact that the gas generator turbine spins separately from the power turbine). This arrangement is used to increase speed and power output flexibility.

Radial Gas Turbines

In 1963, Jan Mowill initiated the development at Kongsberg Vapenfabrikk in Norway. Various successors have made good progress in the refinement of this mechanism. Owing to a configuration that keeps heat away from certain bearings the durability of the machine is improved while the radial turbine is well matched in speed requirement.

Scale Jet Engines

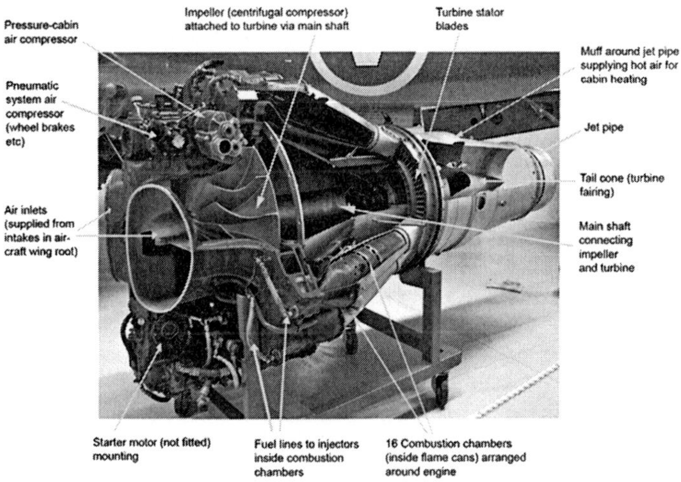

Figure 9: Scale jet engines are scaled down versions of this early full scale engine

Also known as miniature gas turbines or micro-jets.

With this in mind the pioneer of modern Micro-Jets, Kurt Schreckling, produced one of the world's first Micro-Turbines, the FD3/67. This engine can produce up to 22 newtons of thrust, and can be built by most mechanically minded people with basic engineering tools, such as a metal lathe.

Microturbines

Also known as:

- Turbo alternators
- Turbogenerator

Microturbines are touted to become widespread in distributed power and combined heat and power applications. They are one of the most promising technologies for powering hybrid electric vehicles. They range from hand held units producing less than a kilowatt, to commercial sized systems that produce tens or hundreds of kilowatts. Basic principles of microturbine are based on micro combustion.

Part of their claimed success is said to be due to advances in electronics, which allows unattended operation and interfacing with the commercial power grid. Electronic power switching technology eliminates the need for the generator to be synchronised with the power grid. This allows the generator to be integrated with the turbine shaft, and to double as the starter motor.

Microturbine systems have many claimed advantages over reciprocating engine generators, such as higher power-to-weight ratio, low emissions and few, or just one, moving part. Advantages are that microturbines may be designed with foil bearings and air-cooling operating without lubricating oil, coolants or other hazardous materials. Nevertheless reciprocating engines overall are still cheaper when all factors are considered. Microturbines also have a further advantage of having the majority of the waste heat contained in the relatively high temperature exhaust making it simpler to capture, whereas the waste heat of reciprocating engines is split between its exhaust and cooling system.

However, reciprocating engine generators are quicker to respond to changes in output power requirement and are usually slightly more efficient, although the efficiency of microturbines is increasing. Microturbines also lose more efficiency at low power levels than reciprocating engines.

When used in extended range electric vehicles the static efficiency drawback is irrelevant, since the gas turbine can be run at or near maximum power, driving an alternator to produce electricity either for the wheel motors, or for the batteries, as appropriate to speed and battery state. The batteries act as a "buffer" (energy storage) in delivering the required amount of power to the wheel motors, rendering throttle response of the GT completely irrelevant.

There is, moreover, no need for a significant or variable-speed gearbox; turning an alternator at comparatively high speeds allows for a smaller and lighter alternator than would otherwise be the case. The superior power-to-weight ratio of the gas turbine and its fixed speed gearbox, allows for a much lighter prime mover than those in such hybrids as the Toyota Prius (which utilised a 1.8 litre petrol engine) or the Chevrolet Volt (which utilises a 1.4 litre petrol engine). This in turn allows a heavier weight of batteries to be carried. The weight can be made up of more batteries, which allows for a longer electric-only range. Alternatively, the vehicle can use heavier types of batteries such as lead acid batteries (which are cheaper to buy) or safer types of batteries such as Lithium-Iron-Phosphate.

When gas turbines are used in extended-range electric vehicles, like those planned by Land-Rover/Range-Rover in conjunction with Bladon, or by Jaguar also in partnership with Bladon, the very poor throttling response (their high moment of rotational inertia) does not matter, because the gas turbine, which may be spinning at 100,000 rpm, is not directly, mechanically connected to the wheels. It was this poor throttling response that so bedevilled the 1960 Rover gas turbine-powered prototype motor car, which did not have the advantage of an intermediate electric drive train.

Gas turbines accept most commercial fuels, such as petrol, natural gas, propane, diesel, and kerosene as well as renewable fuels such as E85, biodiesel and biogas. However, when running on kerosene or diesel, starting sometimes requires the assistance of a more volatile product such as propane gas - although the new kero-start technology can allow even microturbines fuelled on kerosene to start without propane.

Microturbine designs usually consist of a single stage radial compressor, a single stage radial turbine and a recuperator. Recuperators are difficult to design and manufacture because they operate under high pressure and temperature differentials. Exhaust heat can be used for water heating, space heating, drying processes or absorption chillers, which create cold for air conditioning from heat energy instead of electric energy. Typical microturbine efficiencies are 25 to 35%. When in a combined heat and power cogeneration system, efficiencies of greater than 80% are commonly achieved.

MIT started its millimetre size turbine engine project in the middle of the 1990s when Professor of Aeronautics and Astronautics Alan H. Epstein considered the possibility of creating a personal turbine which will be able to meet all the demands of a modern person's electrical needs, just as a large turbine can meet the electricity demands of a small city.

Problems have occurred with heat dissipation and high-speed bearing in these new microturbines. Moreover, their expected efficiency is a very low 5-6%. According to Professor Epstein, current commercial Li-ion rechargeable batteries deliver about 120-150 W·h/kg. MIT's millimetre size turbine will deliver 500-700 W·h/kg in the near term, rising to 1200-1500 W"'h/kg in the longer term.

External Combustion

Most gas turbines are internal combustion engines but it is also possible to manufacture an external combustion gas turbine which is, effectively, a turbine version of a hot air engine. Those systems are usually indicated as EFGT (Externally Fired Gas Turbine) or IFGT (Indirectly Fired Gas Turbine).

External combustion has been used for the purpose of using pulverised coal or finely ground biomass (such as sawdust) as a fuel. In the indirect system, a heat exchanger is used and only clean air with no combustion products travels through the power turbine. The thermal efficiency is lower in the indirect type of external combustion; however, the turbine blades are not subjected to combustion products and much lower quality (and therefore cheaper) fuels are able to be used. Indirectly fired systems are now commercially available. BTOLA, an Australian based company is now marketing 250 kW - 2MW units.

Gas Turbines in Surface Vehicles

Gas turbines are often used on ships, locomotives, helicopters, tanks, and to a lesser extent, on cars, buses, and motorcycles.

A key advantage of jets and turboprops for aeroplane propulsion - their superior performance at high altitude compared to piston engines, particularly naturally aspirated ones - is irrelevant in automobile applications. Their power-to-weight advantage, though less critical than for aircraft, is still important.

Gas turbines offer a high-powered engine in a very small and light package. However, they are not as responsive and efficient as small piston engines over the wide range of RPMs and powers needed in vehicle applications. In series hybrid vehicles, as the driving electric motors are mechanically detached from the electricity generating engine, the responsiveness, poor performance at low speed and low efficiency at low output problems are much less important. The turbine can be run at optimum speed for its power output, and batteries and ultracapacitors can supply power as needed, with the engine cycled on and off to run it only at high efficiency. The emergence of the continuously variable transmission may also alleviate the responsiveness problem.

Turbines have historically been more expensive to produce than piston engines, though this is partly because piston engines have been mass-produced in huge quantities for decades, while small gas turbine engines are rarities; however, turbines are mass-produced in the closely related form of the turbocharger.

The turbocharger is basically a compact and simple free shaft radial gas turbine which is driven by the piston engine's exhaust gas. The centripetal turbine wheel drives a centrifugal compressor wheel through a common rotating shaft. This wheel supercharges the engine air intake to a degree that can be controlled by means of a wastegate or by dynamically modifying the turbine housing's geometry (as in a VGT turbocharger). It mainly serves as a power recovery device which converts a great deal of otherwise wasted thermal and kinetic energy into engine boost.

Turbocompound engines (actually employed on some trucks) are fitted with blow down turbines which are similar in design and appearance to a turbocharger except for the turbine shaft being mechanically or hydraulically connected to the engine's crankshaft instead of to a centrifugal compressor, thus providing additional power instead of boost. While the turbocharger is a pressure turbine, a power recovery turbine is a velocity one.

Passenger Road Vehicles (Cars, Bikes, and Buses)

A number of experiments have been conducted with gas turbine powered automobiles, the largest by Chrysler. More recently, there has been some interest in the use of turbine engines for hybrid electric cars. For instance, a consortium led by micro gas turbine company Bladon Jets has secured investment from the Technology Strategy Board to develop an Ultra Lightweight Range Extender (ULRE) for next generation electric vehicles. The objective of the consortium, which includes luxury car maker Jaguar Land Rover and leading electrical machine company SR Drives, is to produce the world's first commercially viable - and environmentally friendly - gas turbine generator designed specifically for automotive applications. The common turbocharger for gas or diesel engines is also a turbine derivative.

Concept Cars

The first serious mention investigation of using a gas turbine in cars, was in 1946 when two engineers, Robert Kafka and Robert Engerstein of Carney Associates, a New York engineering firm, came up with the concept where a unique compact turbine engine design would provide power for a rear wheel drive car. After an article appeared in *Popular Science*, there was no further work, beyond the paper stage.

In 1950, designer F.R. Bell and Chief Engineer Maurice Wilks from British car manufacturers Rover unveiled the first car powered with a gas turbine engine. The two-seater JET1 had the engine positioned behind the seats, air intake grilles on either side of the car, and exhaust outlets on the top of the tail. During tests, the car reached top speeds of 140 km/h (87 mph), at a turbine speed of 50,000 rpm. The car ran on petrol, paraffin (kerosene) or diesel oil, but fuel consumption problems proved insurmountable for a production car. It is on display at the London Science Museum.

American car manufacturer Chrysler demonstrated several prototype gas turbine-powered cars from the early 1950s through the early 1980s. Chrysler built fifty Chrysler Turbine Cars in 1963 and conducted the only consumer trial of gas turbine-powered cars. Their turbines employed unique rotating recuperator that significantly increased efficiency. Chrysler put many cars into the hands of consumers, making this effort a small-scale near-production run.

The original General Motors Firebird was a series of concept cars developed for the 1953, 1956 and 1959 Motorama auto shows, powered by gas turbines.

Toyota demonstrated several gas turbine powered concept cars such as the Century gas turbine hybrid in 1975, the Sports 800 Gas Turbine Hybrid in 1979 and the GTV in 1985. No production vehicles were made. The GT24 engine was exhibited in 1977 without a vehicle.

The fictional Batmobile is often said to be powered by a gas turbine or a jet engine. The 1960s television show vehicle was said to be powered by a turbine engine, with a parachute braking system. For the 1989 *Batman* film, the production department built a working turbine vehicle for the Batmobile prop. Its fuel capacity, however, was reportedly only enough for 15 seconds of use at a time.

In the early 1990s Volvo introduced the Volvo Environmental Concept Car(ECC) which was a gas turbine powered hybrid car.

In 1993 General Motors introduced the first commercial gas turbine powered hybrid vehicle—as a limited production run of the EV-1 series hybrid. A Williams International 40 kW turbine drove an alternator which powered the battery-electric powertrain. The turbine design included a recuperator. Later on in 2006 GM went into the EcoJet concept car project with Jay Leno.

At the 2010 Paris Motor Show Jaguar demonstrated its Jaguar C-X75 concept car. This electrically powered supercar has a top speed of 204 mph (328 km/h) and can go from 0 to 62 mph (0 to 100 km/h) in 3.4 seconds. It uses Lithium-ion batteries to power 4 electric motors which combine to produce some 780 bhp. It will do around 100 miles on a single charge of the batteries but in addition it uses a pair of Bladon Micro Gas Turbines to re-charge the batteries extending the range to some 560 miles.

Racing Cars

Rover and the BRM Formula One team joined forces to produce the Rover-BRM, a gas turbine powered coupe, which entered the 1963 24 Hours of Le Mans, driven by Graham Hill and Richie Ginther. It averaged 107.8 mph (173 km/h) and had a top speed of 142 mph (229 km/h). American Ray Heppenstall joined Howmet Corporation and McKee Engineering together to develop their own gas turbine sports car in 1968, the Howmet TX, which ran several American and European events, including two wins, and also participated in the 1968 24 Hours of Le Mans. The cars used Continental gas turbines, which eventually set six FIA land speed records for turbine-powered cars.

For open wheel racing, 1967's revolutionary STP-Paxton Turbocar fielded by racing and entrepreneurial legend Andy Granatelli and driven by Parnelli Jones nearly won the Indianapolis 500; the Pratt & Whitney ST6B-62 powered turbine car was almost a lap ahead of the second place car when a gearbox bearing failed just three laps from the finish line.

The next year the STP Lotus 56 turbine car won the Indianapolis 500 pole position even though new rules restricted the air intake dramatically. In 1971 Lotus principal Colin Chapman introduced the Lotus 56B F1 car, powered by a Pratt & Whitney STN 6/76 gas turbine. Chapman had a reputation of building radical championship-winning cars, but had to abandon the project because there were too many problems with turbo lag.

Buses

The arrival of the Capstone Microturbine has led to several hybrid bus designs, starting with HEV-1 by AVS of Chattanooga, Tennessee in 1999, and closely followed by Ebus and ISE Research in California, and DesignLine Corporation in New Zealand (and later the United States). AVS turbine hybrids were plagued with reliability and quality control problems, resulting in liquidation of AVS in 2003. The most successful design by Designline is now operated in 5 cities in 6 countries, with over 30 buses in operation worldwide, and order for several hundred being delivered to Baltimore , and NYC.

Brescia Italy is using serial hybrid buses powered by microturbines on routes through the historical sections of the city.

Motorcycles

The MTT Turbine SUPERBIKE appeared in 2000 (hence the designation of Y2K Superbike by MTT) and is the first production motorcycle powered by a turbine engine - specifically, a Rolls-Royce Allison model 250 turboshaft engine, producing about 283 kW (380 bhp).

Speed-tested to 365 km/h or 227 mph (according to some stories, the testing team ran out of road during the test), it holds the Guinness World Records for most powerful production motorcycle and most expensive production motorcycle, with a price tag of US$185,000.

Trains

Several locomotive classes have been powered by gas turbines, the most recent incarnation being Bombardier's JetTrain.

Tanks

Figure 10: Marines from 1st Tank Battalion load a Honeywell AGT1500 multi-fuel turbine back into the tank at Camp Coyote, Kuwait, February 2003.

The German Army's development division, the Heereswaffenamt (Army Ordnance Board), studied a number of gas turbine engines for use in tanks starting in mid-1944. The first gas turbine engines used for armoured fighting vehicle GT 101 was installed in the Panther tank.

The second use of a gas turbine in an armoured fighting vehicle was in 1954 when a unit, PU2979, specifically developed for tanks by C. A. Parsons & Co., was installed and trialled in a British Conqueror tank. The Stridsvagn 103 was developed in the 1950s and was the first mass produced main battle tank to use a turbine engine. Since then, gas turbine engines have been used as APUs in some tanks and as main powerplants in Soviet/Russian T-80s and U.S. M1 Abrams tanks, among others.

They are lighter and smaller than diesels at the same sustained power output but the models installed to date are less fuel efficient than the equivalent diesel, especially at idle, requiring more fuel to achieve the same combat range.

Successive models of M1 have addressed this problem with battery packs or secondary generators to power the tank's systems while stationary, saving fuel by reducing the need to idle the main turbine.

T-80s can mount three large external fuel drums to extend their range. Russia has stopped production of the T-80 in favour of the diesel-powered T-90 (based on the T-72), while Ukraine has developed the diesel-powered T-80UD and T-84 with nearly the power of the gas-turbine tank.

The French Leclerc MBT's diesel powerplant features the "Hyperbar" hybrid supercharging system, where the engine's turbocharger is completely replaced with a small gas turbine which also works as an assisted diesel exhaust turbocharger, enabling engine RPM-independent boost level control and a higher peak boost pressure to be reached (than with ordinary turbochargers).

This system allows a smaller displacement and lighter engine to be used as the tank's powerplant and effectively removes turbo lag. This special gas turbine/turbocharger can also work independently from the main engine as an ordinary APU.

A turbine is theoretically more reliable and easier to maintain than a piston engine, since it has a simpler construction with fewer moving parts but in practice turbine parts experience a higher wear rate due to their higher working speeds.

The turbine blades are highly sensitive to dust and fine sand, so that in desert operations air filters have to be fitted and changed several times daily. An improperly fitted filter, or a bullet or shell fragment that punctures the filter, can damage the engine. Piston engines (especially if turbocharged) also need well-maintained filters, but they are more resilient if the filter does fail.

Like most modern diesel engines used in tanks, gas turbines are usually multi-fuel engines.

Marine Applications—Naval

Figure 11: The Gas turbine from MGB 2009

Gas turbines are used in many naval vessels, where they are valued for their high power-to-weight ratio and their ships' resulting acceleration and ability to get underway quickly.

The first gas-turbine-powered naval vessel was the Royal Navy's Motor Gun Boat *MGB 2009* (formerly *MGB 509*) converted in 1947. Metropolitan-Vickers developed the "Beryl" engine equipping an existing F2/3 jet engine with a power turbine. As the test was successful, the Fast Patrol Boats *Bold Pioneer* and *Bold Pathfinder* built in 1953 were the first ships created specifically for gas turbine propulsion.

The first large scale, gas-turbine powered ships, were the Royal Navy's Type 81 (Tribal class) frigates, the first of which (HMS *Ashanti*) was commissioned in 1961.

The Germany Navy launched the first Koln class frigate in 1961 with 2 GTs from BBC in the worlds first combined diesel and gas propulsion system.

The Danish Navy produced 6 Søløven class torpedo boats (in service from 1965 to 1990) which had 3 Rolls Royce Marine Gas Turbines rated at 9,510 kW (12,750 shp) combined, plus two General Motors Diesel engines, rated at 340 kW (460 shp), for better fuel economy at slower speeds. And they also produced 10 Willemoes Class Torpedo / Guided Missile boats (in service from 1974 to 2000) which had 3 Rolls Royce Marine Proteus Gas Turbines also rated at 9,510 kW (12,750 shp), same as the Søløven class boats, and 2 General Motors Diesel Engines, rated at 600 kW (800 shp), also for improved fuel economy at slow speeds.

The Swedish Navy produced 6 Spica class torpedoboats between 1966 and 1967 powered by 3 Bristol Siddeley Proteus 1282, each delivering 3,210 kW (4,300 shp). They were later joined by 12 upgraded Norrkoping class ships, still with the same engines. With their aft torpedo tubes replaced by antishipping missiles they served as missile boats until the last was retired in 2005.

The Finnish Navy issued two Turunmaa class corvettes, Turunmaa and Karjala, in 1968. They were equipped with one 16,410 kW (22,000 shp) Rolls-Royce Olympus TMB3 gas turbine and three Wärtsilä marine diesels for slower speeds. They were the fastest vessels in the Finnish Navy; they regularly achieved 35 knot speeds, with 37.3 knots on sea trials. The Turunmaas were paid off in 2002. *Karjala* is today a museum ship in Turku, and *Turunmaa* serves as a floating machine shop and training ship for Satakunta Polytechnical College.

The next series of major naval vessels were the four Canadian Iroquois class helicopter carrying destroyers first commissioned in 1972. They used 2 ft-4 main propulsion engines, 2 ft-12 cruise engines and 3 Solar Saturn 750 kW generators.

The first U.S. gas-turbine powered ships were the U.S. Coast Guard's *Hamilton*-class High Endurance Cutters the first of which (USCGC *Hamilton*) commissioned in 1967. Since then, they have powered the U.S. Navy's *Perry*-class frigates, *Spruance*-class and *Arleigh Burke*-class destroyers, and *Ticonderoga*-class guided missile cruisers. USS *Makin Island*, a modified *Wasp*-class amphibious assault ship, is to be the Navy's first amphibious assault ship powered by gas turbines. The marine gas turbine operates in a more corrosive atmosphere due to presence of sea salt in air and fuel and use of cheaper fuels.

Non-military Maritime

Gas turbines have been used experimentally to power seagoing commercial vessels since about 1949 (Anglo Saxon Petroleum oil tanker "Auris").

The United States Maritime Commission were looking for options to update WWII Liberty ships and heavy duty gas turbines were one of those selected. In 1956 the "John Sergeant" was lengthened and installed with a General Electric 6600 SHP HD gas turbine, reduction gearing and a variable pitch propeller. It operated for 9700 hours using residual fuel for 7000 hours. The success of this trial opened the way for more development by GE on the use of HD gas turbines for marine use with heavy fuels. The "John Sergeant" was scrapped in 1972 at Portsmouth PA. Boeing launched its first passenger-carrying waterjet-propelled hydrofoil Boeing 929, in April 1974. Those ships were powered by twin Allison gas turbines of the KF-501 series.

Between 1970 and 1982, Seatrain Container Lines operated a scheduled container service across the North Atlantic with four container ships of 26,000 tonnes deadweight tonnage (DWT). Those ships were powered by twin Pratt & Whitney gas turbines of the FT 4 series. The four ships in the class were named "Euroliner", "Eurofreighter", "Asialiner" and "Asiafreighter". They operated a transatlantic container service between ports on the eastern seaboard of the United States and ports in north west Europe. Following the dramatic OPEC(Organisation of the Petroleum Exporting Countries) price increases of the mid-nineteen seventies, operations were constrained by rising fuel costs.

Some modification of the engine systems on those ships was undertaken to permit the burning of a lower grade of fuel (i.e., marine diesel). The modifications were partially successful. It was proved that particular fuel could be used in a marine gas turbine but, savings made were less than anticipated due to increased maintenance requirements. After 1982 the ships were sold, then re-engined with more economical diesel engines. Because the new engines were much larger, there was a consequential loss of some cargo space.

The first passenger ferry to use a gas turbine was the GTS *Finnjet*, built in 1977 and powered by two Pratt & Whitney FT 4C-1 DLF turbines, generating 55 MW and propelling the ship to a speed of 31 knots. However, the Finnjet also illustrated the shortcomings of gas turbine propulsion in commercial craft, as high fuel prices made operating her unprofitable. After four years of service additional diesel engines were installed on the ship to reduce running costs during the off-season.

Another example of commercial usage of gas turbines in a passenger ship is Stena Line's HSS class fastcraft ferries. HSS 1500-class *Stena Explorer*, *Stena Voyager* and *Stena Discovery* vessels use COGAG setups of twin GE LM2500 plus GE LM1600 power for a total of 68 MW. The slightly smaller HSS 900-class *Stena Charisma*, uses twin ABB–STAL GT35 turbines rated at 34,000 kW gross. The *Stena Discovery* was withdrawn from service in 2007, another victim of too high fuel costs.

In July 2000 the Millennium became the first cruise ship to be propelled by gas turbines, in a Combined Gas and Steam Turbine configuration. The liner RMS Queen Mary 2 uses a Combined Diesel and Gas Turbine configuration.

In marine racing applications the 2010 C5000 Mystic catamaran Miss Gieco uses two Lycoming t-55 turbines for its power system.

Advances in Technology

Gas turbine technology has steadily advanced since its inception and continues to evolve; research is active in producing ever smaller gas turbines. Computer design, specifically CFD and finite element analysis along with material advances, has allowed higher compression ratios and temperatures, more efficient combustion and better cooling of engine parts.

On the emissions side, the challenge in technology is increasing turbine inlet temperature while reducing peak flame temperature to achieve lower NOx emissions to cope with the latest regulations. In May 2011, Mitsubishi Heavy Industries achieved a turbine inlet temperature of 1,600 °C on a 320 megawatt gas turbine, 460 MW in gas turbine combined-cycle power generation applications in which gross thermal efficiency exceeds 60%.

Additionally, compliant foil bearings were commercially introduced to gas turbines in the 1990s. They can withstand over a hundred thousand start/stop cycles and eliminated the need for an oil system. On another front, microelectronics and power switching technology have enabled commercially viable micro turbines for distributed and vehicle power.

Advantages and Disadvantages of Gas Turbine Engines

Advantages of Gas Turbine Engines

- Very high power-to-weight ratio, compared to reciprocating engines;
- Smaller than most reciprocating engines of the same power rating.
- Moves in one direction only, with far less vibration than a reciprocating engine.
- Fewer moving parts than reciprocating engines.
- Low operating pressures.
- High operation speeds.
- Low lubricating oil cost and consumption.
- Can run on a wide variety of fuels.

Disadvantages of Gas Turbine Engines

- Cost is very high
- Less efficient than reciprocating engines at idle speed
- Longer startup than reciprocating engines
- Less responsive to changes in power demand compared to reciprocating engines

Wind Turbine

Figure 12: Offshore wind farm using 5MW turbines REpower 5M in the North Sea off Belgium

A wind turbine is a device that converts kinetic energy from the wind into mechanical energy. If the mechanical energy is used to produce electricity, the device may be called a wind generator or wind charger. If the mechanical energy is used to drive machinery, such as for grinding grain or pumping water, the device is called a windmill or wind pump. Developed for over a millennium, today's wind turbines are manufactured in a range of vertical and horizontal axis types. The smallest turbines are used for applications such as battery charging or auxiliary power on sailing boats; while large grid-connected arrays of turbines are becoming an increasingly large source of commercial electric power.

History

Figure 13: James Blyth's electricity generating wind turbine photographed in 1891

Windmills were used in Persia (present-day Iran) as early as 200 B.C. The windwheel of Heron of Alexandria marks one of the first known instances of wind powering a machine in history. However, the first known practical windmills were built in Sistan, a region between Afghanistan and Iran, from the 7th century. These "Panemone" were vertical axle windmills, which had long vertical driveshafts with rectangular blades. Made of six to twelve sails covered in reed matting or cloth material, these windmills were used to grind grain or draw up water, and were used in the gristmilling and sugarcane industries.

Windmills first appeared in Europe during the middle ages. The first historical records of their use in England date to the 11th or 12th centuries and there are reports of German crusaders taking their windmill-making skills to Syria around 1190. By the 14th century, Dutch windmills were in use to drain areas of the Rhine delta.

Figure 14: The first automatically operated wind turbine, built in Cleveland in 1887 by Charles F. Brush. It was 60 feet (18 m) tall, weighed 4 tons (3.6 metric tonnes) and powered a 12kW generator.

The first electricity generating wind turbine, was a battery charging machine installed in July 1887 by Scottish academic James Blyth to light his holiday home in Marykirk, Scotland. Some months later American inventor Charles F Brush built the first automatically operated wind turbine for electricity production in Cleveland, Ohio. Although Blyth's turbine was considered uneconomical in the United Kingdom electricity generation by wind turbines was more cost effective in countries with widely scattered populations.

In Denmark by 1900, there were about 2500 windmills for mechanical loads such as pumps and mills, producing an estimated combined peak power of about 30 MW. The largest machines were on 24-metre (79 ft) towers with four-bladed 23-metre (75 ft) diametre rotors. By 1908 there were 72 wind-driven electric generators operating in the US from 5 kW to 25 kW. Around the time of World War I, American windmill makers were producing 100,000 farm windmills each year, mostly for water-pumping. By the 1930s, wind generators for electricity were common on farms, mostly in the United States where distribution systems had not yet been installed. In this period, high-tensile steel was cheap, and the generators were placed atop prefabricated open steel lattice towers.

A forerunner of modern horizontal-axis wind generators was in service at Yalta, USSR in 1931. This was a 100 kW generator on a 30-metre (98 ft) tower, connected to the local 6.3 kV distribution system. It was reported to have an annual capacity factor of 32 per cent, not much different from current wind machines. In the fall of 1941, the first megawatt-class wind turbine was synchronised to a utility grid in Vermont. The Smith-Putnam wind turbine only ran for 1,100 hours before suffering a critical failure. The unit was not repaired because of shortage of materials during the war.

The first utility grid-connected wind turbine to operate in the U.K. was built by John Brown & Company in 1951 in the Orkney Islands.

Resources

A quantitative measure of the wind energy available at any location is called the Wind Power Density (WPD) It is a calculation of the mean annual power available per square metre of swept area of a turbine, and is tabulated for different heights above ground. Calculation of wind power density includes the effect of wind velocity and air density. Colour-coded maps are prepared for a particular area described, for example, as "Mean Annual Power Density at 50 Metres."

In the United States, the results of the above calculation are included in an index developed by the U.S. National Renewable Energy Lab and referred to as "NREL CLASS." The larger the WPD calculation, the higher it is rated by class. Classes range from Class 1 (200 watts/square metre or less at 50 metres altitude) to Class 7 (800 to 2000 watts/square metre). Commercial wind farms generally are sited in Class 3 or higher areas, although isolated points in an otherwise Class 1 area may be practical to exploit.

Types

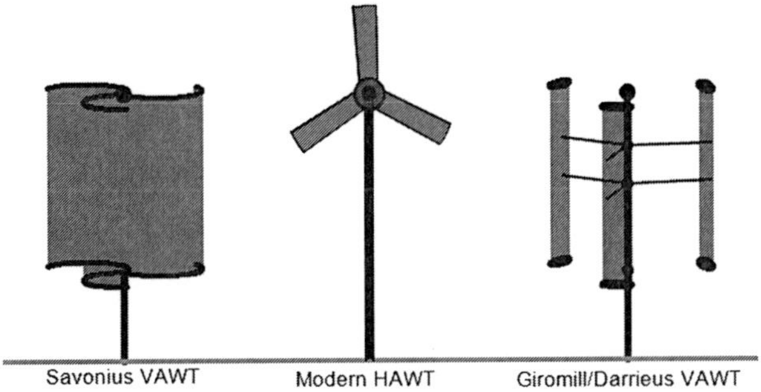

Figure 15: The three primary types: VAWT Savonius, HAWT towered; VAWT Darrieus as they appear in operation.

Wind turbines can rotate about either a horizontal or a vertical axis, the former being both older and more common.

Horizontal Axis

Figure 16: Components of a horizontal axis wind turbine (gearbox, rotor shaft and brake assembly) being lifted into position

Horizontal-axis wind turbines (HAWT) have the main rotor shaft and electrical generator at the top of a tower, and must be pointed into the wind. Small turbines are pointed by a simple wind vane, while large turbines generally use a wind sensor coupled with a servo motor. Most have a gearbox, which turns the slow rotation of the blades into a quicker rotation that is more suitable to drive an electrical generator.

Since a tower produces turbulence behind it, the turbine is usually positioned upwind of its supporting tower. Turbine blades are made stiff to prevent the blades from being pushed into the tower by high winds. Additionally, the blades are placed a considerable distance in front of the tower and are sometimes tilted forward into the wind a small amount.

Downwind machines have been built, despite the problem of turbulence (mast wake), because they don't need an additional mechanism for keeping them in line with the wind, and because in high winds the blades can be allowed to bend which reduces their swept area and thus their wind resistance. Since cyclical (that is repetitive) turbulence may lead to fatigue failures, most HAWTs are of upwind design.

Figure 17: 11 x 7,5 MW E126 Estinnes Windfarm, Belgium, July 2010, one month before completion, with unique 2 part blades.

Modern Wind Turbines

Figure 18: Turbine blade convoy passing through Edenfield in the UK

Turbines used in wind farms for commercial production of electric power are usually three-bladed and pointed into the wind by computer-controlled motors. These have high tip speeds of over 320 kilometres per hour (200 mph), high efficiency, and low torque ripple, which contribute to good reliability. The blades are usually coloured light gray to blend in with the clouds and range in length from 20 to 40 metres (66 to 130 ft) or more.

The tubular steel towers range from 60 to 90 metres (200 to 300 ft) tall. The blades rotate at 10-22 revolutions per minute. At 22 rotations per minute the tip speed exceeds 90 metres per second (300 ft/s). A gear box is commonly used for stepping up the speed of the generator, although designs may also use direct drive of an annular generator. Some models operate at constant speed, but more energy can be collected by variable-speed turbines which use a solid-state power converter to interface to the transmission system. All turbines are equipped with protective features to avoid damage at high wind speeds, by feathering the blades into the wind which ceases their rotation, supplemented by brakes.

Vertical Axis Design

Vertical-axis wind turbines (or VAWTs) have the main rotor shaft arranged vertically. Key advantages of this arrangement are that the turbine does not need to be pointed into the wind to be effective. This is an advantage on sites where the wind direction is highly variable, for example when integrated into buildings.

The key disadvantages include the low rotational speed with the consequential higher torque and hence higher cost of the drive train, the inherently lower power coefficient, the 360 degree rotation of the aerofoil within the wind flow during each cycle and hence the highly dynamic loading on the blade, the pulsating torque generated by some rotor designs on the drive train, and the difficulty of modelling the wind flow accurately and hence the challenges of analysing and designing the rotor prior to fabricating a prototype.

With a vertical axis, the generator and gearbox can be placed near the ground, using a direct drive from the rotor assembly to the ground-based gearbox, hence improving accessibility for maintenance.

When a turbine is mounted on a rooftop, the building generally redirects wind over the roof and this can double the wind speed at the turbine. If the height of the rooftop mounted turbine tower is approximately 50% of the building height, this is near the optimum for maximum wind energy and minimum wind turbulence. It should be borne in mind that wind speeds within the built environment are generally much lower than at exposed rural sites.

Another type of vertical axis is the Parallel turbine similar to the crossflow fan or centrifugal fan it uses the Ground effect. Vertical axis turbines of this type have been tried for many years The Magenn WindKite blimp uses this configuration as well, chosen because of the ease of running.

Subtypes

Figure 19: Darrieus wind turbine of 30 m in the Magdalen Islands

Darrieus Wind Turbine

"Eggbeater" turbines, or Darrieus turbines, were named after the French inventor, Georges Darrieus. They have good efficiency, but produce large torque ripple and cyclical stress on the tower, which contributes to poor reliability. They also generally require some external power source, or an additional Savonius rotor to start turning, because the starting torque is very low. The torque ripple is reduced by using three or more blades which results in greater solidity of the rotor. Solidity is measured by blade area divided by the rotor area. Newer Darrieus type turbines are not held up by guy-wires but have an external superstructure connected to the top bearing.

Giromill

A subtype of Darrieus turbine with straight, as opposed to curved, blades. The cycloturbine variety has variable pitch to reduce the torque pulsation and is self-starting. The advantages of variable pitch are: high starting torque; a wide, relatively flat torque curve; a lower blade speed ratio; a higher coefficient of performance; more efficient operation in turbulent winds; and a lower blade speed ratio which lowers blade bending stresses. Straight, V, or curved blades may be used.

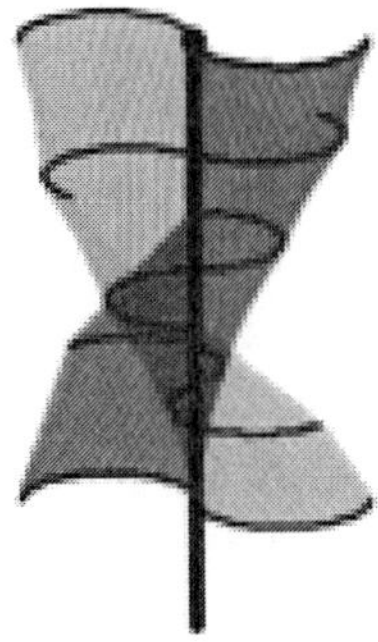

Figure 20: Twisted Savonius

Savonius Wind Turbine

These are drag-type devices with two (or more) scoops that are used in anemometres, *Flettner* vents (commonly seen on bus and van roofs), and in some high-reliability low-efficiency power turbines. They are always self-starting if there are at least three scoops.

Twisted Savonius

Twisted Savonius is a modified savonius, with long helical scoops to give a smooth torque, this is mostly used as roof windturbine or on some boats (like the Hornblower Hybrid).

Turbine Design and Construction

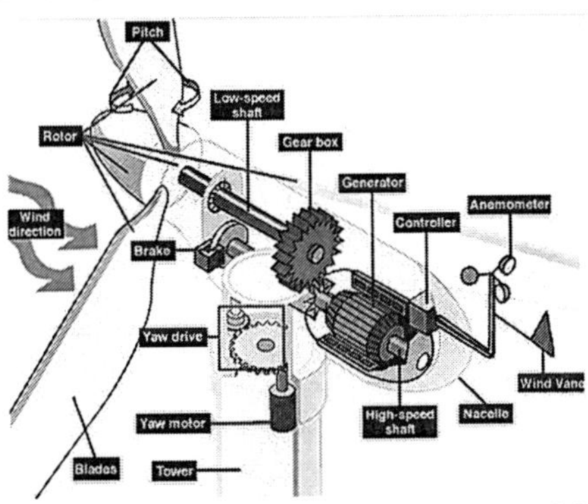

Figure 21: Components of a horizontal-axis wind turbine

Wind turbines are designed to exploit the wind energy that exists at a location. Aerodynamic modelling is used to determine the optimum tower height, control systems, number of blades and blade shape.

Wind turbines convert wind energy to electricity for distribution. Conventional horizontal axis turbines can be divided into three components.

- The rotor component, which is approximately 20% of the wind turbine cost, includes the blades for converting wind energy to low speed rotational energy.
- The generator component, which is approximately 34% of the wind turbine cost, includes the electrical generator, the control electronics, and most likely a gearbox (e.g. planetary gearbox, adjustable-speed drive or continuously variable transmission) component for converting the low speed incoming rotation to high speed rotation suitable for generating electricity.
- The structural support component, which is approximately 15% of the wind turbine cost, includes the tower and rotor yaw mechanism.

A 1.5 MW wind turbine of a type frequently seen in the United States has a tower 80 metres high. The rotor assembly (blades and hub) weighs 48,000 pounds (22,000 kg). The nacelle, which contains the generator component, weighs 115,000 pounds (52,000 kg). The concrete base for the tower is constructed using 58,000 pounds (26,000 kg) of reinforcing steel and contains 250 cubic yards (190 cubic metres) of concrete. The base is 50 feet (15 m) in diametre and 8 feet (2.4 m) thick near the centre.

Unconventional Wind Turbines

One E-66 wind turbine at Windpark Holtriem, Germany, carries an observation deck, open for visitors. Another turbine of the same type, with an observation deck, is located in Swaffham, England. Airborne wind turbines have been investigated many times but have yet to produce significant energy. Conceptually, wind turbines may also be used in conjunction with a large vertical solar updraft tower to extract the energy due to air heated by the sun. Wind turbines which utilise the Magnus effect have been developed. The Ram air turbine is a specialist form of small turbine that is fitted to some aircraft. When deployed, the RAT is spun by the airstream going past the aircraft and can provide power for the most essential systems if there is a loss of all on–board electrical power.

Small Wind Turbines

Small wind turbines may be as small as a fifty-watt generator for boat or caravan use. The US Department of Energy's National Renewable Energy Laboratory (NREL) defines small wind turbines as those smaller than or equal to 100 kilowatts. Small units often have direct drive generators, direct current output, aeroelastic blades, lifetime bearings and use a vane to point into the wind.

Wind Turbine Spacing

On most horizontal windturbine farms, a spacing of about 6-10 times the rotor diametre is often upheld. However, for large wind farms distances of about 15 rotor diametres should be more economically optimal, taking into account typical wind turbine and land costs.

This conclusion has been reached by research conducted by Charles Meneveau of the Johns Hopkins University and Johan Meyers of Leuven University in Belgium, based on computer simulations that take into account the detailed interactions among wind turbines (wakes) as well as with the entire turbulent atmospheric boundary layer. Moreover, recent research by John Dabiri of Caltech suggests that vertical wind turbines may be placed much more closely together so long as an alternating pattern of rotation is created allowing blades of neighbouring turbines to move in the same direction as they approach one another .

Record-holding Turbines

Largest Capacity: The Enercon E-126 has a rated capacity of 7.58 MW, has an overall height of 198 m (650 ft), a diametre of 126 m (413 ft), and is the world's largest-capacity wind turbine since its introduction in 2007.

At least four companies are working on the development of a 10MW turbine:

- American Superconductor
- Wind Power Ltd are developing a 10 MW VAWT, the Aerogenerator X
- Sway AS announced the proposed development of a prototype 10 MW wind turbine with a height of 162.5 m (533 ft) and a rotor diametre of 145 m (475 ft).
- Astralux Ltd are developing vertical axis magneto levitated 10 MW turbine with 230 m height and 260 m rotor diametre
- Clipper Windpower were developing the Britannia 10 MW HAWT, but terminated the project due to financial challenges.

Largest Swept Area

The turbine with the largest swept area is a prototype installed by Gamesa at Jaulín, Zaragoza, Spain in 2009. The G10X – 4.5 MW has a rotor diametre of 128m.

Tallest

The tallest wind turbine is Fuhrländer Wind Turbine Laasow. Its axis is 160 metres above ground and its rotor tips can reach a height of 205 metres. It is the only wind turbine in the world taller than 200 metres.

Largest Vertical-axis

Le Nordais wind farm in Cap-Chat, Quebec has a vertical axis wind turbine (VAWT) named Eole, which is the world's largest at 110 m. It has a nameplate capacity of 3.8MW.

Most Southerly

The turbines currently operating closest to the South Pole are three *Enercon E-33* in Antarctica, powering New Zealand's Scott Base and the United States' McMurdo Station since December 2009 although a modified HR3 turbine from Northern Power Systems operated at the Amundsen-Scott South Pole Station in 1997 and 1998. In March 2010 CITEDEF designed, built and installed a wind turbine in Argentine Marambio Base.

Most Productive

Four turbines at Rønland wind farm in Denmark share the record for the most productive wind turbines, with each having generated 63.2 GWh by June 2010

Highest-situated

The world's highest-situated wind turbine is made by DeWind installed by the Seawind Group and located in the Andes, Argentina around 4,100 metres (13,500 ft) above sea level. The site uses a type D8.2 - 2000 kW / 50 Hz turbine. This turbine has a new drive train concept with a special torque converter (WinDrive) made by Voith and a synchronous generator. The WKA was put into operation in December 2007 and has supplied the Veladero mine of Barrick Gold with electricity since then.

Largest Floating Wind Turbine

The world's largest—and also the first operational deep-water *large-capacity*—floating wind turbine is the 2.3 MW Hywind currently operating 10 kilometres (6.2 mi) offshore in 220-metre-deep water, southwest of Karmøy, Norway. The turbine began operating in September 2009 and utilises a Siemens 2.3 MW turbine

Chapter 6

High-altitude Wind Power

High-altitude wind power (HAWP) has been imagined as a source of useful energy since 1833 with John Etzler's vision of capturing the power of winds high in the sky by use of tether and cable technology. An atlas of the high-altitude wind power resource has been prepared for all points on earth.

Various mechanisms are proposed for capturing the kinetic energy of winds such as kites, kytoons, aerostats, gliders, gliders with turbines for regenerative soaring, sailplanes with turbines, or other airfoils, including multiple-point building- or terrain-enabled holdings. Once the mechanical energy is derived from the wind's kinetic energy, then many options are available for using that mechanical energy: direct traction, conversion to electricity aloft or at ground station, conversion to laser or microwave for power beaming to other aircraft or ground receivers.

Energy generated by a high-altitude system may be used aloft or sent to the ground surface by conducting cables, mechanical force through a tether, rotation of endless line loop, movement of changed chemicals, flow of high-pressure gases, flow of low-pressure gases, or laser or microwave power beams.

High-altitude Wind for Power Purposes

Winds at higher altitudes become steadier, more persistent, and of higher velocity. Because power available in wind increases as the cube of velocity (the velocity-cubed law), assuming other parameters remaining the same, doubling a wind's velocity gives 2x2x2=8 times the power; tripling the velocity gives 3 x 3 x 3 = 27 times the available power. With steadier and more predictable winds, high-altitude wind has an advantage over wind near the ground. Being able to locate HAWP to effective altitudes and using the vertical dimension of airspace for wind farming brings further advantage using high-altitude winds for generating energy.

High-altitude wind generators can be adjusted in height and position to maximise energy return, which is impractical with fixed tower-mounted wind generators.

In each range of altitudes there are altitude-specific concerns being addressed by researchers and developers. As altitude increases, tethers increase in length, the temperature of the air changes, and vulnerability to atmospheric lightning changes. With increasing altitude, exposure to liabilities increase, costs increase, turbulence exposure changes, likelihood of having the system fly in more than one directional strata of winds increases, and the costs of operation changes. HAWP systems that are flown must climb through all intermediate altitudes up to final working altitudes—being at first a low- and then a high- altitude device.

Methods of Capturing Kinetic Energy of High-altitude Winds

Energy can be captured from the wind by kites, kytoons, tethered gliders, tethered sailplanes, aerostats (spherical as well as shaped kytoons), bladed turbines, airfoils, airfoil matrices, balloons, parachutes, drogues, variable drogues, spiral airfoils, Darrieus turbines, Magnus-effect VAWT blimps, multiple-rotor complexes, fabric Jalbert-parafoil kites, uni-blade turbines, flipwings, tethers, bridles, string loops, wafting blades, undulating forms, piezoelectric materials, and more.

When a scheme's purpose is to propel ships and boats, the objects tether-placed in the wind will tend to have most of the captured energy be in useful tension in the main tether. The aloft working bodies will be operated to maintain useful tension even while the ship is moving. This is the method for powerkiting sports. This sector of HAWP is the most installed method. The folklore is that Benjamin Franklin used the traction method of HAWP. George Pocock was a leader in tugging vehicles by traction.

Controls

HAWP aircraft need to be controlled. Solutions in built systems have control mechanisms variously situated. Some systems are passive, or active, or a mix. When a kite steering unit (KSU) is lofted, the KSU may be robotic and self-contained; a KSU may be operated from the ground via radio-control by a live human operator or by smart computer programs. Some systems have built sensors in the aircraft body that report parameters like position, relative position to other parts. Kite control units (KCU) have involved more than steering; tether reeling speeds and directions can be adjusted in response to tether tensions and needs of the system during a power-generating phase or return-non-power-generating phase. Kite control parts vary widely.

Methods of Converting Mechanical Energy to Other Forms of Energy

The mechanical energy of the device may be converted to heat, sound, electricity, light, tension, pushes, pulls, laser, microwave, chemical changes, or compression of gases. Traction is a big direct use of the mechanical energy as in tugging cargo ships and kiteboarders. The methods of getting the mechanical energy from the wind's kinetic energy are several. Lighter-than-air (LTA) moored aerostats are employed as lifters of turbines. Heavier-than-air (HTA) tethered airfoils are being used as lifters or turbines themselves. Combinations of LTA and HTA devices in one system are being built and flown to capture HAWP. Even a family of free-flight airborne devices are represented in the literature that capture the kinetic energy of high-altitude winds (beginning with a description in 1967 by Richard Miller in book Without Visible Means of Support) and a contemporary patent application by Dale C. Kramer, soaring sailplane competitor, inventor.

Electric Generator Position in a HAWP System

Electricity generation is just one of the optional choices of using captured mechanical energy; however, this option dominates the focus of professionals aiming to supply large amounts of energy to commerce and utilities. A long array of secondary options include tugging water turbines, pumping of water, or compressing air or hydrogen. The position of the electric generator is a distinguishing feature among systems. Flying the generator aloft is done in a variety of ways. Keeping the generator at the mooring region is another large design option. The option in one system of a generator aloft and at the ground station has been used where a small generator operates electronic devices aloft while the ground generator is the big worker to make electricity for significant loads.

Aerostat-based HAWP

One method of keeping working HAWP systems aloft is to use buoyant aerostats whether or not the electric generator is lifted or left on the ground. The aerostats are usually, but not always, shaped to achieve a kiting lifting effect. Recharging leaked lifting gas receives various solutions.

- W. R. Benoit US Patent 4350897 Lighter than air wind energy conversion system by William R. Benoit, filed Oct 24,1980, and issued: Sep 21, 1982.

- The TWIND system is based on the use of a sail surface elevated by the climbing force of an aerostatic balloon connected to the ground by a cable used also for energy transmission. The wind present at high altitudes creates a horizontal push on the sail which in its movement transmits this energy to the ground via the connecting cable. At the end of its movement forward, the sail surface is reduced allowing it to move upwind with reduced energy waste.
- The Magenn aerostat is a vertical-axis wind turbine held with its axis horizontal by bridling the axis traverse to the wind so that Magnus-effect lift obtains during autorotation; the electricity is generated with end-hub generators.
- The LTA Windpower PowerShip uses lift from both an aerostat and wings. It operates close to neutral buoyancy and doesn't require a winch. Power is generated by turbines with the propellers on the trailing edge of the wings. The system is designed to be able to take off and land unattended.
- Airbine proposes to lift wind turbines into the air by use of aerostats; the electricity would return to ground loads by way of conductive tether.
- Airship power turbine by William J. Mouton, Jr., and David F. Thompson: Their system integrated the turbine within the central portion of a near-toroidal aerostat, like putting a turbine in the hole of an aerostat donut.

Non-airborne HAWP

Conceptually, two adjacent mountains (natural or terrain-enabled) or artificial buildings or towers (urban or artificial) could have a wind turbine suspended between them by use of cables. When HAWP is cabled between two mountain tops across a valley, the HAWP device is not airborne, but borne up by the cable system. No such systems are known to be in use, though patents teach these methods. When non-cabled bridges are the foundation for holding wind turbines high above the ground, then these are grouped with conventional towered turbines and are outside the intent of HAWP where the tethering an airborne system is foundational.

HAWP Safety

Lightning, aircraft traffic, emergency procedures, system inspections, visibility marking of system parts and its tethers, electrical safety, runaway-wing procedures, over-powering controls, appropriate mooring, and more form the safety environment for HAWP systems.

Challenges of HAWP as an Emerging Industry

There have been several periods of high interest in HAWP before the contemporary activity. The first period had a high focus on pulling carriages over the lands and capturing atmospheric electricity and lightning for human use. The second period was in the 1970s and 1980s when research and investment flourished; a drop in oil price resulted in no significant installations of HAWP. Return on investment (ROI) has been the key parameter; that ROI remains in focus in the current development activity while in the background is the renewable and sustainable energy movement supporting wind power of any kind; but HAWP must compete on ROI with conventional towered solutions.

HAWP Patents

Patents for kite and kytoon systems that work with energy obtained from upper winds are one of the significant tools being employed to advance the nascent HAWP industry.

Timeline for HAWP

Early centuries of kiting demonstrated that the kite is a rotary engine that rotates its tether part about its mooring point and causes hands and arms to move because of the energy captured from higher winds into the mechanical device. The tension in the lofted devices performs the work of lifting and pulling body parts and things. Airborne wind energy (AWE) for HAWP was birthed thousands of years ago; naming what happened and developing the implied potentials of tethered aircraft for doing special works is what is occurring in AWE HAWP. What is "low" for some workers is "high" for others.

- 1796 George Pocock (inventor) used traction mode to travel in vehicles over land roads.
- 1827 George Pocock's book 'The Aeropleustic Art' or 'Navigation in the Air by the Use of Kites or Buoyant Sails' was published. The book was to be republished again several times. The Charvolant or Kite Carriage was described. Importantly Pocock described use of kites for land and sea travel.
- 1833 John Adolphus Etzler saw HAWP blossoming at least for traction.
- 1864? Book's chapter *Kite-Ship* well describes key dynamics of HAWP used for tugging ships by kites. John Gay's: or Work for Boys. Chapter XVIII in the Summer volume.

- 1943 Stanley Biszak instructed using potential energy in free-flight for converting ambient winds impacting turbine to drive electric generator to charge batteries.
- 1967 Richard Miller, former editor of Soaring magazine, published book Without Visible Means of Support that describes the feasibility of free-flight coupled non-ground-moored kites to capture differences in wind strata to travel across continents; such HAWP is the subject of Dale C. Kramer's contemporary patent application.
- 1973? Hermann Oberth In the appendix of his book Primer for Those Who Would Govern there are sketches and a photograph of a model of the *Kite Power Station* from the Oberth Museum.
- 1977 April 3, 1977, invention declared. On September 21, 1979, Douglas Selsam notarised his kite-lifted endless chain of airfoils HAWP system, generic type that would later show in Dutch astronaut Wubbo Ockels' device called LadderMill described in a patent of 1997. Douglas Selsam conceived his *Auto-oriented Wind Harnessing Buoyant Aerial Tramway* on April 3, 1977. On the Selsam notarised disclosure of invention was placed a date of Sept. 20, while the notary placed the final signing on Sept. 21, 1979. notes and drawings.
- 1979 Professor Bryan Roberts begins giromill gyrocopter-type HAWP wind generator development.
- 1986 Bryan Roberts' AWE HAWP rotor generates electricity and lifts itself in tethered flight.
- 1992 *Free Rotor* WO/1992020917 Free Rotor by JACK, Colin, Humphry, Bruce (one man). Colin Jack. Colin Bruce. Multi-rotors are treated. Faired tethers are recognised. 1992.
- 2001 September 6. *Method of Utilisation a Flow Energy and Power Installation* by physicist Alexander Bolonkin details ground generator of huge AWECS systems.
- 2001 Founding of SkySails company. As leader in ship traction HAWP, in 2010 they note having just less than 10 installations.
- 2002 Drachen Foundation awarded a grant to David Lang so that he would explore the use of kites to generate practical power.
- 2004 Physicist Alexander Bolonkin presented a strong analysis toward groundgen AWECS.

- 2004 Drachen Foundation had David Lang, former NASA expert, aerospace consultant, conduct a survey-study of kite-based power generation. He published a summary of methods in his view.
- 2005 HAWP conference held at AeroVironment, Pasadena, CA; attendees: Paul MacCready, Dave Lang, Joe Hadzicki, Scott Skinner. The long meeting was videotaped; the tape is in the archives of the Drachen Foundation, open to researchers.
- 2006 on October 19: Jong Chul Kim, filed for international patent protection for "Electric power generation system using hydro turbine tracted by paraglider." His further studies has been for at-sea production of hydrogen and other chemicals using HAWP tethering kite systems.
- 2006 Kite Sailing Symposium, Seattle, Washington, September 28–30, 2006.
- 2006 Dr. Paul MacCready, American aeronautical engineer, Inventor of the Year Award in 1981, publishes an influential article *In Using Kites to Tap Power of Wind.*
- 2006 January 30: First International Workshop on Modelling and Optimisation of Power Generating Kites KITE-OPT 07 held in Belgium featured keynote speaker astronaut Prof. Wubbo Ockels.
- 2006 September: A 40 kW prototype of high-altitude wind genereator is tested in Italy by Kite Gen Research Massimo Ippolito.
- 2007 Dave Santos demonstrated lofted Portland KiteMotor at alt-energy West Coast Climate Convergence; the generation success was announced in HIPFiSH Columbia-Pacific's Alternative Monthly (Aug.-Sept./2007). KiteMotor Growing Pains
- 2008 Dave Santos demonstrates flipwings for AWE passive-control generator systems.
- 2009 January 28, University of Texas hosted *Airborne Wind Energy Seminar* in its Aerospace Engineering Department. Keynote presenter: Dave Santos of KiteLab.
- 2009 Dave Santos demonstrated passive control of kite-lifted working dynamic kite operating a string tripod to transfer kite-gained mechanical energy to a ground-based generator via

pulley and crank; this was not a reel-method scheme.

- 2009 Towered WECS companies open departments for AWE HAWP: Two ground-hugging towered wind-turbine companies open AWE HAWP departments: SpiralAirfoil Airborne and Selsam.
- 2009 A first international industry association was founded: Airborne Wind Energy Industry Association
- 2009 August. Dr. Hong Zhang, Kyle Fitzpatrick, and other students demonstrate working generation of electricity from powerkiting in preparation for further academic HAWP studies at Rowan University, New Jersey.
- 2009 Nov. 5-6 : HAWP conference in Chico, CA and Oroville, CA. Scores of HAWP and AWE companies and inventors attended. Systems demonstrated: KiteLab's lifted bladed turbine with generator aloft, Selsam multi-rotor torque-tube hybrid with groundstationed generator, SkyMill Energy autogiro RC-controlled reel-in-out method.
- 2009 Dec. 9 : TU Delft University holds Kite Dynamics Symposium 09 for HAWP as central focus.
- 2010 January 13. An announcement that a large investment will be made to construct a commercial utility-scale HAWP project in the city of Foshan, China.
- 2010 March 1–3: Three HAWP companies had booths at ARPA-E Technology Showcase, Gaylord Convention Centre, Washington, DC. Investment and showcasing by HAWP entities: Joby Energy, Sky Windpower, Makani Power directed visitors to each other's booths.
- 2010 March 6. Eight airborne wind energy technology entity leaders in USA held a telephone conference and decided without objection to propose a modification to bill HR 3165; the primary line-item proposed modification entered to a congressional member was the inclusion of "airborne wind energy technology."
- 2010 April 24. Multiple chaotic systems effectively captured for useful work: Full demonstration of working mini-AWECS involving Double Pendulum and flapping kite wing-mill under lifter kite to drive clockworks system's charging lithium battery by KiteLab of Ilwaco, Washington, USA. Designer and builder:

Dave Santos.

- 2010 Another international organisation has been formed: Airborne Wind Energy Consortium
- 2010 May: USA's NASA firms attention on AWECS by establishing an AWT data web site: NASA Wind Energy Airborne Harvesting Stystem Study. The NASA 2010 Innovation Fund selected Airborne Wind Turbine Energy Harvesting UAV Systems; project leader is Dr. Mark D. Moore, Aeronautics Systems.
- 2010, June 16 An international partnership between Dutch and Norwegian entities provides second-round financing for classic reel-in HAWP method using tethered powerplane system. Airborne wind energy technology firm Ampyx Power partners with Byte and Statkraft.
- 16 June 2010 A unique treasure of 700+ drawings by airborne wind energy techologist Dave Santos made from 2006 and into 2009 were made available to the public. Seven-hundred pages of drawings made from 2006 into 2009 by hand and mind of Dave Santos Alternate full name: David Santos Gorena-Guinn, roboticist, artist, founder of KiteLab, master instructor for KitePilotSchool, and chief technical officer for Airborne Wind Energy Industry Association (AWEIA).
- 2010 July 9: World's First Gondola Turbine Array Demo by KiteLab.
- 2010 September 29–30 Airborne Wind Energy Conference 2010. This complements the 2009 November first major international conference.
- *2010 October:* The Wayne German Prize for Contributions to Airborne Wind Energy stated its second year awardee.
- *2010 November 25:* CONOPS are being drafted by KiteLab, Ilwaco, WA, USA, and also NASA AWT project leader for submission to the USA's FAA toward obtaining response for potentially having airborne wind energy conversion systems as an integral part of aviation and air space.
- *2010 December 15. IPP:* Capturing Wind Energy Through Airborne Platforms (IPP :: Innovative Partnerships Program NASA)

- *2011 January 15:* The 10th Annual Windless Kite Festival in Long Beach, WA (USA): Dave Santos successfully demonstated keeping a kite flying circling in no wind via use of tri-tether rig. Such proves a method for AWECS station-keeping when wind fails to keep the HAWP craft flying. The arrangement is a reverse of what Santos priorly demonstrated where kites can drive ground generator through a tri-tether rig.
- *2011 August 25:* The 1st report on the high-altitude wind energy technologies and markets is published by GL Garrad Hassan. It is the first of it's kind and purely focuses on high-altitude wind energy systems. The Market Status Report on High Altitude Wind Energy provides an detailed overview on wind energy resources, different systems and technologies, involved companies and the market potential. The market part focuses on the offshore application of high-altitude wind energy.

Airborne Wind Turbine

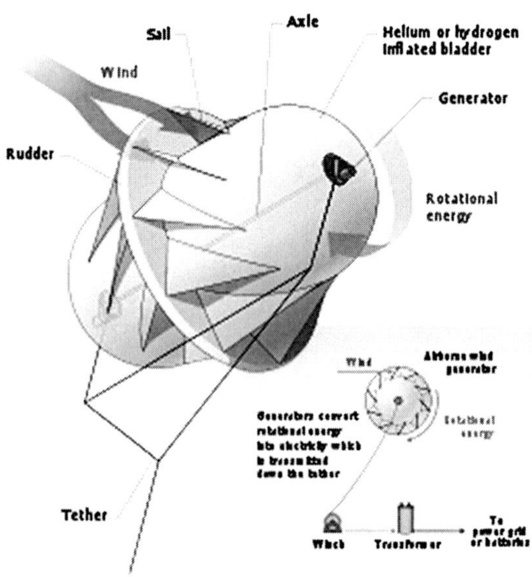

Figure 1: Airborne wind generator of Savonius style

An airborne wind turbine is a design concept for a wind turbine that is supported in the air without a tower. Airborne wind turbines may operate in low or high altitudes; they are part of a wider class of airborne wind energy systems (AWE) addressed by high-altitude wind power. When the generator is on the ground, then the tethered aircraft need not carry the generator mass or have a conductive tether. When the generator is aloft, then a conductive tether would be used to transmit energy to the ground or used aloft or beamed to receivers using microwave or laser. Airborne turbine systems would have the advantage of tapping an almost constant wind, without requirements for slip rings or yaw mechanism, and without the expense of tower construction. Kites and 'helicopters' come down when there is insufficient wind; kytoons and blimps resolve the matter. Also, bad weather such as lightning or thunderstorms, could temporarily suspend use of the machines, probably requiring them to be brought back down to the ground and covered. Some schemes require a long power cable and, if the turbine is high enough, an aircraft exclusion zone. As of 2011, no commercial airborne wind turbines are in regular operation.

Aerodynamic Variety

An aerodynamic airborne wind power system relies on the wind for support.

Bryan Roberts, a professor of engineering at the University of Technology, in Sydney, Australia, has proposed a helicopter-like craft which flies to 15,000 feet (4,600 m) altitude and stays there, held aloft by wings that generate lift from the wind, and held in place by a cable to a ground anchor. According to its designers, while some of the energy in the wind would be 'lost' on lift, the constant and potent winds would allow it to generate constant electricity. Since the winds usually blow horizontally, the turbines would be at an angle from the horizontal, catching winds while still generating lift. Deployment could be done by feeding electricity to the turbines, which would turn them into electric motors, lifting the structure into the sky.

The Dutch ex-astronaut and physicist Wubbo Ockels, working with the Delft University of Technology in the Netherlands, has designed and demonstrated an airborne wind turbine he calls a "Laddermill". It consists of an endless loop of kites. The kites lift one end of the endless loop, (the "ladder") up, and the released energy is used to drive an electric generator.

A Sept'09 paper from Carbon Tracking Ltd., Ireland has shown the capacity factor of a kite using ground-based generation to be 52.2%, which compares favourably with terrestrial wind-farm capacity factors of 30%.

A team from Worcester Polytechnic Institute in the United States has developed a smaller-scale with an estimated output of about 1 kW. It uses a kiteboarding kite to induce a rocking motion in a pivoting beam. The Kitegen uses a prototype vertical-axis wind turbine. It is an innovative plan (still in the construction phase) that consists of one wind farm with a vertical spin axis, and employs kites to exploit high-altitude winds.

The Kite Wind Generator (KWG) or Kitegen is claimed to eliminate all the static and dynamic problems that prevent the increase of the power (in terms of dimensions) obtainable from the traditional horizontal-axis wind turbine generators. Generating equipment would remain on the ground, and only the airfoils are supported by the wind. Such a wind power plant would be capable of producing the energy equivalent to a nuclear plant, while using an area of few square kilometres, without occupying it exclusively. (The majority of this area can still be used for agriculture, or navigation in the case of an offshore installation.)

Dave Santos, of KiteLab in Ilwaco, Washington, has been advancing single-surface wingmills to generate useful electricity with a ground-based generator.

The Rotokite is developed from Gianni Vergnano's idea. It uses aerodynamic profiles similar to kites that have been rotated on their own axis, emulating the performance of a propeller. The use of the rotation principle simplifies the problem of checking the flight of the kites and eliminates the difficulties due to the lengths of cables, enabling the production of wind energy at low cost. The Heli Wind Power is a project of Gianni Vergnano that uses a tethered kite.

In August 2011 the German company SkySails, producer of kites for ship propulsion, announced a kite-based wind power system for on- and offshore applications that is supposed to be "30% cheaper than current offshore solutions".

Aerostat Variety

An aerostat-type wind power system relies at least in part on buoyancy to support the wind-collecting elements. Aerostats vary in their designs and resulting lift-over-drag aerodynamic characteristic; the kiting effect of higher lift-over-drag shapes for the aerostat can effectively keep an airborne turbine aloft; a variety of such kiting balloons were made famous in the kytoon by Domina C. Jalbert.

Balloons can be added to the mix to keep systems up without wind, but balloons leak slowly and have to be resupplied with lifting gas, possibly patched as well. Very large, sun heated balloons may solve the helium or hydrogen leakage problems.

An Ontario based company called Magenn is developing a turbine called the Magenn Air Rotor System (MARS). The 1,000-foot (300 m)-wide MARS system uses a horizontal rotor in a helium suspended apparatus which is tethered to a transformer on the ground. Magenn claims that their technology provides high torque, low starting speeds, and superior overall efficiency thanks to its ability to deploy higher in comparison to non-aerial solutions. The first prototypes were built by TCOM in April 2008. No production units have been delivered.

Figure 2: Concept drawing of the Twind technology.

The Twind Technology concept uses a pair of captive balloons at an altitude of 800 metres. The tether cables transmit force to a rotating platform on the ground. Each balloon has a sail connected to it. The two balloons move alternately, the balloon with the sail open moves downwind and draws the other balloon upwind, and then the motion reverses. The tether cable can be used to turn the shaft of a generator to produce electrical energy or perform other works (grinding, sawing, pumping).

Estimated Costs

Sky Windpower estimates that its technology will be capable of producing electricity for \$0.02 per KWh, while a system of raising a kite to a high altitude while turning a generator on the ground, and then changing its shape so that it can be drawn back down with less energy than it produced on the way up, has been estimated to be capable of producing electricity for \$0.01 per KWh - both numbers being significantly lower than the current price of non-subsidized electricity.

Kite Applications

The kite is used to do certain things; one kite or many kites are applied to achieve certain purposes, objectives, or tasks, that is:

applications. Humans have applied the kite to bring perceived benefits during peace and war alike. New applications for the kite continue to be found. Only some innovative applications appear in national patents; others are communicated in newspapers, magazines, books, and Internet pages. Air kites, water kites, bi-media kites, fluid kites, gas kites, kytoons, paravanes, soil kites, solid kites, and plasma kites have niche applications that are furthering the interests of humans. Non-human-made kites have applications; some spiders make use of kiting.

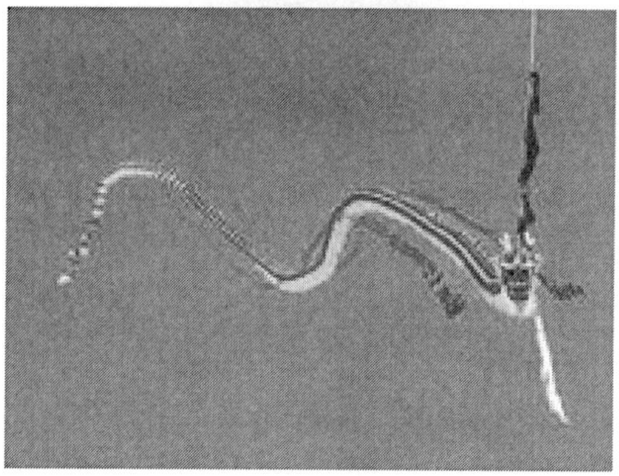

Figure 3: Chinese dragon kite more than one hundred feet long which flew in the annual Berkeley, California, kite festival in 2000. It is a kite-train of hundreds of linked circles (with outriggers ending in feathers for balance). The dragon's head is a bamboo frame with painted silk covering. Application: entertainment, recreation, event promotion, city or institution promotion.

Kites for Transportation

Long-distance travel across land , ice, and sea started centuries ago, but today significant tasks of moving people and goods from point A to point B are occurring; this is so in great part from the advances in kites and kite systems designs and technology, better understanding of winds, and use of computers and GPS. In 1889 kite sailing was carefully instructed via controlling large kite systems towing boats: Aerial Apparatus

Free-flight cross-country hang gliding kites both in the hang glider style and the paraglider style are permitting trips of hundreds of miles; records are recorded by the FAI. George Pocock (inventor) was an early pioneer in kites for transportation. NASA continues to explore free-flying kites for delivering goods to earth surface and non-earth planet surfaces, including Mars. The severe projects for using very large kites to sail cargo ships is currently underway; KiteSail(tm) and KiteShip (tm) along with a series of patents and improvements in control of large ship-carried kite systems aim to save significant fuel. Kiting one's kayak is getting a significant following A. Kinsman - kite kayaking tutorial..

Kite Aerial Photography (KAP)

Figure 4: William Abner Eddy patented a kite. He used kites for aerial photography (KAP).

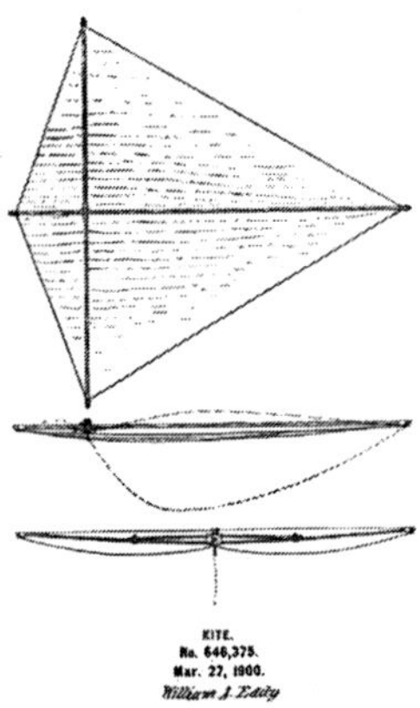

Figure 5: US646375 Willian A. Eddy kite patent image.

Figure 6: Photo of the ruins of the 1906 San Francisco Earthquake by George Lawrence of Chicago. Photographed using a kite system to loft a panoramic camera.

William A. Eddy of Eddy-kite fame lifted cameras to take photographs of cities and landscapes. Today KAP is the hobby of many enthusiasts, is a tool for surveying land and animals, and a mode for artistic expression.

Professor Charles Benton illustrates how KAP can grow in one's life. Scott Haefner has one of the most extensive collection of KAP photographs; he shares his technology. San Francisco. Those who do KAP are called *kapers* KAP.

Kites for Teaching Schooling Subjects

The kite is frequently the vehicle for teaching aerodynamics, mathematics, art, history, culture, materials, cooperation, physical education, and problem solving.

Kites as Subject of Publications

International organisations, national organisations, local kite clubs, kite shops, and individuals have kite and kiting as the focus of their publications. Subscribership and buyer levels vary. Kite pamphlets and books out-of-print become prized collectors' items.

Large Kites Being Applied Over the Oceans

The German company SkySails has developed ship-pulling kites as a supplemental power source for cargo ships, first tested in January 2008 on the ship MS Beluga Skysails. Trials on this 55 m ship have shown that, in favourable winds, the kite increases fuel-efficiency by up to 30%.

This system is planned to be in full commercial production late 2008. Kites are also available as an auxiliary sail or emergency spinnaker for sailing boats. Self-launching Parafoil kites are attached to the mast.

MS Beluga Skysails is the world's first commercial container cargo ship which is partially powered by a giant computer-controlled kite (160 m^2. or 1,722 sq ft.). The kite could reduce fuel consumption by 20%. It was launched 17 December 2007 and is set to leave the northern German port of Bremerhaven to Guanta, Venezuela at 1700 local time (1600 GMT), January 22, 2008.

Stephan Wrage, managing director of SkySails GmbH announced: "During the next few months we will finally be able to prove that our technology works in practice and significantly reduces fuel consumption and emissions."

Verena Frank, project manager at Beluga Shipping GmbH, SkySails GmbH's partner further stated that "the project's core concept was using wind energy as auxiliary propulsion power and using wind as a free of charge energy". Kite yachting may have started with Benjamin Franklin's pond pull. English Channel crossing.

Scientific Kiting

Kites have also been used for scientific purposes, such as Benjamin Franklin's famous (but dangerous) experiment proving that lightning is electricity. Kites had an historical role in lifting scientific instruments to measure atmospheric conditions for weather forecasting.

Kites Used to Test Designs Meant to Become Powered Aircraft

Figure 7: Wright kite with aim to get a powered aircraft.

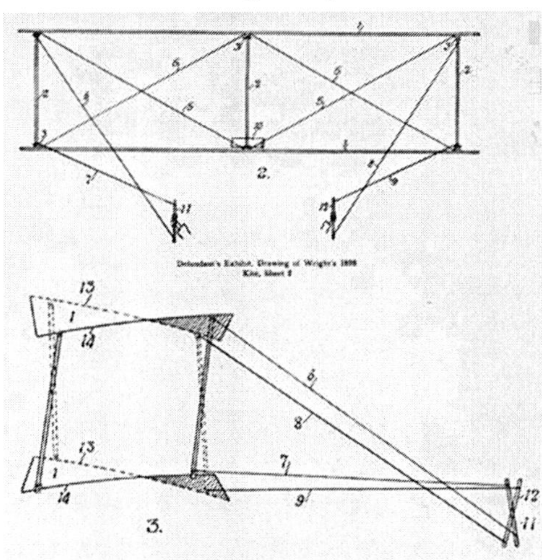

Figure 8: Early quad-line steerable kite by Wright Brothers as they aim for powered flight.

Kites were the precursors to aircraft, and were instrumental in the development of early flying craft. Alexander Graham Bell experimented with very large man-lifting kites, as did the Wright brothers and Lawrence Hargrave.

Conductor carried up by a kite in the sky can lead to a high voltage shock, which can endanger people and equipment.

Kites for Extreme Sports

Kite boarding, kite surfing, kite buggying, kite buggy jumping, kite landboarding, freestyle kite landboarding, snow kiting or snowkiting, downhill speed kiting, hang gliding, and kite high jumping are among the extreme kite sports wherein competitions are held.

Kites for Competitive Stunt-kiting Sport

Peter Powell's emphatic forwarding of two-line stunt kites or sport kites help to move stunt kiting into a popular activity as well as a competitive sport. Also, the parafoil stunt kites feed the same sporting activity. Events for kites of more than two control lines are very evident.

The four-line Revolution kite has been setting new standards in precision flying. Informal field competition and formal sport competition support a stunt-kite industry; seemingly endless refinements to the kite wing and kite line along with accessories continue unabated.

Kites for Sport Kite Fighting

A kite has two essential parts: wing and tether line. In kite fighting, the kite line plays a huge part in the activity. Sport kite fighting is perhaps 2000 year old; participation worldwide is high. North American Kite Fighter Association (NAFKA) Trawl-board and paravane innovator

In *Kite Fishing by the Salt-Water Natives of Mala or Malaita Island*, British Solomon Islands T. W. Edge-Partington, leaf kites are described. The sago palm or ivory nut tree has leaves from which natives of Mala or Malaita Island made kites for fishing.

Kites for Advertising Messages and Logos

Kites can also be used as light-effect carrier, e.g., by carrying light-sticks or battery-powered light effects. Promotional kites: Companies buy large quantities of kites that feature their advertisement. Messages are sometimes displayed by lighting systems that are built into the kite system.

Kites for Commercial Fishing

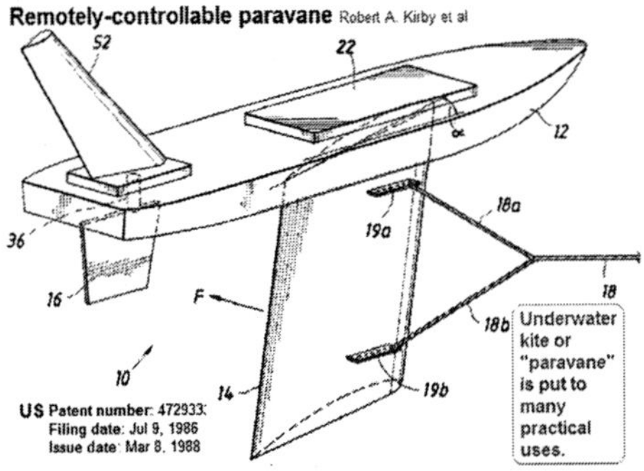

Figure 9: Water kite

Net-spreading underwater kites and kite vanes aid the control of large fishing nets. Remotely-controllable paravane Robert A. Kirby et al.

Kites for Recreational, Sport, and Subsistence Fishing

There are several ways kites are used in recreatonal and sport fishing. Lofting drop lines is one, but things don't stop there. Net-spreading underwater kites, soil kites (kiting achors), kiting bait, control-kite trolling of bait, recreational kiting during fishing sessions, aerial photography of fishing environment using kites, and out and back cycles of trolling bait using a kite.

Recreational fishing, commercial fishing, and scientific and military uses of depressors of tow lines use water kiting to accomplish the effects wanted.

The Use of Kites for Fishing—George Webster wrote comprehensively on kite fishing. Jetty/Pier Fishing.; _ Paravanes for Sportfishing. A plan view of a Solomon islander's leaf fishing kite is shown in a photograph held by the Pitt-Rivers Museum is viewable at *Natural History Magazine* online; Pick from the Past, *Natural History*, April 1957: "Go Fly a Kite".

Kites Used to Decorate Homes and Businesses

Kites stilled are hung decoratively in rooms of homes and businesses to set the tone of a home or selling environment.

Kites Used to Sell Kites

Many kite stores fly kites regularly so that people will see the kites; one of the final purposes is for the store to profit from the flying of the kites. Kites are necessary to increase sales of kites.

Kites Used for Generating Electricity from Rotation

Both air and hydro kites are used to generate electricity; the kite is set in the stream of air or water; various schemes are used to extract some of the stream's energy for converting that energy to electricity.

A major research and development project called Makani Power, based in California and funded by Google.org, is investigating the use of kites in harnessing high altitude wind currents to generate electricity.

Video links for generating electricity by using kites:

- Back and forth taking kite generates electricity:
- *MagGenn (Magnus-effect kite wind generator):* a Magnus-effect rotating kite electricity generation system: Magenn Power, Inc.;
- KiteGen (kite generator). Italian company is working on several methods of generating electricity from kite systems. One key method is the "stem" system. High altitude wind power: an era of abundance?
- Makani Power's 10kW test platform autonomously generating power:

Military Applications

Kites have been used for military uses in the past for signalling, for delivery of munitions, for free-flight kiting payloads from aircraft to ground positions, for kiting troops to points where they could parachute to destinations, for underwater kiting via paravanes to perform various underwater duties, for lifting payloads from one point to another, for raising rescue signals from rafts or stressed areas, for raising communications antenna, and for observation by lifting an observer above the field of battle, and by using kite aerial photography. Barrage kites have been used in both open frame kites and kytoon types to defence against enemy aircraft.

Kim Yu-Sin (or Kim Yushin), a Korean general, in 637 C.E. rallied his troops to defeat rebels by kite lofting a burning ball. Kites were also used by Admiral Yi of the Joseon (1392-1910) Dynasty of Korea.

During the Japanese invasions of Korea (1592-1598), Admiral Yi commanded his navy with kites. His kites had specific markings directing his fleet to perform his order. Admiral Yi was said to have over 300 such kites. The war eventually resulted in a Chinese and Korean victory; the kites played a minor role in the war's conclusion.

In more modern times the British navy also used kites to haul human lookouts high into the air to see over the horizon and possibly the enemy ships, for example with the kite developed by Samuel Franklin Cody.

Barrage kites were used to protect London as well as the Pacific coast of the United States during the last century. Kites and kytoons were used for lofting communications antenna. Submarines lofted observers in rotary kites. The Rogallo parawing kite and the Jalbert parafoil kite were used for governable parachutes (free-flying kites) to deliver troops and supplies.

Science and Meteorology

Kites have been used for scientific purposes, such as Benjamin Franklin's famous experiment proving that lightning is electricity. Kites were the precursors to aircraft, and were instrumental in the development of early flying craft.

Alexander Graham Bell experimented with very large man-lifting kites, as did the Wright brothers and Lawrence Hargrave. Kites had an historical role in lifting scientific instruments to measure atmospheric conditions for weather forecasting. The Weather Doctor Almanac.

Collecting kited spiders with kited nets: Kites are used to take samples of upper air and to collect things found in the upper air. The spiders that kite to disperse (so-called ballooning (spider)s) have been found in nets raised to upper air for collecting ; the method is noted carefully in *Spider Ballooning: Development and Evaluation of Field Trapping Methods (Araneae)* Balloon kite of the so-called ballooning spiderlings; the spiders' kite is not a balloon.

Radio Aerials and Light Beacons

Kites can be used for radio purposes, by kites carrying antennas for MF, LF or VLF-transmitters. This method was used for the reception station of the first transatlantic transmission by Marconi. Captive balloons may be more convenient for such experiments, because kite carried antennas require a lot of wind, which may be not always possible with heavy equipment and a ground conductor.

It must be taken into account during experiments, that a conductor carried by a kite can lead to a high voltage toward ground, which can endanger people and equipment, if suitable precautions (grounding through resistors or a parallel resonant-circuit tuned to transmission frequency) are not taken.

Figure 10: Antenna raising in 1915. The kite balloon ship HMS CANNING anchored off Salonika with kite balloon aloft, November 1915.

Figure 11: Kytoon equipped ship. US Navy photo. A kite balloon has been deployed from the USS Arizona. The kite balloon has a two man crew.

Kites for Lifting Antennas: Kites can be used to carry light effects such as lightsticks or battery powered lights.

Kite Traction

Kites can be used to pull people and vehicles downwind. Efficient foil-type kites such as power kites can also be used to sail upwind under the same principles as used by other sailing craft, provided that lateral forces on the ground or in the water are redirected as with the keels, centre boards, wheels and ice blades of traditional sailing craft. In the last two decades several kite sailing sports have become popular, such as kite buggying and kite surfing.

Kite sailing opens several possibilities not available in traditional sailing:

- Wind speeds are greater at higher altitudes
- Kites may be manoeuvred dynamically which increases the force available dramatically
- There is no need for mechanical structures to withstand bending forces; vehicles or hulls can be very light or dispensed with all together

The German company SkySails has developed ship-pulling kites as a supplemental power source for cargo ships, first tested in January 2008 on the ship MS Beluga Skysails. Trials on this 55 m ship have shown that, in favourable winds, the kite reduces fuel consumption by up to 30%. This system is planned to be in full commercial production late. Kites are available as an auxiliary sail or emergency spinnaker for sailing boats. Self-launching Parafoil kites are attached to the mast. MS Beluga Skysails is the world's first commercial container cargo ship partially powered by a giant computer-controlled kite (160 m^2. or 1,722 sq ft.). The kite could reduce fuel consumption by 20%. It was launched on 17 December 2007 and was set to leave the northern German port of Bremerhaven to Guanta, Venezuela on January 22, 2008. Stephan Wrage, managing director of SkySails GmbH announced: "During the next few months we will finally be able to prove that our technology works in practice and significantly reduces fuel consumption and emissions." Verena Frank, project manager at Beluga Shipping GmbH, SkySails GmbH's partner further stated that "the project's core concept was using wind energy as auxiliary propulsion power and using wind as a free of charge energy". Anne Quemere has crossed solo the Atlantic Ocean using power kites. And on November 4, 2008, she started her bid to cross the Pacific Ocean under power kites as sailing powerto pull her in her small boat called *Oceankite*.

Power Generation

Main article: High altitude wind power. Stakeholders of the nascent industry of generating power and saving fuel by use of large kite systems by 2010 was over 500 persons. A conceptual research and development project by Makani Power, based in California and funded by Google, is investigating the use of kites in harnessing high altitude wind currents to generate electricity. Magenn Power's Floating Wind Generators Nearing Deployment and KiteGen.

Non-man-made Kite Applications

There are natural kites that play a part in shaping what happens on earth. Some leaves kite to relieve wind pressures, pump fluids, and to disconnect annually to fertilize the soils. Poet Marvin Bell recognised leaves are kites in his *Nightworks: Poems 1962-2000.* The leaf wafts kitingly in the wind held by the tethered leaf stem; when it is fall time, the leaf stem has a de-mooring disconnect process; the wind then easily interracts with the leaf to cause it to fly off the trees and into a gliding fall to the ground. There is a following of kite makers that bridle leaves to fly them again as kites.

Billions of spiders use kiting to travel or disperse; others use kiting to build bridge lines for their webs. Spiders hanging in the moving air on their silks are deflected to various points where they make anchor points for their web building. Carol Frost, biology researcher of the University of Alberta, Canada, observed kiting in spiderlings.

Power Kite

Figure 12: Kitesurfing in strong onshore winds off the north shore of O»ahu in Hawaii. Note the windsurfer catching the wave break.

A power kite or traction kite is a large kite designed to provide significant pull to the user.

They come in three main forms: foils, leading edge inflatables and supported leading edge. There are also rigid-framed kites and soft single skin kites. There are several different control systems used with these kites which have two to five lines and a bar or handles. Most foil kites are made for use on land as they are "opened celled" so air flows in and out easily, this can be used as a safety feature to depower the kite. There are a few exceptions for this with new foils that use closed cell technology; they float on the water while you relaunch.

Power kites are generally used in conjunction with a vehicle or board, such as in:

- kitesurfing on a kiteboard
- kite buggying on a purpose-built 3-wheeled cart
- kite landboarding on an all-terrain/mountain/land board
- kite skating on all-terrain roller skates
- snowkiting on skis or snowboards
- kite jumping, using a power kite for jumping above the ground
- man lifting, where a harnessed kite flier is moored to the ground or one or more people to provide tension and lift
- kiteboating, on a boat

Research is also under way in the use of kites to generate electric power to be fed into a power grid. Kites can be used to reach high altitude winds such as a jet stream, which are always present, even if ground level winds available to wind turbines are absent.

Kites of related design are used for sailing, including speed sailing. *Jacob's Ladder*, a kite-powered boat, set the C-Class world sailing speed record with a speed of 25 knots (46 km/h) in 1982, a record that stood for six years. A kiteboard was the first sailing craft to exceed a speed of 50 knots (93 km/h) in October 2008. Power kites range in size from 1.5 to 50 m (16 to 540 sq^2 ft). All kites are made for specific purposes: some for water, land, power or manoeuvrability.

Floating wind Turbine

Figure 13: The world's first full-scale floating wind turbine, Hywind, being assembled in the Amøy Fjord near Stavanger, Norway, before deployment in the North Sea.

A floating wind turbine is an offshore wind turbine mounted on a floating structure that allows the turbine to generate electricity in water depths where bottom-mounted towers are not feasible. The wind can be stronger and steadier over water due to the absence of topographic features that may disrupt wind flow. The relocation of wind farms into the sea can reduce visual pollution if the windmills are sited more than 12 miles (19 km) offshore, provide better accommodation of fishing and shipping lanes, and allow silting near heavily developed coastal cities. Floating wind parks are wind farms that site several floating wind turbines closely together to take advantage of common infrastructure such as power transmission facilities.

History

The concept for "large-scale offshore floating wind turbines was introduced by Professor William E. Heronemus at the University of Massachusetts in 1972. It was not until the mid 1990s, after the commercial wind industry was well established, that the topic was taken up again by the mainstream research community." As of 2003, existing offshore fixed-bottom wind turbine technology deployments had been limited to water depths of 30 metres. Worldwide deep-water wind resources are extremely abundant in subsea areas with depths up to 600 metres, which are thought to best facilitate transmission of the generated electric power to shore communities.

Operational Deep-water Platforms

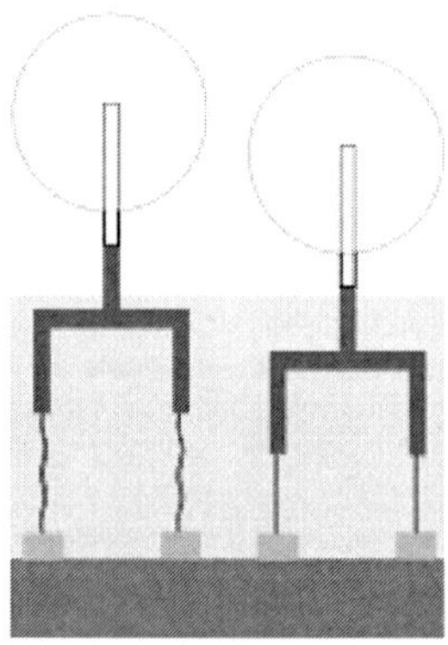

Figure 14: A tension leg mooring system as used by Blue H: left-hand tower-bearing structure (grey) is free floating, the right-hand structure is pulled by the tensioned cables (red) down towards the seabed anchors (light-grey)

As of 2011, there have been only three operational floating wind turbines used to farm wind energy. Blue H deployed the first 80 kW floating wind turbine 113 kilometres (70 mi) off the coast of Italy in December, 2007. It was then decommissioned at the end of 2008 after completing a planned test year of gathering operational data. SeaTwirl deployed their first floating grid connected wind turbine off the coast of Sweden in August, 2011. There new design is storing energy as a flywheel and by transporting water it can be charged and uncharged even after the wind has stopped blowing. The first large-capacity, 2.3 megawatt floating wind turbine is Hywind, which became operational in the North Sea off of Norway in September 2009, and is still operational as of October 2010.

Blue H Technologies

Blue H Technologies of the Netherlands operated the first floating wind turbine, a prototype deep-water platform with an 80-kilowatt turbine off Puglia, southeast Italy in 2008. Installed 21 km off the coast in waters 113 metres deep in order to gather test data on wind and sea conditions, the small prototype unit was decommissioned at the end of 2008. Blue H has plans to build a 38-unit deepwater wind farm at the same location. The Blue H technology utilised a tension-leg platform design and a two-bladed turbine. The two-bladed design can have a "much larger chord, which allows a higher tip speed than those of three-bladers. The resulting increased background noise of the two-blade rotor is not a limiting factor for offshore sites." As of 2009, Blue H was building a full-scale commercial 2.4 MW_e unit in Brindisi, Italy which it expected to deploy at the same site of the prototype in the southern Adriatic Sea in 2010. This is the first unit in the planned 90 MW *Tricase* offshore wind farm, located more than 20 km off the Puglia coast line.

Hywind

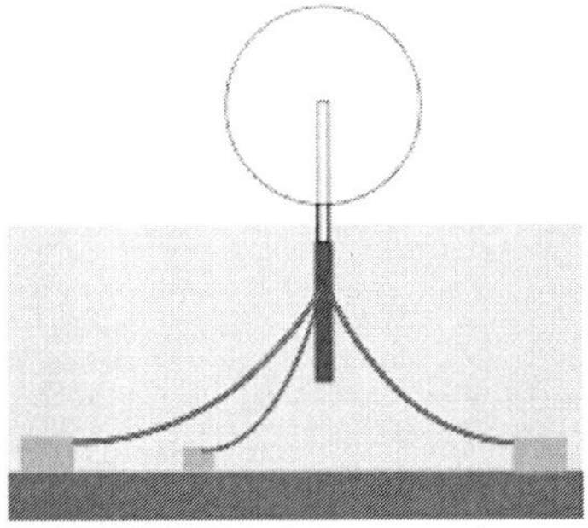

Figure 15: A single floating cylindrical spar buoy moored by catenary cables. Hywind uses a *ballasted catenary* layout that adds 60 tonne weights hanging from the midpoint of each anchor cable to provide additional tension.

The world's first operational deep-water floating *large-capacity* wind turbine is the Hywind, in the North Sea off Norway. The Hywind was towed out to sea in early June 2009.

The 2.3 megawatt turbine was constructed by Siemens Wind Power and mounted on a floating tower with a 100 metre deep draft. The float tower was constructed by Technip. Statoil says that floating wind turbines are still immature and commercialisation is distant.

The installation is owned by Statoil and will be tested for two years. After assembly in the calmer waters of Amøy Fjord near Stavanger, Norway, the 120-metre-tall tower with a 2.3 MW turbine was towed 10 km offshore into 220-metre-deep water, 10 km southwest of Karmøy, on 6 June 2009 for a two year test deployment."

Alexandra Beck Gjorv of Statoil said, " The experiment should help move offshore wind farms out of sight ... The global market for such turbines is potentially enormous, depending on how low we can press costs." The unit became operational in the summer of 2009. Hywind was inaugurated on 8 September 2009. As of October 2010, after a full year of operation, the Hywind turbine is still operating and generating electricity for the Norwegian grid, and still is as of February 2011.

The turbine cost 400 million kroner (around US$62 million) to build and deploy. The 13-kilometre (8.1 mi) long submarine power transmission cable was installed in July, 2009 and system test including rotor blades and initial power transmission was conducted shortly thereafter. The installation is expected to generate about 9 GW·h of electricity annually. The SWATH (Small Waterplane Area Twin Hull), a new class of offshore wind turbine service boat, will be tested at Hywind.

Hywind delivered 7.3 GWh in 2010, and survived 11 metre waves with seemingly no wear. As of June 2011, additional pilot Hywind installations are planned in the US and in the North Sea off the coast of Scotland.

Topologies

Platform topologies can be classified into:

- *single-turbine-floater* (one wind turbine mounted on a floating structure)
- *multiple turbine floaters* (multiple wind turbines mounted on a floating structure)

Engineering Considerations

Undersea mooring of floating wind turbines are accomplished with three principal mooring systems. Two common types of engineered design for anchoring floating structures include tension-leg and catenary loose mooring systems. *Tension leg mooring systems* have vertical tethers under tension providing large restoring moments in pitch and roll. *Catenary mooring systems* provide station keeping for an offshore structure yet provide little stiffness at low tensions." A third form of mooring system is the *ballasted catenary* configuration, created by adding multiple-tonne weights hanging from the midsection of each anchor cable in order to provide additional cable tension and therefore increase stiffness of the above-water floating structure.

Economics

"Technically, the [theoretical] feasibility of deepwater [floating] wind turbines is not questioned as long-term survivability of floating structures has already been successfully demonstrated by the marine and offshore oil industries over many decades. However, the economics that allowed the deployment of thousands of offshore oil rigs have yet to be demonstrated for floating wind turbine platforms. For deepwater wind turbines, a floating structure will replace pile-driven monopoles or conventional concrete bases that are commonly used as foundations for shallow water and land-based turbines.

The floating structure must provide enough buoyancy to support the weight of the turbine and to restrain pitch, roll and heave motions within acceptable limits. The capital costs for the wind turbine itself will not be significantly higher than current marinised turbine costs in shallow water. Therefore, the economics of deepwater wind turbines will be determined primarily by the additional costs of the floating structure and power distribution system, which are offset by higher offshore winds and close proximity to large load centres (e.g. shorter transmission runs)."

As of 2009 however, the economic feasibility of shallow-water offshore wind technologies is more completely understood. With empirical data obtained from fixed-bottom installations off many countries for over a decade now, representative costs are well understood. Shallow-water turbines cost between 2.4 and 3 million United States dollars per megawatt to install, according to the World Energy Council.

As of 2009, the practical feasibility and per-unit economics of deep-water, floating-turbine offshore wind is yet to be seen. Initial deployment of single full-capacity turbines in deep-water locations began only in 2009.

As of October 2010, new feasibility studies are supporting that floating turbines are becoming both technically and economically viable in the UK and global energy markets. "The higher up-front costs associated with developing floating wind turbines would be offset by the fact that they would be able to access areas of deep water off the coastlne of the UK where winds are stronger and reliable."

The recent Offshore Valuation study conducted in the UK has confirmed that using just one third of the UK's wind, wave and tidal resource could generate energy equivalent to 1 billion barrels of oil per year; the same as North Sea oil and gas production. Some of the primary challenges are the coordination needed to develop transmission lines.

Floating Design Concepts

Figure 16:

Ideol

Ideol is a French company that has patented a new floating platform concept specifically designed for offshore wind.

While the floater concept and patents are not yet publicly disclosed, the company is communicating on its web site mobility solution to reduce wake losses in an offshore wind farm by repositioning the floating turbines depending on the wind direction. The company has patented a mechanical solution to move the floater along its mooring lines and has developed a software to optimise in real-time the farm layout. Eliminating wake losses allows to increase significantly the power production as well as to reduce the long-term components failures. According to publicly released information, Ideol has a construction and installation cost of around 1M Euro per MW. As such, the company intends to offer an alternative to fixed foundations starting from 40 m water depth.

OffshoreWind.biz reported that the company will build a 5 MW floating prototype off the European coast in 2013.

WindFloat

WindFloat is a floating foundation for offshore wind turbines designed and patented by Principle Power. It is to be tested in autumn 2011 off the coast of Portugal with a Vestas V80 2MW wind turbine.

The foundation attempts to improve dynamic stability at shallow draft by dampening wave and turbine induced motion utilising a tri-column triangular platform with the wind turbine positioned on only one of the three columns. The triangular platform is then "moored with 4 lines, 2 of which are connected to the column stabilising the turbine, thus creating an asymmetric" mooring to increase stability and reduce motion.

As the wind shifts direction and changes the loads on the turbine and foundation, pumps will shift ballast water between foundation chambers. The project is managed by the joint venture WindPlus (led by electricity provider Energias de Portugal). Vestas turbines will be the standard for the project. Construction cost is expected to be below $30 million, and funded by the project partners and Fundo de Apoio a Inovacao. This technology could allow wind turbines to be sited in offshore areas that were previously considered inaccessible, areas having water depth exceeding 50 metres and more powerful wind resources than shallow-water offshore wind farms typically encounter.

Nautica Windpower

Figure 17: Nautica Windpower's AFT design features a downwind two-bladed rotor with passive wind alignment to reduce costs

Nautica Windpower uses a patented technology aimed at reducing system weight, complexity and costs for deep water sites. Scale model tests in open water have been conducted and structural dynamics modelling is under development for a multi-megawatt design. Nautica Windpower's Advanced Floating Turbine (AFT) uses a single mooring line and a downwind two-bladed rotor configuration that is deflection tolerant and aligns itself with the wind without an active yaw system. Two-bladed, downwind turbine designs that can accommodate flexibility in the blades will potentially prolong blade lifetime, diminish structural system loads and reduce offshore maintenance needs, yielding lower lifecycle costs.

OC3-Hywind

The International Energy Agency (IEA), under the auspices of their *Offshore Code Comparison Collaboration* (OC3) initiative, has completed high-level design and simulation modelling of the *OC-3 Hywind* system, a 5-MW wind turbine installed on a floating spar buoy, moored with catenary mooring lines, in water depth of 320 metres. The spar buoy platform would extend 120 metres below the surface and the mass of such a system, including ballast would exceed 7.4 million kg.

DeepWind

Risø and 11 international partners started a 4-year program called DeepWind in October 2010 to create and test economical floating Vertical Axis Wind Turbines up to 20MW. The program is supported with €3m through EUs Seventh Framework Programme. Partners include TUDelft, SINTEF, Statoil and United States National Renewable Energy Laboratory.

VertiWind—Proposals—Floating Wind Farms

As of September 2011, Japan plans to build a pilot floating wind farm, with six 2-megawatt turbines, off the Fukushima coast of northeast Japan where the recent disaster has created a scarcity of electric power. After the evaluation phase is complete in 2016, "Japan plans to build as many as 80 floating wind turbines off Fukushima by 2020." The cost is expected to be in the range of 10-20 billion Yen over five years to build the first six floating wind turbines. Some foreign companies also plan to bid on the 1 GW large floating wind farm that Japan hopes to build by 2020.

As of November 2011, Statoil plans to build a multi-turbine project in Scottish waters utilising the Hywind design.

The US State of Maine solicited proposals in September 2010 to build the world's first floating, commercial wind farm. The RFP is seeking proposals for 25 MW of deep-water offshore wind capacity to supply power for 20-year long-term contract period via grid-connected floating wind turbines in the Gulf of Maine. Successful bidders must enter into long-term power supply contracts with either Central Maine Power Company (CMP), Bangor Hydro-Electric Company (BHE), or Maine Public Service Company (MPS). Proposals were due by May 2011. Some vendors who could bid on the proposed project have expressed concerns about dealing with the United States regulatory environment. Since the proposed site is in Federal waters, developers would need a permit from the Minerals Management Service, "which took more than seven years to approve a yet-to-be-built, shallow-water wind project off Cape Cod," and is also the agency under fire in June 2010 for lax oversight of deepwater oil drilling in Federal waters. "Uncertainty over regulatory hurdles in the United States ... is 'the Achilles heel' for Maine's ambitions for deepwater wind."

Wind Power

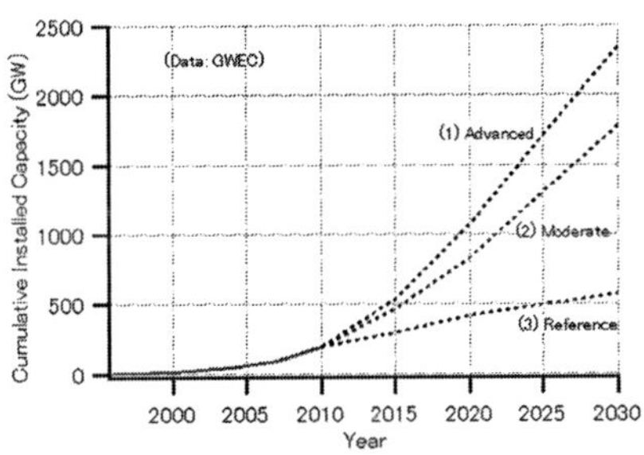

Figure 18: *Wind power:* worldwide installed capacity forecast

Figure 19: Burbo Bank Offshore Wind Farm, at the entrance to the River Mersey in North West England

Wind power is the conversion of wind energy into a useful form of energy, such as using wind turbines to make electricity, windmills for mechanical power, windpumps for water pumping or drainage, or sails to propel ships.

The total amount of economically extractable power available from the wind is considerably more than present human power use from all sources. At the end of 2010, worldwide nameplate capacity of wind-powered generators was 197 gigawatts (GW). Wind power now has the capacity to generate 430 TWh annually, which is about 2.5% of worldwide electricity usage. Over the past five years the average annual growth in new installations has been 27.6 percent. Wind power market penetration is expected to reach 3.35 percent by 2013 and 8 percent by 2018. Several countries have already achieved relatively high levels of wind power penetration, such as 21% of stationary electricity production in Denmark, 18% in Portugal, 16% in Spain, 14% in Ireland and 9% in Germany in 2010. As of 2011, 83 countries around the world are using wind power on a commercial basis.

A large wind farm may consist of several hundred individual wind turbines which are connected to the electric power transmission network. Offshore wind power can harness the better wind speeds that are available offshore compared to on land, so offshore wind power's contribution in terms of electricity supplied is higher. Small onshore wind facilities are used to provide electricity to isolated locations and utility companies increasingly buy back surplus electricity produced by small domestic wind turbines.

Although a variable source of power, the intermittency of wind seldom creates problems when using wind power to supply up to 20% of total electricity demand, but as the proportion rises, increased costs, a need to use storage such as pumped-storage hydroelectricity, upgrade the grid, or a lowered ability to supplant conventional production may occur. Power management techniques such as excess capacity, storage, dispatchable backing supply (usually natural gas), exporting and importing power to neighbouring areas or reducing demand when wind production is low, can mitigate these problems.

Wind power, as an alternative to fossil fuels, is plentiful, renewable, widely distributed, clean, produces no greenhouse gas emissions during operation, and uses little land. In operation, the overall cost per unit of energy produced is similar to the cost for new coal and natural gas installations. The construction of wind farms is not universally welcomed, but any effects on the environment from wind power are generally much less problematic than those of any other power source.

History

Humans have been using wind power for at least 5,500 years to propel sailboats and sailing ships. Windmills have been used for irrigation pumping and for milling grain since the 7th century AD in what is now Afghanistan, India, Iran and Pakistan.

In the US, the development of the "water-pumping windmill" was the major factor in allowing the farming and ranching of vast areas otherwise devoid of readily accessible water. Windpumps contributed to the expansion of rail transport systems throughout the world, by pumping water from water wells for the steam locomotives. The multi-bladed wind turbine atop a lattice tower made of wood or steel was, for many years, a fixture of the landscape throughout rural America. When fitted with generators and battery banks, small wind machines provided electricity to isolated farms.

In July 1887, a Scottish academic, Professor James Blyth, undertook wind power experiments that culminated in a UK patent in 1891. In the US, Charles F. Brush produced electricity using a wind powered machine, starting in the winter of 1887-1888, which powered his home and laboratory until about 1900. In the 1890s, the Danish scientist and inventor Poul la Cour constructed wind turbines to generate electricity, which was then used to produce hydrogen. These were the first of what was to become the modern form of wind turbine.

Small wind turbines for lighting of isolated rural buildings were widespread in the first part of the 20th century. Larger units intended for connection to a distribution network were tried at several locations including Balaklava USSR in 1931 and in a 1.25 megawatt (MW) experimental unit in Vermont in 1941.

In the 1970s, U.S. industries teamed with NASA in a research program which created the NASA wind turbines, developing and testing many of the features of modern utility-scale turbines.

The modern wind power industry began in 1979 with the serial production of wind turbines by Danish manufacturers Kuriant, Vestas, Nordtank, and Bonus. These early turbines were small by today's standards, with capacities of 20–30 kW each. Since then, they have increased greatly in size, with the Enercon E-126 capable of delivering up to 7 MW, while wind turbine production has expanded to many countries.

Today, wind generated energy is the fastest growing source of renewable energy. Wind power is expected to grow worldwide in the twenty-first century.

Wind Energy

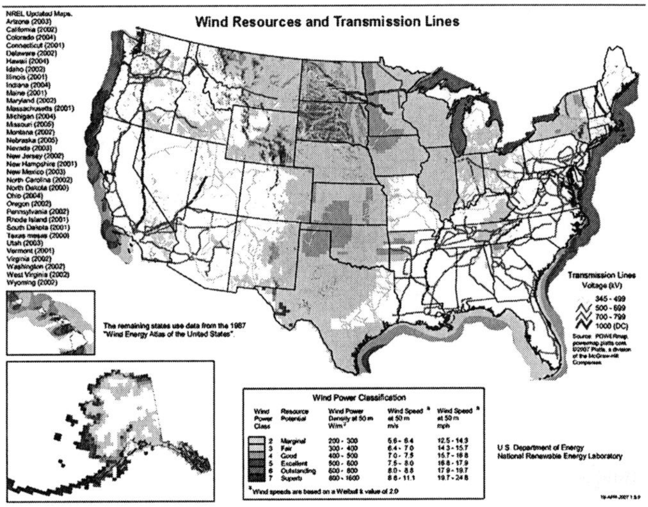

Figure 20: Map of available wind power for the United States. Colour codes indicate wind power density class.

The Earth is unevenly heated by the sun, such that the poles receive less energy from the sun than the equator; along with this, dry land heats up (and cools down) more quickly than the seas do. The differential heating drives a global atmospheric convection system reaching from the Earth's surface to the stratosphere which acts as a virtual ceiling. Most of the energy stored in these wind movements can be found at high altitudes where continuous wind speeds of over 160 km/h (99 mph) occur. Eventually, the wind energy is converted through friction into diffuse heat throughout the Earth's surface and the atmosphere.

The total amount of economically extractable power available from the wind is considerably more than present human power use from all sources. The most comprehensive study as of 2005 found the potential of wind power on land and near-shore to be 72 TW, equivalent to 54,000 MToE (million tons of oil equivalent) per year, or over five times the world's current energy use in all forms. The potential takes into account only locations with mean annual wind speeds e" 6.9 m/s at 80 m. The study assumes six 1.5 megawatt, 77 m diametre turbines per square kilometre on roughly 13% of the total global land area (though that land would also be available for other compatible uses such as farming). The authors acknowledge that many practical barriers would need to be overcome to reach this theoretical capacity.

Others authors disagree with the bottom-up methodology and cites problems with such methods which can be "violating the first principle of energy conservation". The principle is that the amount of energy which can be extracted from wind power can actually exceed the power currently present in the lower atmosphere using such bottom-up analyses. (i.e. There is 100 TW of total power in the lower 200m of the entire atmosphere and somes studies go well over that limit.) Their results show 1 TWe for the limit of wind power energy, which is much lower than previous estimates.

The practical limit to exploitation of wind power will be set by economic and environmental factors, since the resource available is far larger than any practical means to develop it.

Distribution of Wind Speed

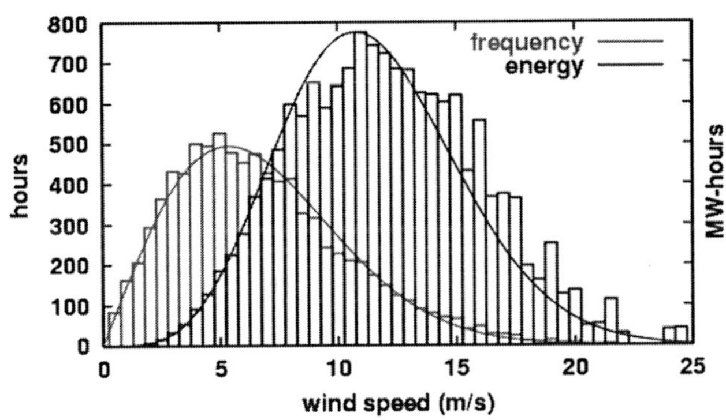

Figure 21: Distribution of wind speed (red) and energy (blue) for all of 2002 at the Lee Ranch facility in Colorado. The histogram shows measured data, while the curve is the Rayleigh model distribution for the same average wind speed.

The strength of wind varies, and an average value for a given location does not alone indicate the amount of energy a wind turbine could produce there. To assess the frequency of wind speeds at a particular location, a probability distribution function is often fit to the observed data. Different locations will have different wind speed distributions. The Weibull model closely mirrors the actual distribution of hourly wind speeds at many locations. The Weibull factor is often close to 2 and therefore a Rayleigh distribution can be used as a less accurate, but simpler model.

Wind Farms

A wind farm is a group of wind turbines in the same location used for production of electric power. A large wind farm may consist of several hundred individual wind turbines, and cover an extended area of hundreds of square miles, but the land between the turbines may be used for agricultural or other purposes. A wind farm may also be located offshore.

Many of the largest operational onshore wind farms are located in the US. As of November 2010, the Roscoe Wind Farm is the largest onshore wind farm in the world at 781.5 MW, followed by the Horse Hollow Wind Energy Centre (735.5 MW).

As of November 2010, the Thanet Wind Farm in the UK is the largest offshore wind farm in the world at 300 MW, followed by Horns Rev II (209 MW) in Denmark.

There are many large wind farms under construction and these include BARD Offshore 1 (400 MW), Clyde Wind Farm (548 MW), Greater Gabbard wind farm (500 MW), Lincs Wind Farm (270 MW), London Array (1000 MW), Lower Snake River Wind Project (343 MW), Macarthur Wind Farm (420 MW), Shepherds Flat Wind Farm (845 MW), Sheringham Shoal (317 MW), and the Walney Wind Farm (367 MW).

Wind Power Usage

Worldwide there are now many thousands of wind turbines operating, with a total nameplate capacity of 194,400 MW. World wind generation capacity more than quadrupled between 2000 and 2006, doubling about every three years.

The United States pioneered wind farms and led the world in installed capacity in the 1980s and into the 1990s. In 1997 German installed capacity surpassed the U.S., a position it held until passed by the U.S. in 2008. China rapidly expanded its wind installations in the late 2000s and passed the U.S. in 2010 to become the world leader.

Europe accounted for 48% of the world total in 2009. In 2010, Spain became Europe's leading producer of wind energy, achieving 42,976 GWh. However, Germany holds the first place in Europe in terms of installed capacity, with a total of 27,215 MW at December 31, 2010.

Wind power accounts for approximately 21% of electricity use in Denmark, 18% in Portugal, 16% in Spain, 14% in the Republic of Ireland, and 9% in Germany.

Table 1: Top 10 countries by nameplate windpower capacity (2010)

Country	*Windpower Capacity (MW)*
China	44,733
United States	40,180
Germany	27,215
Spain	20,676
India	13,066
Italy	5,797
France	5,660
United Kingdom	5,204
Canada	4,008
Denmark	3,734

Table 2: Top 10 EU countries by windpower electricity production (December 2010)

Country	*Windpower Electricity Production (GWh)*
Spain	42,976
Germany	35,500
United Kingdom	11,440
France	9,600
Portugal	8,852
Denmark	7,808
Netherlands	3,972
Sweden	3,500
Ireland	3,473
Greece	2,200
Austria	2,100

Growth Trends

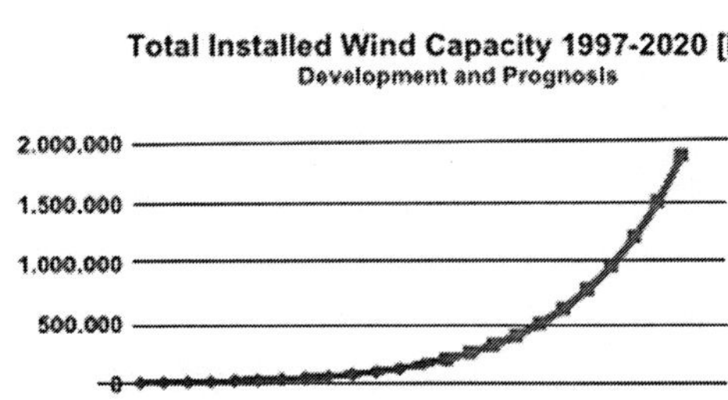

Figure 22: Worldwide installed capacity 1997–2020 [MW], developments and prognosis. Data source: WWEA

In 2010, more than half of all new wind power was added outside of the traditional markets in Europe and North America. This was largely from new construction in China, which accounted for nearly half the new wind installations (16.5 GW).

Global Wind Energy Council (GWEC) figures show that 2007 recorded an increase of installed capacity of 20 GW, taking the total installed wind energy capacity to 94 GW, up from 74 GW in 2006. Despite constraints facing supply chains for wind turbines, the annual market for wind continued to increase at an estimated rate of 37%, following 32% growth in 2006.

In terms of economic value, the wind energy sector has become one of the important players in the energy markets, with the total value of new generating equipment installed in 2007 reaching €25 billion, or US$36 billion.

Although the wind power industry was affected by the global financial crisis in 2009 and 2010, a BTM Consult five year forecast up to 2013 projects substantial growth. Over the past five years the average growth in new installations has been 27.6 percent each year. In the forecast to 2013 the expected average annual growth rate is 15.7 percent.

More than 200 GW of new wind power capacity could come on line before the end of 2013. Wind power market penetration is expected to reach 3.35 percent by 2013 and 8 percent by 2018.

Offshore Wind Power

Offshore wind power refers to the construction of wind farms in bodies of water to generate electricity from wind. Better wind speeds are available offshore compared to on land, so offshore wind power's contribution in terms of electricity supplied is higher.

Siemens and Vestas are the leading turbine suppliers for offshore wind power. DONG Energy, Vattenfall and E.ON are the leading offshore operators.

As of October 2010, 3.16 GW of offshore wind power capacity was operational, mainly in Northern Europe. According to BTM Consult, more than 16 GW of additional capacity will be installed before the end of 2014 and the UK and Germany will become the two leading markets.

Offshore wind power capacity is expected to reach a total of 75 GW worldwide by 2020, with significant contributions from China and the US.

Electricity Generation

Figure 23: Typical components of a wind turbine (gearbox, rotor shaft and brake assembly) being lifted into position

In a wind farm, individual turbines are interconnected with a medium voltage (often 34.5 kV), power collection system and communications network. At a substation, this medium-voltage electric current is increased in voltage with a transformer for connection to the high voltage electric power transmission system.

The surplus power produced by domestic microgenerators can, in some jurisdictions, be fed into the network and sold to the utility company, producing a retail credit for the microgenerators' owners to offset their energy costs.

Grid Management

Induction generators, often used for wind power, require reactive power for excitation so substations used in wind-power collection systems include substantial capacitor banks for power factor correction. Different types of wind turbine generators behave differently during transmission grid disturbances, so extensive modelling of the dynamic electromechanical characteristics of a new wind farm is required by transmission system operators to ensure predictable stable behaviour during system faults.

In particular, induction generators cannot support the system voltage during faults, unlike steam or hydro turbine-driven synchronous generators. Doubly fed machines generally have more desirable properties for grid interconnection. Transmission systems operators will supply a wind farm developer with a *grid code* to specify the requirements for interconnection to the transmission grid. This will include power factor, constancy of frequency and dynamic behaviour of the wind farm turbines during a system fault.

Capacity Factor

Since wind speed is not constant, a wind farm's annual energy production is never as much as the sum of the generator nameplate ratings multiplied by the total hours in a year. The ratio of actual productivity in a year to this theoretical maximum is called the capacity factor. Typical capacity factors are 20–40%, with values at the upper end of the range in particularly favourable sites. For example, a 1 MW turbine with a capacity factor of 35% will not produce 8,760 MW·h in a year (1 × 24 × 365), but only 1 × 0.35 × 24 × 365 = 3,066 MW·h, averaging to 0.35 MW. Online data is available for some locations and the capacity factor can be calculated from the yearly output.

Unlike fuelled generating plants, the capacity factor is affected by several parameters, including the variability of the wind at the site, but also the generator size- having a smaller generator would be cheaper and achieve higher capacity factor, but would make less electricity (and money) in high winds. Conversely a bigger generator would cost more and generate little extra power and, depending on the type, may stall out at low wind speed. Thus an optimum capacity factor can be used, which is usually around 20-35%.

In a 2008 study released by the U.S. Department of Energy's Office of Energy Efficiency and Renewable Energy, the capacity factor achieved by the wind turbine fleet is shown to be increasing as the technology improves. The capacity factor achieved by new wind turbines in 2004 and 2005 reached 36%.

Penetration

Wind energy "penetration" refers to the fraction of energy produced by wind compared with the total available generation capacity. There is no generally accepted "maximum" level of wind penetration. The limit for a particular grid will depend on the existing generating plants, pricing mechanisms, capacity for storage or demand management, and other factors. An interconnected electricity grid will already include reserve generating and transmission capacity to allow for equipment failures; this reserve capacity can also serve to regulate for the varying power generation by wind plants. Studies have indicated that 20% of the total electrical energy consumption may be incorporated with minimal difficulty. These studies have been for locations with geographically dispersed wind farms, some degree of dispatchable energy, or hydropower with storage capacity, demand management, and interconnection to a large grid area export of electricity when needed. Beyond this level, there are few technical limits, but the economic implications become more significant. Electrical utilities continue to study the effects of large (20% or more) scale penetration of wind generation on system stability and economics.

At present, a few grid systems have penetration of wind energy above 5%: Denmark (values over 19%), Spain and Portugal (values over 11%), Germany and the Republic of Ireland (values over 6%). But even with a modest level of penetration, there can be times where wind power provides a substantial percentage of the power on a grid. For example, in the morning hours of 8 November 2009, wind energy produced covered more than half the electricity demand in Spain, setting a new record.

Variability and Intermittency

Electricity generated from wind power can be highly variable at several different timescales: from hour to hour, daily, and seasonally. Annual variation also exists, but is not as significant. Related to variability is the short-term (hourly or daily) predictability of wind plant output. Like other electricity sources, wind energy must be "scheduled". Wind power forecasting methods are used, but predictability of wind plant output remains low for short-term operation.

Because instantaneous electrical generation and consumption must remain in balance to maintain grid stability, this variability can present substantial challenges to incorporating large amounts of wind power into a grid system. Intermittency and the non-dispatchable nature of wind energy production can raise costs for regulation, incremental operating reserve, and (at high penetration levels) could require an increase in the already existing energy demand management, load shedding, or storage solutions or system interconnection with HVDC cables. At low levels of wind penetration, fluctuations in load and allowance for failure of large generating units requires reserve capacity that can also regulate for variability of wind generation. Wind power can be replaced by other power stations during low wind periods. Transmission networks must already cope with outages of generation plant and daily changes in electrical demand. Systems with large wind capacity components may need more spinning reserve (plants operating at less than full load).

Pumped-storage hydroelectricity or other forms of grid energy storage can store energy developed by high-wind periods and release it when needed. Stored energy increases the economic value of wind energy since it can be shifted to displace higher cost generation during peak demand periods.

The potential revenue from this arbitrage can offset the cost and losses of storage; the cost of storage may add 25% to the cost of any wind energy stored, but it is not envisaged that this would apply to a large proportion of wind energy generated. For example, in the UK, the 2 GW Dinorwig pumped storage plant evens out electrical demand peaks, and allows base-load suppliers to run their plant more efficiently. Although pumped storage power systems are only about 75% efficient, and have high installation costs, their low running costs and ability to reduce the required electrical base-load can save both fuel and total electrical generation costs.

In particular geographic regions, peak wind speeds may not coincide with peak demand for electrical power. In the US states of California and Texas, for example, hot days in summer may have low wind speed and high electrical demand due to air conditioning. Some utilities subsidize the purchase of geothermal heat pumps by their customers, to reduce electricity demand during the summer months by making air conditioning up to 70% more efficient; widespread adoption of this technology would better match electricity demand to wind availability in areas with hot summers and low summer winds. Another option is to interconnect widely dispersed geographic areas with an HVDC "Super grid". In the US it is estimated that to upgrade the transmission system to take in planned or potential renewables would cost at least $60 billion.

In the UK, demand for electricity is higher in winter than in summer, and so are wind speeds. Solar power tends to be complementary to wind. On daily to weekly timescales, high pressure areas tend to bring clear skies and low surface winds, whereas low pressure areas tend to be windier and cloudier. On seasonal timescales, solar energy typically peaks in summer, whereas in many areas wind energy is lower in summer and higher in winter. Thus the intermittencies of wind and solar power tend to cancel each other somewhat. The Institute for Solar Energy Supply Technology of the University of Kassel pilot-tested a combined power plant linking solar, wind, biogas and hydrostorage to provide load-following power around the clock, entirely from renewable sources.

A report on Denmark's wind power noted that their wind power network provided less than 1% of average demand 54 days during the year 2002. Wind power advocates argue that these periods of low wind can be dealt with by simply restarting existing power stations that have been held in readiness or interlinking with HVDC. Electrical grids with slow-responding thermal power plants and without ties to networks with hydroelectric generation may have to limit the use of wind power.

Conversely, on particularly windy days, even with penetration levels of 16%, wind power generation can surpass all other electricity sources in a country. In Spain, on November 8, 2009 wind power production reached the highest percentage of electricity production till then, with wind farms covering 53% of the total demand.

Three reports on the wind variability in the UK issued in 2009, generally agree that variability of wind needs to be taken into account, but it does not make the grid unmanageable; and the additional costs, which are modest, can be quantified. A 2006 International Energy Agency forum presented costs for managing intermittency as a function of wind-energy's share of total capacity for several countries, as shown:

Table 3: Increase in system operation costs, Euros per MW·h, for 10% and 20% wind share

	10%	*20%*
Germany	2.5	3.2
Denmark	0.4	0.8
Finland	0.3	1.5
Norway	0.1	0.3
Sweden	0.3	0.7

Capacity Credit and Fuel Saving

Many commentators concentrate on whether or not wind has any "capacity credit" without defining what they mean by this and its relevance. Wind does have a capacity credit, using a widely accepted and meaningful definition, equal to about 20% of its rated output. This means that reserve capacity on a system equal in MW to 20% of added wind could be retired when such wind is added without affecting system security or robustness. But the precise value is irrelevant since the main value of wind (in the UK, worth 5 times the capacity credit value) is its fuel and CO2 savings.

According to a 2007 Stanford University study published in the Journal of Applied Meteorology and Climatology, interconnecting ten or more wind farms can allow an average of 33% of the total energy produced to be used as reliable, baseload electric power, as long as minimum criteria are met for wind speed and turbine height.

Economics

Cost Trends: Wind power has low ongoing costs, but a moderate capital cost. The estimated average cost per unit incorporates the cost of construction of the turbine and transmission facilities, borrowed funds, return to investors (including cost of risk), estimated annual production, and other components, averaged over the projected useful life of the equipment, which may be in excess of twenty years. Energy cost estimates are highly dependent on these assumptions so published cost figures can differ substantially.

A 2011 report from the American Wind Energy Association stated, "Wind's costs have dropped over the past two years, in the range of 5 to 6 cents per kilowatt-hour recently.... about 2 cents cheaper than coal-fired electricity, and more projects were financed through debt arrangements than tax equity structures last year.... winning more mainstream acceptance from Wall Street's banks.... Equipment makers can also deliver products in the same year that they are ordered instead of waiting up to three years as was the case in previous cycles.... 5,600 MW of new installed capacity is under construction in the United States, more than double the number at this point in 2010. Thirty-five percent of all new power generation built in the United States since 2005 has come from wind, more than new gas and coal plants combined, as power providers are increasingly enticed to wind as a convenient hedge against unpredictable commodity price moves."

A British Wind Energy Association report gives an average generation cost of onshore wind power of around 3.2 pence (between US 5 and 6 cents) per kW·h (2005). Cost per unit of energy produced was estimated in 2006 to be comparable to the cost of new generating capacity in the US for coal and natural gas: wind cost was estimated at $55.80 per MW·h, coal at $53.10/MW·h and natural gas at $52.50. Similar comparative results with natural gas were obtained in a governmental study in the UK in 2011. Other sources in various studies have estimated wind to be more expensive than other sources. A 2009 study on wind power in Spain by Gabriel Calzada Alvarez Universidad Rey Juan Carlos concluded that each installed MW of wind power led to the loss of 4.27 jobs, by raising energy costs and driving away electricity-intensive businesses. The U.S. Department of Energy found the study to be seriously flawed, and the conclusion unsupported. The presence of wind energy, even when subsidised, can reduce costs for consumers (€5 billion/yr in Germany) by reducing the marginal price by minimising the use of expensive 'peaker plants'.

The marginal cost of wind energy once a plant is constructed is usually less than 1 cent per kW·h. In 2004, wind energy cost a fifth of what it did in the 1980s, and some expected that downward trend to continue as larger multi-megawatt turbines were mass-produced. However, capital costs have increased. For example, in the United States, installed cost increased in 2009 to $2,120 per kilowatt of nameplate capacity, compared with $1,950 in 2008, a 9% increase. Not as many facilities can produce large modern turbines and their towers and foundations, so constraints develop in the supply of turbines resulting in higher costs.

Incentives

Wind energy in many jurisdictions receives financial or other support to encourage its development. Wind energy benefits from subsidies in many jurisdictions, either to increase its attractiveness, or to compensate for subsidies received by other forms of production which have significant negative externalities.

In the US, wind power receives a tax credit for each kW·h produced; at 1.9 cents per kW·h in 2006, the credit has a yearly inflationary adjustment. Another tax benefit is accelerated depreciation. Many American states also provide incentives, such as exemption from property tax, mandated purchases, and additional markets for "green credits".

Countries such as Canada and Germany also provide incentives for wind turbine construction, such as tax credits or minimum purchase prices for wind generation, with assured grid access (sometimes referred to as feed-in tariffs). These feed-in tariffs are typically set well above average electricity prices. The Energy Improvement and Extension Act of 2008 contains extensions of credits for wind, including microturbines.

Secondary market forces also provide incentives for businesses to use wind-generated power, even if there is a premium price for the electricity. For example, socially responsible manufacturers pay utility companies a premium that goes to subsidize and build new wind power infrastructure. Companies use wind-generated power, and in return they can claim that they are undertaking strong "green" efforts. In the US the organisation Green-e monitors business compliance with these renewable energy credits.

Full Costs and Lobbying

A House of Lords Select Committee report (2008) on renewable energy in the UK reported a "concern over the prospective role of wind generated and other intermittent sources of electricity in the UK, in the absence of a break-through in electricity storage technology or the integration of the UK grid with that of continental Europe".

Commenting on the EU's 2020 renewable energy target, Helm is critical of how the costs of wind power are cited by lobbyists. Helm also says that wind's problem of intermittent supply will probably lead to another dash-for-gas or dash-for-coal in Europe, possibly with a negative impact on energy security. In the US, the wind power industry has recently increased its lobbying efforts considerably, spending about \$5 million in 2009 after years of relative obscurity in Washington. By comparison, the US nuclear industry alone spent over \$650 million on its lobbying efforts during a single ten year period ending in 2008.

Environmental Effects

Compared to the environmental impact of traditional energy sources, the environmental impact of wind power is relatively minor. Wind power consumes no fuel, and emits no air pollution, unlike fossil fuel power sources. The energy consumed to manufacture and transport the materials used to build a wind power plant is equal to the new energy produced by the plant within a few months. While a wind farm may cover a large area of land, many land uses such as agriculture are compatible, with only small areas of turbine foundations and infrastructure made unavailable for use.

There are reports of bird and bat mortality at wind turbines as there are around other artificial structures. The scale of the ecological impact may or may not be significant, depending on specific circumstances. Prevention and mitigation of wildlife fatalities, and protection of peat bogs, affect the siting and operation of wind turbines.

A study on wind farm noise reported that people are annoyed by sound from wind turbines at far less sound levels than they are by noises from railroads, aircraft, or road traffic. The study found the percentage of respondents who found noise levels highly annoying rose quickly as sound levels increased above about 37dbA (about the level of a conversation).

Small-scale Wind Power

Small-scale wind power is the name given to wind generation systems with the capacity to produce up to 50 kW of electrical power.

Isolated communities, that may otherwise rely on diesel generators may use wind turbines to displace diesel fuel consumption. Individuals may purchase these systems to reduce or eliminate their dependence on grid electricity for economic or other reasons, or to reduce their carbon footprint. Wind turbines have been used for household electricity generation in conjunction with battery storage over many decades in remote areas.

Grid-connected wind turbines may use grid energy storage, displacing purchased energy with local production when available. Off-grid system users can either adapt to intermittent power or use batteries, photovoltaic or diesel systems to supplement the wind turbine. Equipment such as parking metres or wireless Internet gateways may be powered by a wind turbine that charges a small battery, replacing the need for a connection to the power grid.

In locations near or around a group of high-rise buildings, wind shear generates areas of intense turbulence, especially at street-level. The risks associated with mechanical or catastrophic failure have thus plagued urban wind development in densely populated areas, rendering the costs of insuring urban wind systems prohibitive. Moreover, quantifying the amount of wind in urban areas has been difficult, as little is known about the actual wind resources of towns and cities.

A new Carbon Trust study into the potential of small-scale wind energy has found that small wind turbines could provide up to 1.5 terawatt hours (TW·h) per year of electricity (0.4% of total UK electricity consumption), saving 0.6 million tonnes of carbon dioxide (Mt CO_2) emission savings. This is based on the assumption that 10% of households would install turbines at costs competitive with grid electricity, around 12 pence (US 19 cents) a kW·h.

Distributed generation from renewable resources is increasing as a consequence of the increased awareness of climate change. The electronic interfaces required to connect renewable generation units with the utility system can include additional functions, such as the active filtering to enhance the power quality.

Research and Development

Despite growing worldwide demand for wind energy, present wind technology is not optimised and there are still significant challenges. Most of the research has occurred in industry, and is not always easily shared. According to a research agenda from from a coalition of researchers from universities, industry, and government, supported by the Atkinson Centre for a Sustainable Future, wind energy research requires a drastic transformation. According to the report:

The gains that we are seeking require new innovations in fluid dynamics, control, materials, manufacturing, structures, and electric power distribution, as well of new ways of engaging the public in appreciating and accepting this technology, the associated transmission infrastructure and its effects on reducing climate change. Design and analysis tools need to be developed. Common computer codes need to be shared and refined in an open collegial way that cannot occur in industry. Researchers need to disseminate, debate, and share results openly, accelerating innovation in the subject.

Wind Hybrid Power Systems

Wind hybrid power systems combines wind turbines with other storage and/or generation sources. One of the key issues with wind energy is its intermittent nature. This has led to numerous methods of storing energy.

Wind-hydro System

A wind-hydro system generates electric energy combining wind turbines and pumped storage. The combination has been the subject of long-term discussion, but had not been implemented at a single location as of late 2010. Wind-hydro stations dedicate all, or a significant portion, of their wind power resources to pumping water into pumped storage reservoirs. These reservoirs are an implementation of grid energy storage.

Advantages

Wind and its generation potential is inherently variable. However, when this energy source is used to pump water into reservoirs at an elevation (the principle behind pumped storage), the potential energy of the water is relatively stable and can be used to generate electrical power by releasing it into a hydropower plant when needed. The combination has been described as particularly suited to islands that are not connected to larger grids.

Proposals

During the 1980s, an installation was proposed in the Netherlands. The IJsselmeer would be used as the reservoir, with wind turbines located on its dike. Feasibility studies have been conducted for installations on the island of Ramea (Newfoundland and Labrador) and on the Lower Brule Indian Reservation (South Dakota). An installation at Ikaria Island, Greece, had entered the construction phase as of 2010.

The island of El Hierro is where the first world's first wind-hydro power station is expected to be complete. Current TV called this "a blueprint for a sustainable future on planet Earth". It is designed to cover between 80-100% of the island's power and is set to be operational in 2012.

Wind-hydrogen System

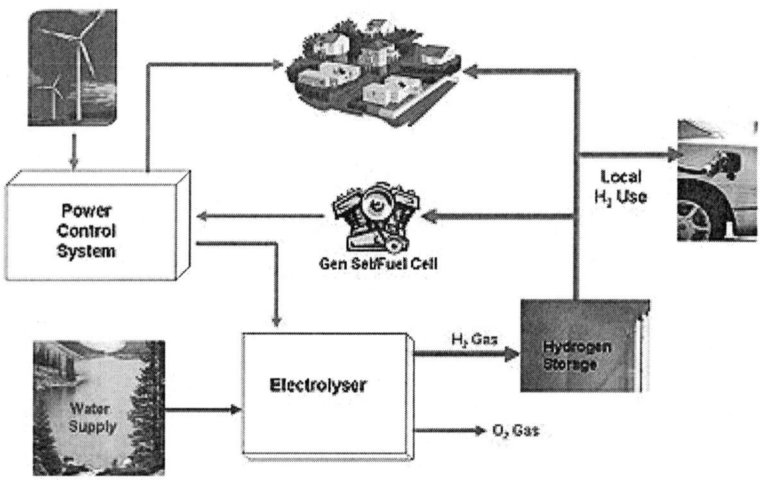

Figure 24.

One method of storing wind energy is the production of hydrogen through the electrolysis of water. This hydrogen is subsequently used to generate electricity during periods when demand can not be matched by wind alone. The energy in the stored hydrogen can be converted into electrical power through fuel cell technology or a combustion engine linked to an electrical generator.

Successfully storing hydrogen has many issues which need to be overcome, such as embrittlement of the materials used in the power system.

This technology is being developed in many countries and has even seen a recent IPO of an Australian firm called Wind Hydrogen that looks to commercialise this technology in both Australia and the UK. Essentially Wind Hydrogen offers a source of domestic and vehicular energy for rural communities where current energy transmission costs are prohibitive. Test sites include:

Community	*Country*	*Wind MW*
Ramea, Newfoundland and Labrador	Newfoundland, Canada	0.3
Prince Edward Island Wind-Hydrogen Village	PEI, Canada	
Lolland	Denmark	
Bismark	North Dakota, USA	
Koluel Kaike	Santa Cruz, Argentina	
Ladymoor Renewable Energy Project (LREP)	Scotland	
Hunterston Hydrogen Project	Scotland	
RES2H2	Greece	0.50
Unst	Scotland	0.03
Utsira	Norway	0.60

Wind-diesel System

A wind-diesel hybrid power system combines diesel generators and wind turbines, usually alongside ancillary equipment such as energy storage, power converters, and various control components, to generate electricity.

They are designed to increase capacity and reduce the cost and environmental impact of electrical generation in remote communities and facilities that are not linked to a power grid. Wind-diesel hybrid systems reduce reliance on diesel fuel, which creates pollution and is costly to transport.

History

Wind-diesel generating systems have been under development and trialled in a number of locations during the latter part of the 20th century. A growing number of viable sites have been developed with increased reliability and minimised technical support costs in remote communities.

Technology

The successful integration.of wind energy with diesel generating sets relies on complex controls to ensure correct sharing of intermittent wind energy and controllable diesel generation to meet the demand of the usually variable load.

The common measure of performance for wind diesel systems is Wind Penetration which is the ratio between Wind Power and Total Power delivered, e.g. 60% wind penetration implies that 60% of the system power comes from the wind. Wind Penetration figures can be either peak or long term. Sites such as Mawson Station, Antarctica, as well as Coral Bay and Bremer Bay in Australia have peak wind penetrations of around 90%.

Technical solutions to the varying wind output include controlling wind output using variable speed wind turbines (e.g. Enercon, Denham, Western Australia), controlling demand such as the heating load (e.g. Mawson), storing energy in a flywheel (e.g. Powercorp, Coral Bay). Some installations are now being converted to wind hydrogen systems such as on Ramea in Canada which is due for completion in 2010.

Wind-compressed Air Systems

At power stations that use compressed air energy storage (CAES), electrical energy is used to compress air and store it in underground facilities such as caverns or abandoned mines. During later periods of high electrical demand, the air is released to power turbines, generally using supplemental natural gas. Power stations that make significant use of CAES are operational in McIntosh, Alabama, Germany, and Japan. System disadvantages include some energy losses in the CAES process; also, the need for supplemental use of fossil fuels such as natural gas means that these systems do not completely make use of renewable energy.

The Iowa Stored Energy Park, projected to begin commercial operation in 2015, will use wind farms in Iowa as an energy source in conjunction with CAES.

Wind-solar Building

The Pearl River Tower in Guangzhou, China, will mix solar panel on its windows and several wind turbines at different stories of its structure, allowing this tower to be energy positive.

Wind-solar Lighting

In several parts of China, there are lighting pylons with combinations of solar panels and wind-turbines at their top. This allows space already used for lighting to be used more efficiently with two complementary energy productions units. Most common models use horizontal axis wind-turbines, but now models are appearing with vertical axis wind-turbines, using a helicoidal shaped, twisted-Savonius system.

Compact Wind Acceleration Turbine

Compact Wind Acceleration Turbines (CWATs) are a class of wind turbine that uses structures to accelerate wind before it enters the wind-generating element. The concept of these structures has been around for decades but has not gained wide acceptance in the marketplace. In 2008, two companies targeting the mid-wind (100 kW-1 MW) marketplace have received funding from venture capital. The first company to receive funding is Optiwind, which received its series A funding in April 2008, and the second company is FloDesign Wind Turbine Corp, which also received its series A funding in April 2008. Optiwind is funded through Charles River Ventures and FloDesign is funded through Kleiner Perkins.

Other CWATs under development include the Windtamer and Enflo turbines.

CWATs are a new acronym that encompasses the class of machines formerly known as DAWTs (diffuser augmented wind turbines). The technologies mentioned above all use diffuser augmentation that is substantially similar to previous designs as the primary means of acceleration.

DAWTs were heavily researched by K. Foreman and Oman of Grumman Aerospace in the 1970s and 1980s and Igra in Israel in the 1970s. At the end of a decade of wind tunnel research and design funded by Grumman, NASA, and the DOE, it was determined that the DAWT system's economics were not sufficient to justify commercialisation. In the 1990s the Grumman technology was licensed to a New Zealand company, Vortec Wind. The attempt to commercialise the Vortec 7 in New Zealand from 1998 to 2002 proved it to be economically untenable when compared to the dominant HAWT (horizontal axis wind turbine) technology.

The historic failure of DAWTs to gain acceptance has to do with the economics of these systems. DAWT systems have not shown improvement in the economics of HAWTs. This lack of advantage has centred around two key factors affecting the development of this type of machine.

The first factor regards power increase and the method of comparison used by DAWT (and more recently CWAT) designers to determine the whether the system is worth developing. Grumman and other attempts to commercialise these machines compare their machines to HAWTs based on a rotor area to rotor area comparison. As Van Bussel of Delft (The Science of Making More Torque from

Wind: Diffuser Experiments and Theory Revisited, G.J.W. van Bussel, Delft, 2007) pointed out, this is an inaccurate comparison and the comparison of power multiples should be made on the basis of the exit area of the diffuser or shroud not the rotor area.

Grumman claimed a 4× increase over an unshrouded turbine based on an acceleration of 1.6 times the ambient wind velocity (An Investigation on Diffuser Augmented Turbines, D.G. Philips, 2003 (reference materials compiled from K.M. Foreman)). A 1.6 acceleration is in fact 2.6 times the power of a HAWT if the ratio of the shrouded rotor to the exit area is 1.6. If however the rotor to exit area ratio is 2.75 (as it was in the Grumman case), the actual power increase over a HAWT with the same swept area as the diffuser exit area is only 1.4× the power. Given that the DAWTs with this ratio have roughly a solidity of 60+% when the blades and the diffuser are accounted for and the solidity of the HAWT is roughly 10%, the cost and amount of material needed to produce the 40% gain outweighs the increase in power.

Second is the structural requirement in terms of resisting overturning and bending in extreme wind events which all wind turbines must be designed for by ISO standard. The DAWT structure has poor drag characteristics. That combined with higher solidity leads to substantially greater structural costs than a HAWT in the support structure, the yaw bearing, and the foundation.

For these reasons, low power increase when referred to exit area and structural costs, DAWTs have not been a solution to improving wind energy at utility scale. The power increases thus far have proved insufficient to offset the structural costs. In small wind applications where structural issues are lessened they may be better than HAWTs if it can be definitively shown that they can improve output by 40% for the same cost.

The recent attempts mentioned above to improve on the DAWT type (the CWATs) appear to have similar issues.

In the case of Optiwind a 3 times increase is claimed for their 150 KW and 300 KW based on the horizontal dimension of their machine. If the power these machines produce is referred to the stack height (the vertical height - the tower height) of their system, it is only roughly 80% of the power produced by a HAWT of that diametre. The solidity and drag properties of the Optiwind are again problematic in extreme wind events and it seems unlikely that the system represents a cost improvement over a HAWT.

Flodesign's MEWT (mixer-ejector wind turbine, another CWAT variation) is differentiated from previous DAWT's by using a lobed two stage diffuser (Grumman and Vortec machine were also two stage, but conical instead of lobed) to equalise the pressure over the exit area of the diffuser.

The theory is that creating a uniform pressure distribution with the lobes and the injection of external flow will prevent boundary layer separation in the diffuser thereby allowing the maximum acceleration through rotor. Werle and Presz's (Flodesign's chief scientists) paper, AAIA technical note Ducted Water/Wind turbines revisited - 2007, details the theory behind their design. Maximum acceleration detailed in their paper is 1.8× the ambient velocity from which they derive that 3-5 times more power is available at the rotor than for an unshrouded turbine. When referred to exit area this multiple drops to between parity and 2.1 times the HAWT power. Flodesign's turbine based on released images and CAD's appears to be substantially similar to the Vortec and Grumman machines except for the lobed inner annulus. This would indicate that it's drag characteristics can be expected to be similar.

Again it is unclear that these machines can outperform HAWTs on either a power output or cost basis. Additionally many CWAT companies seem to misunderstand certain aspects of wind power. There are many claims that because these machines start at lower speeds and stop at higher speeds they can gather even more power. The fact is that there is very little power in winds below 5 metres per second (2% of annual production) and the percentage of annual wind that falls above the HAWT cut out speed of 25 metres per second is less than .05%. Many of these claims including the idea that these machines will use less material than a HAWT for the same rating seem somewhat poorly thought out.

There is one recent DAWT design that appears to have a definitive positive power, if not cost, comparison to HAWTs. This is the Enflo turbine. Based on its rotor:exit ratio and the published power performance this turbine appears to have a confirmed 2 times increase in power output over a HAWT of the diametre of the exit area. It is still unlikely that this machine can scale to larger ratings but based on published data the Enflo appears to be the best performing DAWT/CWAT yet built.

Chapter 7

Unconventional Wind Turbines

As of 2010, the most common type of wind turbine is the three-bladed horizontal-axis wind turbine (HAWT). This article deals with various types of wind turbine that differ from the standard type. Some types have had limited commercial use, while others have only been demonstrated or are only theoretical concepts with no practical applications.

Modified Horizontal

Two Blades: Nearly all modern wind turbines uses rotors with three blades, but some use only two blades. This was the type used at GROWIAN, some other prototypes and several wind turbine types manufactured by Ned Wind. Eemmeerdijk Wind Park uses only 2-blade turbines. Turbines with two blades are manufactured by Nordic Windpower (Model N 1000) and by GC China.

Ducted Rotor

Still something of a research project, the ducted rotor consists of a turbine inside a duct which flares outwards at the back. They are also referred as Diffuser-Augmented Wind Turbines (i.e. DAWT). The main advantage of the ducted rotor is that it can operate in a wide range of winds and generate a higher power per unit of rotor area. Another advantage is that the generator operates at a high rotation rate, so it doesn't require a bulky gearbox, so the mechanical portion can be smaller and lighter. A disadvantage is that (apart from the gearbox) it is more complicated than the unducted rotor and the duct is usually quite heavy, which puts an added load on the tower. The Eolienne Bollee is an example of a DAWT.

Co-axial, Multirotor

Two or more rotors may be mounted to the same driveshaft, with their combined co-rotation together turning the same generator — fresh wind is brought to each rotor by sufficient spacing between rotors combined with an offset angle (alpha) from the wind direction. Wake vorticity is recovered as the top of a wake hits the bottom of the next rotor. Power has been multiplied several times using co-axial, multiple rotors in testing conducted by inventor and researcher Douglas Selsam, for the California Energy Commission in 2004.

The first commercially available co-axial multi-rotor turbine is the patented dual-rotor American Twin Superturbine from Selsam Innovations in California, with 2 propellers separated by 12 feet. It is the most powerful 7-foot-diametre (2.1 m) turbine available, due to this extra rotor.

Counter-Rotating Horizontal-axis

When a system expels or accelerates mass in one direction, the accelerated mass will cause a proportional but opposite force on that system. A single rotor wind turbine causes a significant amount of tangential or rotational air flow to be created by the spinning blades. The energy of this tangential air flow is wasted in a single-rotor propeller design. To use this wasted effort, the placement of a second rotor behind the first takes advantage of the disturbed airflow. Contra-rotation wind energy collection with two rotors, one behind the other, can gain up to 40% more energy from a given swept area as compared with a single rotor. Much work has been done recently on this in the USA.

A patent application dated 1992 exists based on work done with the Trimblemill. Other advantages of contra-rotation include no gear boxes, auto centreing on the wind - no yaw motors/mechanism required. Ability to be point suspended, thus reducing support structure overturning moments theoretical ability to be "grid linked" without electronics, thus giving the possibility of "arrays". Counter-rotating turbines can be used to increase the rotation speed of the electrical generator. As of 2005, no large practical counter-rotating HAWTs are commercially sold. When the counter-rotating turbines are on the same side of the tower, the blades in front are angled forwards slightly so as to avoid hitting the rear ones. If the turbine blades are on opposite sides of the tower, it is best that the blades at the back be smaller than the blades at the front and set to stall at a higher wind speed.

This allows the generator to function at a wider wind speed range than a single-turbine generator for a given tower. To reduce sympathetic vibrations, the two turbines should turn at speeds with few common multiples, for example 7:3 speed ratio. Overall, this is a more complicated design than the single-turbine wind generator, but it taps more of the wind's energy at a wider range of wind speeds. Appa designed and demonstrated a contra-rotor wind turbine in FY 2000–2002 funded by California Energy Commission. This study showed 30 to 40% more power extraction than a comparable single-rotor system. Further, it was observed that the slower the rotor speed, the better the performance. Consequently, Megawatt machines benefit most. This also cancels the gyroscopic forces. There will be an improvement in overall efficiency.

Furling Tail And Twisting Blades

In addition to variable pitch blades, furling tails and twisting blades are other improvements on wind turbines. Similar to the variable pitch blades, they may also greatly increase the efficiency of the turbine and be used in "do-it-yourself" construction

Telescopic Blades

A next step in making improvements to wind turbines may be in the use of telescopic blades. Telescopic blades can change the blade's length, thus increasing or decreasing the turbine's swept area. Telescopic blades make a turbine more productive by increasing the turbine's rotor diametre during low-wind conditions. In high-wind conditions, when the turbine is in need of reducing loads, the blades can be retracted to make the rotor smaller.

Wind-Mill Style

De Nolet is a wind turbine in Rotterdam disguised as a windmill.

Modified Vertical Axis

Aerogenerator: The Aerogenerator is a special design of vertical axis wind turbine which could allow greater energy outputs.

Savonius

The Savonius wind turbine is another special design wind turbine.

Augmented

The "G" Model VAWT Turbine is equipped with three self-positioning *Augmentation And Directioning Wings* (AADW) placed as the outer sections of classical Darrieus blades. The GMWT can increase almost fivefold the efficiency of classical Darrieus Blades: AADWs adjust themselves to the wind direction without any external power. The resulting combination ("G" Model Wind Turbine) works with very low cut-in wind speed, has self starting ability, together with a high capacity factore.

Fuller

The "Fuller" wind turbine is a fully enclosed wind turbine which uses boundary layers instead of blades.

Imagine a stack of CDs on a central shaft with a small air gap in between, the surface tension of moving air as it passes through the small gaps creates friction and using this energy causes the CDs to rotate around the shaft. The Fuller has addition vanes to help direct the air for improved performance, hence it isn't totally bladeless, as has been suggested.

Aerial

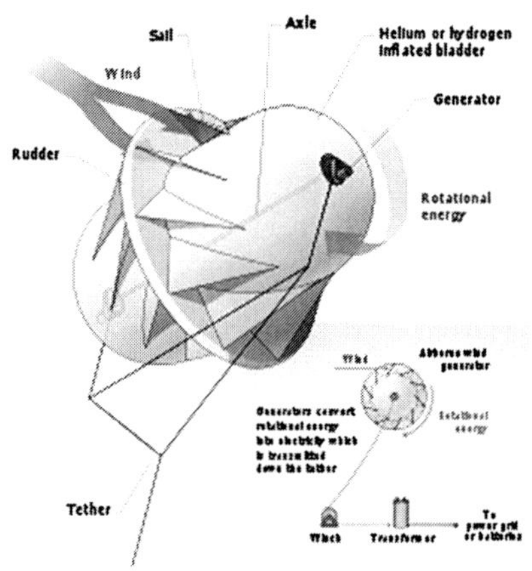

Figure 1: Concept for an airborne wind generator.

It has been suggested that wind turbines could be flown in high-speed winds using high altitude wind power tactics, taking advantage of the steadier winds at high altitudes. A system of automatically controlled tethered kites could also be used to capture energy from high-altitude winds.

H-rotor

This is a vertical axis turbine, but it isn't favoured because of its poor efficiency. One blade is pushed by the wind while the other is being pushed in the opposite direction. Consequently only one blade is working at a time

Wind Belt—Vaneless Ion Wind Generator—Piezoelectric

Another special type of wind turbines are the piezoelectric wind turbines. Turbines with diametres on the scale of 10 centimetres work by flexing piezoelectric crystals as they rotate, sufficient to power small electronic devices.

Traffic-driven

A few proposals call for generating power from the otherwise wasted energy in the draft created by traffic.

Blade Tip Power System (BTPS)

Designed by Imad Mahawili with Honeywell/WindTronics. This design uses many nylon blades and turns a permanent magnet generator inside out. The magnets are on the tips of the blades, and the stator is on the outside of the generator.

Solar Chimney

Wind turbines may also be used in conjunction with a solar collector to extract the energy due to air heated by the Sun and rising through a large vertical Solar updraft tower. Wind turbines are part of experimental wave powered generators where air displaced by waves drives turbines.

Wind Turbines on Public Display

The great majority of wind turbines around the world belong to individuals or corporations who use them to generate electric power or to perform mechanical work. As such, wind turbines are primarily designed to be working devices. However, the large size and height above surroundings of modern industrial wind turbines, combined with their moving rotors, often makes them among the most conspicuous objects in their areas. A few localities have exploited the attention-getting nature of wind turbines by placing them on public display, either with visitor centres around their bases, or with viewing areas farther away. The wind turbines themselves are generally of conventional horizontal-axis, three-bladed design, and generate power to feed electrical grids, but they also serve the unconventional roles of technology demonstration, public relations, and education.

Rooftop Wind-turbines

Wind-turbines can be installed on the top of a roof of a building. This is not as common as may first be assumed. Some examples include Marthalen Landi-Silo in Switzerland, Council House 2 in Melbourne, Australia. Ridgeblade in the UK is like a vertical wind turbine on its side mounted on the apex of a pitched roof. Discovery Tower is an office building in Houston, Texas, that incorporates 10 wind turbines in its architecture.

The Museum of Science in Boston, Massachusetts began constructing a rooftop Wind Turbine Lab in 2009. The Lab will test nine wind turbines from five different manufacturers on the roof of the Museum. Rooftop wind turbines may suffer from turbulence, especially in cities, which reduces power output and accelerates turbine wear. The lab seeks to address the general lack of performance data for urban wind turbines.

Due to the structural limitations of buildings, the limited space in urban areas, and safety considerations, wind turbines mounted on buildings are usually small (with nameplate capacities in the low kilowatts), rather than the megawatt-class wind turbines which are most economical for wind farms. A partial exception is the Bahrain World Trade Centre with three 225 kW wind turbines mounted between twin skyscrapers.

Vertical Axis Wind Turbine

Vertical-axis wind turbines (VAWTs) are a type of wind turbine where the main rotor shaft is set vertically and the main components are located at the base of the turbine.

Among the advantages of this arrangement are that generators and gearboxes can be placed close to the ground, which makes these components easier to service and repair, and that VAWTs do not need to be pointed into the wind. Major drawbacks for the early designs (Savonius, Darrieus and giromill) included the pulsatory torque that can be produced during each revolution and the huge bending moments on the blades. Later designs solved the torque issue by using the helical twist of the blades almost similar to Gorlov's water turbines.

A VAWT tipped sideways, with the axis perpendicular to the wind streamlines, functions similarly. A more general term that includes this option is "transverse axis wind turbine". For example, the original Darrieus patent , includes both options.

Drag-type VAWTs, such as the Savonius rotor, typically operate at lower tipspeed ratios than lift-based VAWTs such as Darrieus rotors and cycloturbines.

General Aerodynamics

The resultant velocity vector, $\vec{W}$, is the vectorial sum of the undisturbed upstream air velocity, $\vec{U}$, and the velocity vector of the advancing blade, $-\vec{\omega}\times\vec{R}$.

$$\vec{W} = \vec{U} + \left(-\vec{\omega}\times\vec{R}\right)$$

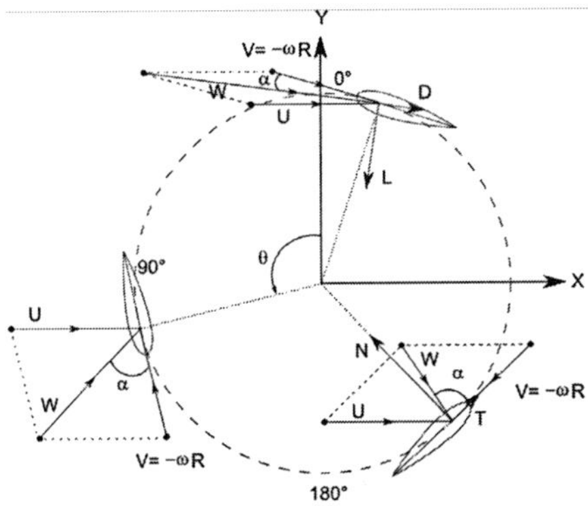

Figure 2: Forces and velocities acting in a Darrieus turbine for various azimuthal positions

Figure 3: Five-kilowatt vertical axis wind turbine

Thus, the oncoming fluid velocity varies, the maximum is found for $\theta = 0^{\circ}$ and the minimum is found for $\theta = 180^{\circ}$, where θ is the azimuthal or orbital blade position. The angle of attack, α, is the angle between the oncoming air speed, W, and the blade's chord. The resultant airflow creates a varying, positive angle of attack to the blade in the upstream zone of the machine, switching sign in the downstream zone of the machine.

From geometrical considerations, the resultant airspeed flow and the angle of attack are calculated as follows:

$$W = U\sqrt{1+2\lambda\cos\theta+\lambda^2}$$

$$\alpha = \tan^{-1}\left(\frac{\sin\theta}{\cos\theta+\lambda}\right)$$

where $\lambda = \frac{\omega R}{U}$ is the tip speed ratio parameter.

The resultant aerodynamic force is decomposed either in lift (F_L) - drag (D) components or normal (N) - tangential (T) components. The forces are considered acting at 1/4 chord from the leading edge (by convention), the pitching moment is determined to resolve the aerodynamic forces. The aeronautical terms lift and drag are, strictly speaking, forces across and along the approaching net relative airflow respectively.

The tangential force is acting along the blade's velocity and, thus, pulling the blade around, and the normal force is acting radially, and, thus, is acting against the bearings. The lift and the drag force are useful when dealing with the aerodynamic behaviour around each blade, i.e. dynamic stall, boundary layer, etc; while when dealing with global performance, fatigue loads, etc., it is more convenient to have a normal-tangential frame. The lift and the drag coefficients are usually normalised by the dynamic pressure of the relative airflow, while the normal and the tangential coefficients are usually normalised by the dynamic pressure of undisturbed upstream fluid velocity.

$$C_L = \frac{F_L}{1/2\,\rho A W^2} \quad ; \quad C_D = \frac{D}{1/2\,\rho A W^2} \quad ; \quad C_T = \frac{T}{1/2\,\rho A U^2} \quad ; \quad C_N = \frac{-}{1}$$

A = Surface Area

The amount of power, P, that can be absorbed by a wind turbine.

$$P = \frac{1}{2} C_p \rho A v^3$$

Where C_p is the power coefficient, ρ is the density of the air, A is the swept area of the turbine, and í is the wind speed.

Advantages of Vertical Axis Wind Turbines

VAWTs offer a number of advantages over traditional horizontal-axis wind turbines (HAWTs). They can be packed closer together in wind farms, allowing more in a given space. This is not because they are smaller, but rather due to the slowing effect on the air that HAWTs have, forcing designers to separate them by ten times their width.

VAWTs are rugged, quiet, omni-directional, and they do not create as much stress on the support structure. They do not require as much wind to generate power, thus allowing them to be closer to the ground. By being closer to the ground they are easily maintained and can be installed on chimneys and similar tall structures.

Disadvantages of Vertical Axis Wind Turbines

Some disadvantages that the VAWTs possess are that they have a tendency to stall under gusty winds. VAWTs have very low starting torque, as well as dynamic stability problems. The VAWTs are sensitive to off-design conditions and have a low installation height limiting to operation to lower wind speed environments.

The blades of a VAWT are prone to fatigue as the blade spins around the central axis. The vertically oriented blades used in early models twisted and bent as they rotated in the wind. This caused the blades to flex and crack. Over time the blades broke apart and sometimes leading to catastrophic failure. Because of these problem, Vertical axis wind turbines have proven less reliable than horizontal-axis wind turbines (HAWTs).

Research programmes (in 2011) have sought to overcome the inefficiencies associated with VAWTs by reconfiguration of turbine placement within wind farms. It is thought that, despite the lower wind-speed environment at low elevations, "the scaling of the physical forces involved predicts that [VAWT] wind farms can be built using less expensive materials, manufacturing processes, and maintenance than is possible with current wind turbines"

Gorlov Helical Turbine

The Gorlov helical turbine (GHT) is a water turbine evolved from the Darrieus turbine design by altering it to have helical blades/foils. It was patented in a series of patents from September 19, 1995 to July 3, 2001 and won 2001 ASME Thomas A. Edison Patent Award. GHT was invented by Professor Alexander M. Gorlov of the Northeastern University.

The physical principles of the GHT work are the same as for its main prototype, the Darrieus turbine, and for the family of similar Vertical axis wind turbines which includes also Turby wind turbine Quietrevolution wind turbine Urban Green Energy. GHT, turby and quietrevolution solved pulsatory torque issues by using the helical twist of the blades.

Fluid Performance

The term "foil" is used to describe the shape of the blade cross-section at a given point, with no distinction for the type of fluid, (thus referring to either a "airfoil" or "hydrofoil"). In the helical design, the blades curve around the axis, which has the effect of evenly distributing the foil sections throughout the rotation cycle, so there is always a foil section at every possible angle of attack. In this way, the sum of the lift and drag forces on each blade do not change abruptly with rotation angle. The turbine generates a smoother torque curve, so there is much less vibration and noise than in the Darrieus design. It also minimises peak stresses in the structure and materials, and facilitates self-starting of the turbine. In testing environments the GHT has been observed to have up to 35% efficiency in energy capture reported by several groups. "Among the other vertical-axis turbine systems, the Davis Hydro turbine, the EnCurrent turbine, and the Gorlov Helical turbine have all undergone scale testing at laboratory or sea. Overall, these technologies represent the current norm of tidal current development."

Turbine Axis-Orientation

The main difference between the Gorlov helical turbine and conventional turbines is the orientation of the axis in relation to current flow. The GHT is a vertical-axis turbine which means the axis is positioned perpendicular to current flow, whereas traditional turbines are horizontal-axis turbines which means the axis is positioned parallel to the flow of the current. Fluid flows, such as wind, will naturally change direction, however they will still remain parallel to the ground. So in all vertical-axis turbines, the flow remains perpendicular to the axis, regardless of the flow direction, and the turbines always rotate in the same direction. This is one of the main advantages of vertical-axis turbines. If the direction of the water flow is fixed, then the Gorlov turbine axis could be verical or horizontal, the only requirement is orthogonality to the flow.

Airfoil / Hydrofoil

The GHT operates under a lift-based concept. The foil sections on the GHT are symmetrical, both top-to-bottom and also from the leading-to-trailing edge. The GHT can actually spin equally well in either direction. The GHT works under the same principle as the Darrieus turbine; that is, it relies upon the movement of the foils in order to change the apparent direction of the flow relative to the foils, and thus change the (apparent) "angle of attack" of the foil.

Environmental Issues

A GHT is proposed for low-head micro hydro installations, when construction of a dam is undesirable. The GHT is an example of damless hydro technology. The technology may potentially offer cost and environmental benefits over dam-based micro-hydro systems. Some advantages of damless hydro are that it eliminates the potential for failure of a dam, which improves public safety. It also eliminates the initial cost of dam engineering, construction and maintenance, reduces the environmental and ecological complications, and potentially simplifies the regulatory issues put into law specifically to mitigate the problems with dams.

In general, a major ecological issue with hydropower installations is their actual and perceived risk to aquatic life. It is claimed that a GHT spins slowly enough that fish can see it soon enough to swim around it. From preliminary tests in 2001, it was claimed that if a fish swims between the slowly moving turbine blades, the fish will not be harmed. Also it would be difficult for a fish to become lodged or stuck in the turbine, because the open spaces between the blades are larger than even the largest fish living in a small river. A fish also would not be tumbled around in a vortex, because the GHT does not create a lot of turbulence, so small objects would be harmlessly swept through with the current.

How it Works

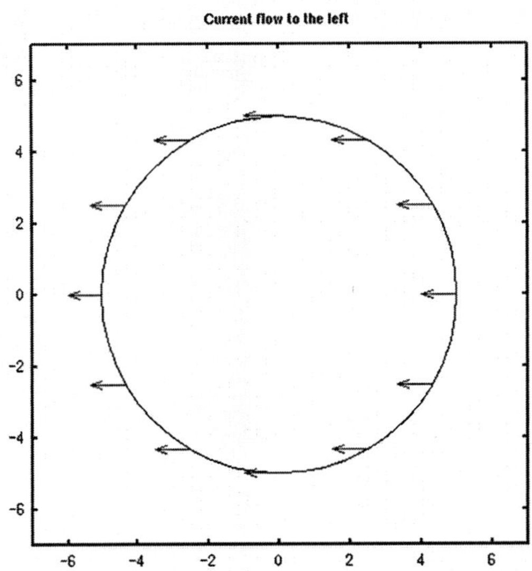

Figure 4: Current flow to the left.

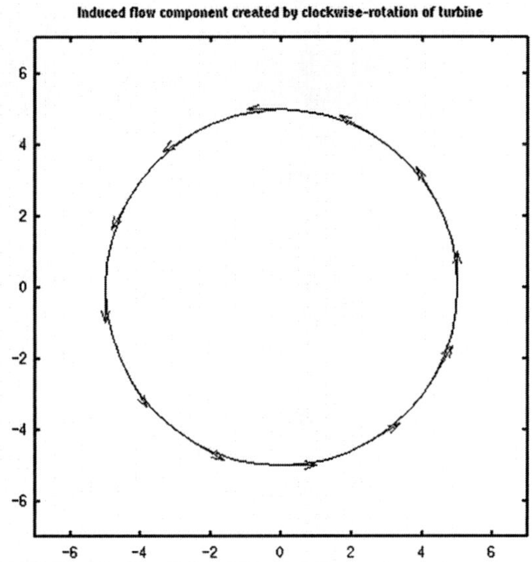

Figure 5: Induced flow component created by clockwise rotation of turbine.

In this Example the Direction of the Fluid Flow is to the Left.

As the turbine rotates, in this case in a clockwise direction, the motion of the foil through the fluid changes the apparent velocity and angle of attack (speed and direction) of the fluid with respect to the frame of reference of the foil. The combined effect of these two flow components (i.e. the vector sum), yields the net total "Apparent flow velocity" as shown in the next figures.

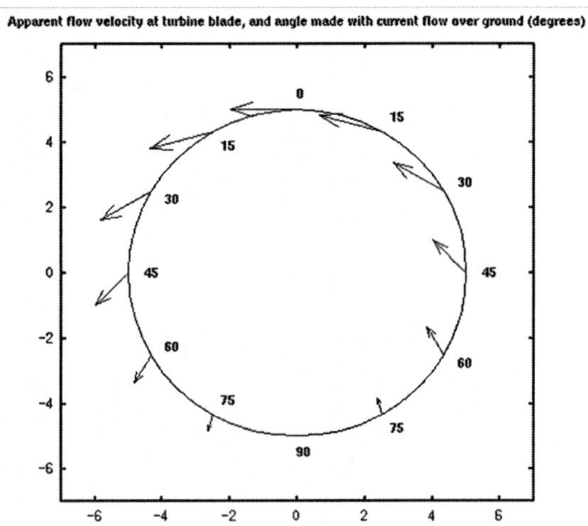

Figure 6: Apparent flow velocity of turbine blade, and angle made with current flow over ground (in degrees).

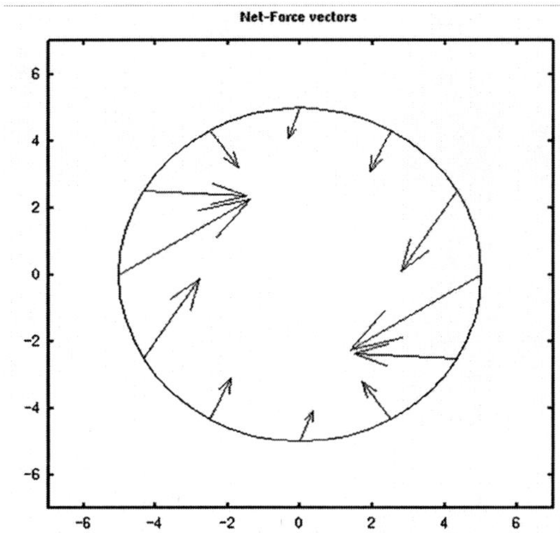

Figure 7: Net force vectors.

The action of this apparent flow on each foil section generates both a lift and drag force, the sum of which is shown in the figures above titled “Net force vectors”. Each of these net force vectors can be split into two orthogonal vectors: a radial component and a tangential component, shown here as “Normal force” and “Axial force” respectively. The normal forces are opposed by the rigidity of the turbine structure and do not impart any rotational force or energy to the turbine. The remaining force component propels the turbine in the clockwise direction, and it is from this torque that energy can be harvested.

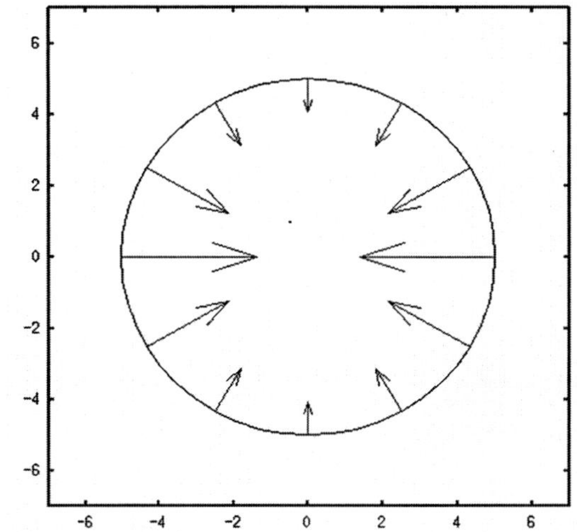

Figure 8: Normal force vectors.

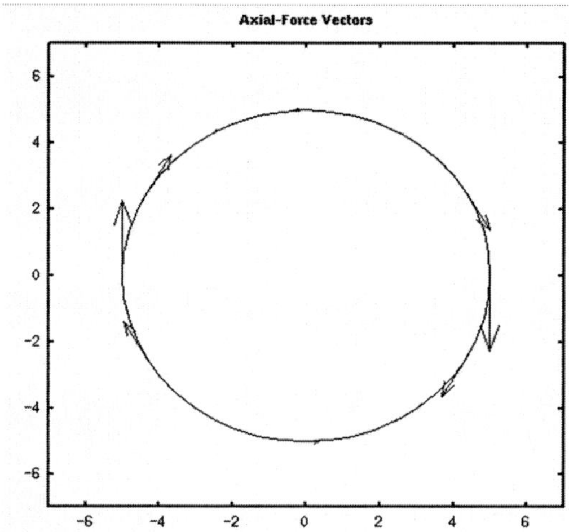

Figure 9: Axial force vectors.

Gas Turbine Locomotive

A Gas turbine locomotive is a locomotive powered by a gas turbine. The majority of gas turbine locomotives have had electric transmission but mechanical transmission has also been used, particularly in the early days. The advantage of using gas turbines is that they have very high power-to-bulk and power-to-weight ratios. The disadvantage is that gas turbines generally have lower thermal efficiency than diesel engines, especially when running at less than full load.

Overview

Where electric transmission is used, the engine is usually a single-shaft machine in which one turbine drives both the compressor and the output shaft. With mechanical transmission, the power turbine must be capable of starting from rest, so a more complex arrangement is necessary. One option is a two-shaft machine, with separate turbines to drive the compressor and the output shaft. Another is to use a separate gas generator, which may be of either rotary or piston type.

Examples

Examples of gas turbine-mechanical locomotives:

- 1933 Nydqvist and Holm, 1-B-1, Sweden
- 1952 Davenport-Bessler Corp., 1-B-1, United States
- 1951 Renault, France, B-B, 1,000 hp (750 kW)
- 1954 Gotaverken, Sweden, 1-C-1, 1,300 hp (970 kW)
- 1958 Renault, France, C-C, 2,000 hp (1,500 kW)

- 1958 Škoda, C-C, Czechoslovakia, 3,200 hp (2,400 kW)
- 1959 British Rail GT3, 2-C-0, 2,700 hp (2,000 kW)

History

Work leading to the emergence of the gas turbine locomotive began in France and Sweden in the 1920s but the first locomotive did not appear until 1933. These early experiments used piston engines as gas generators. This idea has not been widely adopted, but it might be worth re-visiting. High fuel consumption was a major factor in the decline of conventional gas-turbine locomotives and the use of a piston engine as a gas generator would probably give better fuel economy than a turbine-type compressor, especially when running at less than full load.

France

The locomotives were built by Renault and had Pescara free-piston engines as gas generators. Each gas generator consisted of a horizontal, single cylinder, two-stroke diesel engine with opposed pistons. It had no crankshaft and the pistons were returned after each power stroke by compression and expansion of air in a separate cylinder. The exhaust from the diesel engine powered the gas-turbine which drove the wheels through a two-speed gearbox and propeller shafts.

Czechoslovakia

Turbine power was considered for railway traction in the former Czechoslovakia. Two turbine-powered prototypes were built, designated TL 659.001 and .002, featuring C-C wheel arrangement, 3200 hp (2.4 MW) main turbine, helper turbine and Tatra 111 helper diesel engine. The first prototype was finished in February 1958 and was scheduled to be exhibited at Expo '58. This was aborted, because it wasn't ready in time. The first out-of-factory tests were conducted in March 1959 on the Plzeò–Cheb–Sokolov line. On May 15, 1959, the first prototype pulled its heaviest train, 6486 metric tons, but the turbine caught fire only a day later. The engine was never restored and eventually scrapped. The second prototype was built with lessons learned from the first prototype. It left the factory in March 1960 and was the only turbine locomotive to pass the tests for regular service on tracks of the former ÈSD. This engine was tried near Kolín and Plzeò with mixed results. This engine was taken out of service in April 1966 and sold to University of •ilina as an educational instrument. The locomotive was scrapped some time later.

Although these experiments had mixed results, they were the most powerful locomotives with purely mechanical powertrain in the world and also the most powerful independent-traction locomotives in Czechoslovakia.

Sweden

The Power gas locomotive was built by Gotaverken. It had a vertical, five cylinder, two-stroke diesel engine with opposed pistons. There was a single crankshaft connected to both upper and lower pistons. The exhaust from the diesel engine powered the gas turbine which drove the wheels through reduction gearing, jack shaft and side rods.

Free-piston engine

The free-piston engine was patented in 1934 by Raul Pateras Pescara (1890–1966). He was an Argentinian, who worked in Spain and France, and was prominent in helicopter development in the 1919s.

Coal-firing

In the 1940s and 1950s research was done, in both the USA and UK, aimed at building gas turbine locomotives which could run on pulverised coal. The main problem was to avoid erosion of the turbine blades by particles of ash. Some bench testing was done but the projects were abandoned before any complete locomotives were built. The sources for the following information are Robertson and Sampson.

USA

In the USA, the plan was to use a gas turbine similar to an oil-fuelled one and to remove ash particles with filters. Details of the US research (done in 1946) were passed to Britain's London, Midland and Scottish Railway.

UK

On 23 December 1952 the UK Ministry of Fuel and Power placed an order for a coal-fired gas turbine locomotive to be used on British Railways. The locomotive was to be built by the North British Locomotive Company and the turbine would be supplied by C. A. Parsons and Company.

According to Sampson, the plan was to use indirect heating. The pulverised coal would be burned in a combustion chamber and the hot gases passed to a heat exchanger. Here, the heat would be transferred to a separate body of compressed air which would power the turbine.

Essentially, it would have been a hot air engine using a turbine instead of a piston. Robertson shows a diagram which confirms Sampson's information but also refers to problems with erosion of turbine blades by ash. This is strange because, with a conventional shell and tube heat exchanger, there would be no risk of ash entering the turbine circuit.

Working Cycle

There were two separate, but linked, circuits - the combustion circuit and the turbine circuit.

1. Combustion circuit. Pulverised coal and air were mixed and burned in a combustion chamber and the hot gases passed to a heat exchanger where heat was transferred to the compressed air in the turbine circuit. After leaving the heat exchanger the combustion gases entered a boiler to generate steam for train heating.
2. Turbine circuit. Air entered the compressor and was compressed. The compressed air passed to the heat exchanger where it was heated by the combustion gases. The heated compressed air drove two turbines - one to drive the compressor and the other to power the locomotive. The turbine exhaust (which was hot air) then entered the combustion chamber to support the combustion.

Specification

The locomotive was never built but the specification was as follows:

- Wheel arrangement: C-C, later changed to 1A1A-A1A1
- Horsepower: 1,800, later reduced to 1,500
- Weight: 117 tons, later increased to 150 tons

The projected output was:

- Tractive effort,
  - 30,000 lbf (130 kN) at 72 mph (116 km/h)
  - 45,000 lbf (200 kN) at 50 mph (80 km/h)
- Thermal efficiency,
  - 10% at 1/10 load
  - 16% at half load
  - 19% at full load

The transmission was to be mechanical, via a two-speed gearbox, giving a high speed for passenger working and a lower speed for freight. The tractive effort figures, quoted above, look suspiciously high for the specified speeds. It seems more likely that the figures quoted are for starting tractive effort and maximum speed in high gear and low gear respectively.

There is a model of the proposed locomotive at Glasgow Museum of Transport and some records are held at the National Railway Museum.

The British Rail GT3 was a much simpler machine consisting essentially of a standard oil-fired gas turbine mounted on a standard steam locomotive chassis, built as a demonstrator by English Electric in 1961. Its almost crude simplicity enabled it to avoid much of the unreliability which had plagued the complex experimental GTELs 18000 and 18100 in earlier years, but it nevertheless failed to be competitive against conventional traction and was scrapped.

Gas Turbine-Electric Locomotive

A gas turbine - electric locomotive, or GTEL, is a locomotive that uses a gas turbine to drive an electric generator or alternator. The electric current thus produced is used to power traction motors. This type of locomotive was first experimented with during the Second World War, but reached its peak in the 1950s to 1960s. Few locomotives use this system today.

Description

A GTEL uses a turbo-electric drivetrain in which a turboshaft engine drives an electrical generator or alternator via a system of gears. The electrical power is distributed to power the traction motors that drive the locomotive. In overall terms the system is very similar to a conventional diesel-electric, with the large diesel engine replaced with a smaller gas turbine of similar power.

A gas turbine offers some advantages over a piston engine. There are few moving parts, decreasing the need for lubrication and potentially reducing maintenance costs, and the power-to-weight ratio is much higher.

A turbine of a given power output is also physically smaller than an equally powerful piston engine, allowing a locomotive to be very powerful without being inordinately large. However, a turbine's power output and efficiency both drop dramatically with rotational speed, unlike a piston engine, which has a comparatively flat power curve. This makes GTEL systems useful primarily for long-distance high-speed runs.

Union Pacific operated the largest fleet of such locomotives of any railroad in the world, and was the only railroad to use them for hauling freight. Most other GTELs have been built for small passenger trains, and only a few have seen any real success in that role. With a rise in fuel costs (eventually leading to the 1973 oil crisis), gas turbine locomotives became uneconomical to operate, and many were taken out of service. Additionally, Union Pacific's locomotives required more maintenance than originally anticipated, due to fouling of the turbine blades by the Bunker C oil used as fuel.

History

Switzerland: In 1939 the Swiss Federal Railways ordered a GTEL with a 1620 kW (2170 hp) of maximum engine power from Brown Boveri. It was completed in 1941, and then underwent testing before entering regular service. The Am 4/6 was the first gas turbine - electric locomotive. It was intended primarily to work light, fast, passenger trains on routes which normally handle insufficient traffic to justify electrification.

United Kingdom

In 1949 the Brown Boveri completed the BR 18000. It was a 1840 kW (2470 hp) GTEL, ordered by the Great Western Railway and used for express passenger services. In 1951 the Metropolitan-Vickers built the BR 18100 for the British Railways.

It had an aircraft-type gas turbine of 3,000 horsepower (2.2 MW). Maximum speed was 90 miles per hour (140 km/h). In 1961 English Electric built the GT3 locomotive for British Railways. It had a steam locomotive 4-6-0 wheel arrangement and a tender which carried fuel oil for the turbine.

The British Rail APT-E, prototype of the Advanced Passenger Train, was turbine-powered. Like the French TGV, later models were electric instead. This choice was made because British Leyland, the turbine supplier, ceased production of the model used in the APT-E.

France

SNCF The first TGV prototype, TGV 001, was powered by a gas turbine, but steep oil prices prompted the change to overhead electric lines for power delivery.

United States

Union Pacific ran a large fleet of turbine-powered freight locomotives starting in the 1950s. These were widely used on long-haul routes, and were cost-effective despite their poor fuel economy due to their use of "leftover" fuels from the petroleum industry. At their height the railroad estimated that they powered about 10% of Union Pacific's freight trains, a much wider use than any other example of this class.

As other uses were found for these heavier petroleum byproducts, notably for plastics, the cost of the Bunker C fuel increased until the units became too expensive to operate and they were retired from service by 1969.

In April 1950, Westinghouse completed an experimental 4,000 hp (3,000 kW) turbine locomotive, #4000, known as the Blue Goose, with a B-B-B-B wheel arrangement. The locomotive used two 2,000 hp (1,500 kW) turbine engines, was equipped for passenger train heating with a steam generator that utilised the waste exhaust heat of the right hand turbine, and was geared for 100 miles per hour (160 km/h) While it was demonstrated successfully in both freight and passenger service on the PRR, MKT, and CNW, no production orders followed, and it was scrapped in 1953.

In 1997 the Federal Railroad Administration (FRA) solicited proposals to develop high speed locomotives for routes outside the Northeast Corridor where electrification was not economical. Bombardier Ltd, at the Plattsburg, N.Y. plant where the Acela was produced, developed a prototype (JetTrain) which combined a Pratt & Whitney Canada PW100 gas turbine and a diesel engine with a single gearbox powering four traction motors identical to those in Acela.

The diesel provided head end power and low speed traction, with the turbine not being started until after leaving stations. The prototype was completed in June 2000, and safety testing was done at the FRA's Pueblo, CO test track beginning in the summer of 2001. A maximum speed of 156 mph was reached. The prototype was then taken on a tour of potential sites for high speed service, but no service has yet begun.

Canada

In 2002, Bombardier Transportation announced the launch of the JetTrain, a high-speed trainset consisting of tilting carriages and a locomotive powered by a Pratt & Whitney turboshaft engine. Proposals were made to use the trains for Quebec City-Windsor, Orlando-Miami, and in Alberta, Texas, Nevada and the UK.

However, nothing ever came of any of these proposals, and the JetTrain essentially disappeared, being superseded by the Bombardier Zefiro line of conventionally-powered high speed and very high speed trains. The JetTrain no longer appears on any of Bombardier's current web sites or promotional materials, although it can still be found on older web sites bearing the Canadair logos.

Soviet Union

Two gas turbine-electric locomotive types underwent testing in the Soviet Union. The G1-01 freight GTEL was intended to consist of two locomotives of a C-C wheel arrangement, but only one section was built. The test program began in 1959 and lasted into the early 1970s.

The GP1 was a similar design, also with a C-C wheel arrangement, introduced in 1964. Two units were built, GP1-0001 and GP1-0002, which were also used in regular service. Both types had a maximum power output of 2,600 kW (3,500 hp).

Russia

In 2006, Russian Railways introduced the GEM-10 switcher GTEL. The turbine's maximum power output is 1,000 kW (1,300 hp) and it runs on liquefied natural gas.

The GEM-10 has a C-C wheel arrangement. The TGEM10-0001 is a two-unit (cow-calf) switcher GTEL, with a B-B+B-B wheel arrangement, and uses the same turbine and fuel as the GEM-10.

The GT1-001 freight GTEL, introduced in 2007, runs on liquefied natural gas and has a maximum power output of 8,300 kW (11,100 hp). The locomotive has a B-B-B+B-B-B wheel arrangement, and up to three GT1s can be coupled.

On January 23, 2009 the locomotive conducted a test run with a 159 car train weighing 15,000 metric tons. Further heavy-haul tests were conducted in December 2010. In a test run conducted in September 2011 the locomotive pulled 170 freight cars weighing 16,000 metric tons.

Twin-turbo

Twin-turbo refers to a turbocharged engine, in which two turbochargers compress the intake charge. There are two commonly used twin turbo configurations: parallel twin-turbo and sequential twin-turbo. A third kind of twin-turbocharging, staged turbocharging, is used in diesel automobile racing applications.

Parallel Twin-turbo

Figure 10: 3.5 Ford EcoBoost engine (Twin Turbo)

Parallel twin-turbo refers to a turbocharger configuration in which two identical turbochargers function simultaneously, splitting the turbocharging duties equally. Each turbocharger is driven by half of the engine's spent exhaust energy. In most applications, the compressed air from both turbos is combined in a common intake manifold and sent to the individual cylinders. Usually each turbocharger is mounted to its own individual exhaust/turbo manifold, but on inline-type engines both turbochargers can be mounted to a single turbo manifold.

Parallel twin turbos applied to V-shaped engines are usually mounted with one turbo assigned to each cylinder bank, providing packaging symmetry and simplifying plumbing over a single turbo setup. When used on inline engines parallel twin turbos are commonly applied with two smaller turbos, which can provide similar performance with less turbo lag than a single larger turbo.

Some examples of parallel twin-turbo inline engines are Nissan's RB26DETT, BMW's N54 and Volvo's B6284T and B6294T. Some examples of V formation engines with parallel twin-turbos include Mitsubishi's 6A12TT and 6G72TT; Nissan's VG30DETT and VR38DETT; and Audi's 1997-2002 S4 (B5), 1997-2005 A6, and 2003-2004 RS6.

While a parallel twin-turbo set-up theoretically has less turbo lag than a single turbocharger set up, this is not always the case due to many factors. Marginally reduced combined inertial resistance, simplified exhaust plumbing, and the simultaneous spooling of both turbos means that there can still be a noticeable bit of lag, especially in high-flow turbo/high boost applications. Some ways to counter this are to use a light pressure set up with smaller turbos, where the turbos are designed to output less boost but spool earlier. While this setup sacrifices some top end power, it still has less lag than a similar engine with a single turbo set up making the same power. Another system would be the use of variable geometry turbochargers. This system changes the angle of the guide vanes depending on the exhaust pressure, giving the system excellent power throughout the rev range. Once used mainly in turbocharged diesel engines, Chrysler was the first to use it in mass-production gasoline-powered vehicles with the Shelby CSX, debuted in 1989.

It is possible to use parallel operation with more than two turbochargers. Two such examples are the Bugatti EB110 and Bugatti Veyron, both of which run four turbochargers in parallel. The EB110 runs 4 turbos on a 3.5 litre V12 engine, producing 542 hp (404 kW) at 8000 rpm, while the Veyron uses a 16 cylinder engine to generate 1,200 PS (880 kW; 1,200 hp).

Sequential Turbos

Figure 11: BMW 3 Series Diesel turbo setup

Sequential turbos refer to a set-up in which the motor utilises one turbocharger for lower engine speeds, and a second or both turbochargers at higher engine speeds. Typically, larger high-flow turbochargers are not as efficient at low RPMs, resulting in lower intake manifold pressures under these conditions. On the other hand, smaller turbos spool up quickly at low RPMs but cannot supply enough air at higher RPMs. During low to mid engine speeds, when available spent exhaust energy is minimal, only one relatively small turbocharger (called the primary turbocharger) is active. During this period, all of the engine's exhaust energy is directed to the primary turbocharger only, providing the small turbo's benefits of a lower boost threshold, minimal turbo lag, and increased power output at low engine speeds. As RPM increases, the secondary turbocharger is partially activated in order to pre-spool prior to its full utilisation. Once a preset engine speed or boost pressure is attained, valves controlling compressor and turbine flow through the secondary turbocharger are opened completely. (The primary turbocharger is deactivated at this point in some applications.) In this way a full twin-turbocharger setup provides the benefits associated with a large turbo, including maximum power output, without the disadvantage of increased turbo lag.

Sequential turbocharger systems provide a way to decrease turbo lag without compromising ultimate boost output and engine power. Examples of cars with a sequential twin-turbo setup include the 1986-1988 Porsche 959, the 1990-1995 Eunos Cosmo JC, 1992-2002 Mazda RX-7 Turbo (FD3S), the 1993-1998 Toyota Supra Turbo (JZA8x), the 1994-2005 JDM Subaru Legacy GT, GT-B & B4 RSK, and the 2008-2010 Ford Super Duty 6.4L diesel.

Staged Turbos

A sequential turbo can also refer to a system where the output pressure must be greater than can be provided by a single turbo, commonly called a staged twin turbo system. In this case, multiple similarly sized turbochargers are used in sequence, but both operate constantly. The first turbo boosts pressure as much as possible (for example to three times the intake pressure). Subsequent turbos takes the charge from the previous stage and compresses it further (for example to an additional three times intake pressure, for a total boost of nine times atmospheric pressure). This configuration is commonly found on piston engine aircraft which usually do not need to rapidly raise and lower engine speed (and thus where turbo lag is not a primary design consideration), and where the intake pressure is quite low due to low atmospheric pressure at altitude, requiring a very high pressure ratio. High-performance diesel engines also sometimes use this configuration, since diesel engines do not suffer from pre-ignition issues and can use significantly higher boost pressure than Otto cycle engines.

Advantages in Diesel Emissions

While spark ignition engines have fallen out of favour of the sequential turbo design, many diesel companies now make engines with sequential turbos in order to reduce emissions. Caterpillar Inc. ACERT engines utilise sequential turbos as well as International on some of their newest engines.

Chapter 8

Fossil-fuel Power Station

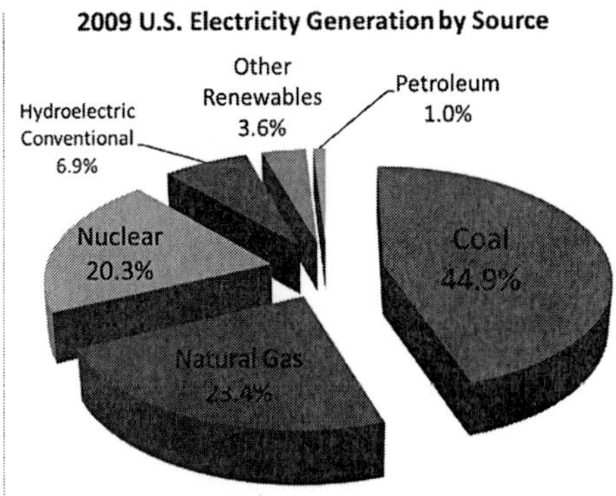

Figure 1: Sources of electricity in the U.S. in 2009.

A fossil-fuel power station is a power station that burns fossil fuels such as coal, natural gas or petroleum (oil) to produce electricity. Central station fossil-fuel power plants are designed on a large scale for continuous operation. In many countries, such plants provide most of the electrical energy used.

Fossil fuel power stations (except for MHD generators) have some kind of rotating machinery to convert the heat energy of combustion into mechanical energy, which then operate an electrical generator. The prime mover may be a steam turbine, a gas turbine or, in small isolated plants, a reciprocating internal combustion engine. All plants use the drop between the high pressure and temperature of the steam or combusting fuel and the lower pressure of the atmosphere or condensing vapour in the steam turbine.

Byproducts of power thermal plant operation need to be considered in both the design and operation. Waste heat due to the finite efficiency of the power cycle must be released to the atmosphere, using a cooling tower, or river or lake water as a cooling medium. The flue gas from combustion of the fossil fuels is discharged to the air; this contains carbon dioxide and water vapour, as well as other substances such as nitrogen, nitrogen oxides, sulphur oxides, and (in the case of coal-fired plants) fly ash, mercury and traces of other metals. Solid waste ash from coal-fired boilers must also be removed. Some coal ash can be recycled for building materials.

Fossil fuelled power stations are major emitters of CO_2, a greenhouse gas (GHG) which according to a consensus of scientific organisations is a contributor to global warming observed over the last 100 years. Brown coal emits 3 times as much CO_2 as natural gas, black coal emits twice as much CO_2 per unit of electric energy. Carbon capture and storage of emissions are not expected to be available on a commercial economically viable basis until 2025. A recent study indicates that sulphur emissions from fossil fuelled power stations in China have caused a 10-year lull in global warming (1998-2008)

Basic Concepts

In a fossil fuel power plant the chemical energy stored in fossil fuels such as coal, fuel oil, natural gas or oil shale) and oxygen of the air is converted successively into thermal energy, mechanical energy and, finally, electrical energy for continuous use and distribution across a wide geographic area. Each fossil fuel power plant is a highly complex, custom-designed system. Construction costs, as of 2004, run to US$1,300 per kilowatt, or $650 million for a 500 MWe unit. Multiple generating units may be built at a single site for more efficient use of land, natural resources and labour. Most thermal power stations in the world use fossil fuel, outnumbering nuclear, geothermal, biomass, or solar thermal plants.

Heat Into Mechanical Energy

The second law of thermodynamics states that any closed-loop cycle can only convert a fraction of the heat produced during combustion into mechanical work. The rest of the heat, called waste heat, must be released into a cooler environment during the return portion of the cycle. The fraction of heat released into a cooler medium must be equal or larger than the ratio of absolute temperatures of the cooling system (environment) and the heat source (combustion furnace). Raising the furnace temperature improves the efficiency but complicates the design, primarily by the selection of alloys used for construction, making the furnace more expensive. The waste heat cannot be converted into mechanical energy without an even cooler cooling system. However, it may be used in cogeneration plants to heat buildings, produce hot water, or to heat materials on an industrial scale, such as in some oil refineries, plants, and chemical synthesis plants.

Typical thermal efficiency for electrical generators in the industry is around 33% for coal and oil-fired plants, and up to 50% for combined-cycle gas-fired plants. Plants designed to achieve peak efficiency while operating at capacity will be less efficient when operating off-design (i.e. temperatures too low.)

The worlds most efficient Coal-fired Power Station is Vattenfall's 400MWe, Unit 3 at Nordjylland Power Station.

The Carnot cycle is the theoretically most efficient closed thermodynamic cycle for conversion of heat energy into useful work, and practical fossil-fuel stations cannot exceed this limit. In principle, fuel cells do not have the same thermodynamic limits as they are not heat engines.

Fuel Transport and Delivery

Figure 2: Big Bend Coal Power Station in Apollo Beach, Florida in the United States.

Figure 3: Coal fired power plants provide about 46% of consumed electricity in the United States. This is the Castle Gate Plant near Helper, Utah.

Coal is delivered by highway truck, rail, barge, collier ship or coal slurry pipeline. Some plants are even built near coal mines and coal is delivered by conveyors. A large coal train called a "unit train" may be two kilometres (over a mile) long, containing 100 cars with 100 short tons of coal in each one, for a total load of 10,000 tons. A large plant under full load requires at least one coal delivery this size every day. Plants may get as many as three to five trains a day, especially in "peak season" during the hottest Summer and/or coldest Winter months (depending on local climate) when power consumption is high. A large thermal power plant such as the one in Nanticoke, Ontario stores several million metric tons of coal for winter use when the lakes are frozen.

Modern unloaders use rotary dump devices, which eliminate problems with coal freezing in bottom dump cars. The unloader includes a train positioner arm that pulls the entire train to position each car over a coal hopper. The dumper clamps an individual car against a platform that swivels the car upside down to dump the coal. Swiveling couplers enable the entire operation to occur while the cars are still coupled together. Unloading a unit train takes about three hours.

Shorter trains may use railcars with an "air-dump", which relies on air pressure from the engine plus a "hot shoe" on each car. This "hot shoe" when it comes into contact with a "hot rail" at the unloading trestle, shoots an electric charge through the air dump apparatus and causes the doors on the bottom of the car to open, dumping the coal through the opening in the trestle. Unloading one of these trains takes anywhere from an hour to an hour and a half. Older unloaders may still use manually operated bottom-dump rail cars and a "shaker" attached to dump the coal. Generating stations adjacent to a mine may receive coal by conveyor belt or massive diesel-electric-drive trucks.

A collier (cargo ship carrying coal) may hold 40,000 long tons of coal and takes several days to unload. Some colliers carry their own conveying equipment to unload their own bunkers; others depend on equipment at the plant. Colliers are large, seaworthy, self-powered ships. For transporting coal in calmer waters, such as rivers and lakes, flat-bottomed vessels called barges are often used. Barges are usually unpowered and must be moved by tugboats or towboats.

For start up or auxiliary purposes, the plant may use fuel oil as well. Fuel oil can be delivered to plants by pipeline, tanker, tank car or truck. Oil is stored in vertical cylindrical steel tanks with capacities as high as 90,000 barrels (14,000 m)' worth. The heavier no. 5 "bunker" and no. 6 fuels are typically steam-heated before pumping in cold climates. Plants fuelled by natural gas are usually built adjacent to gas transport pipelines or have dedicated gas pipelines extended to them.

Fuel Processing

Coal is prepared for use by crushing the rough coal to pieces less than 2 inches (5 cm) in size. The coal is then transported from the storage yard to in-plant storage silos by rubberized conveyor belts at rates up to 4,000 short tons per hour. In plants that burn pulverised coal, silos feed coal pulverisers (coal mills) that take the larger 2-inch (51 mm) pieces, grind them to the consistency of face powder, sort them, and mix them with primary combustion air which transports the coal to the furnace and preheats the coal to drive off excess moisture content. A 500 MWe plant may have six such pulverisers, five of which can supply coal to the furnace at 250 tons per hour under full load. In plants that do not burn pulverised coal, the larger 2-inch (51 mm) pieces may be directly fed into the silos which then feed either mechanical distributors that drop the coal on a travelling grate or the cyclone burners, a specific kind of combustor that can efficiently burn larger pieces of fuel.

Steam-Electric

Most electric power made from fossil fuel is produced by thermal power stations. Reciprocating steam engines fell out of use rapidly after the first steam turbines were introduced around 1906.

Gas Turbine Plants

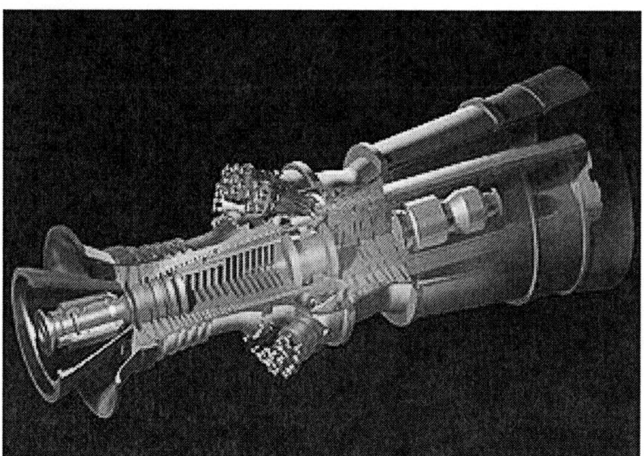

Figure 4: 480 megawatt GE H series power generation gas turbine

Figure 5: Currant Creek Power Plant near Mona, Utah is a natural gas fired electrical plant.

One type of fossil fuel power plant uses a gas turbine in conjunction with a heat recovery steam generator (HRSG). It is referred to as a combined cycle power plant because it combines the Brayton cycle of the gas turbine with the Rankine cycle of the HRSG. The thermal efficiency of these plants has reached a record heat rate of 5690 Btu/(kW·h), or just under 60%, at a facility in Baglan Bay, Wales.

The turbines are fuelled either with natural gas, syngas or fuel oil. While more efficient and faster to construct (a 1,000 MW plant may be completed in as little as 18 months from start of construction), the economics of such plants is heavily influenced by the volatile cost of fuel, normally natural gas. The combined cycle plants are designed in a variety of configurations composed of the number of gas turbines followed by the steam turbine. For example, a 3-1 combined cycle facility has three gas turbines tied to one steam turbine. The configurations range from (1-1), (2-1), (3-1), (4-1), (5-1), to (6-1)

Simple-cycle or open cycle gas turbine plants, without a steam cycle, are sometimes installed as emergency or peaking capacity; their thermal efficiency is much lower. The high running cost per hour is offset by the low capital cost and the intention to run such units only a few hundred hours per year. Other gas turbine plants are installed in stages, with an open cycle gas turbine the first stage and additional turbines or conversion to a closed cycle part of future project plans.

Reciprocating Engines

Diesel engine generator sets are often used for prime power in communities not connected to a widespread power grid. Emergency (standby) power systems may use reciprocating internal combustion engines operated by fuel oil or natural gas. Standby generators may serve as emergency power for a factory or data centre, or may also be operated in parallel with the local utility system to reduce peak power demand charge from the utility.

Diesel engines can produce strong torque at relatively low rotational speeds, which is generally desirable when driving an alternator, but diesel fuel in long-term storage can be subject to problems resulting from water accumulation and chemical decomposition. Rarely-used generator sets may correspondingly be installed as natural gas or LPG to minimise the fuel system maintenance requirements.

Spark-ignition internal combustion engines operating on gasoline (petrol), propane, or LPG are commonly used as portable temporary power sources for construction work, emergency power, or recreational uses.

Reciprocating external combustion engines such as the Stirling engine can be run on a variety of fossil fuels, as well as renewable fuels or industrial waste heat. Installations of Stirling engines for power production are relatively uncommon.

Environmental Impacts

Figure 6: The Mohave Power Station, a 1,580 MW coal power station near Laughlin, Nevada, out of service since 2005 due to environmental restrictions

The world's power demands are expected to rise 60% by 2030. With the worldwide total of active coal plants over 50,000 and rising, the International Energy Agency (IEA) estimates that fossil fuels will account for 85% of the energy market by 2030.

World organisations and international agencies, like the IEA, are concerned about the environmental impact of burning fossil fuels, and coal in particular. The combustion of coal contributes the most to acid rain and air pollution, and has been connected with global warming. Due to the chemical composition of coal there are difficulties in removing impurities from the solid fuel prior to its combustion. Modern day coal power plants pollute less than older designs due to new "scrubber" technologies that filter the exhaust air in smoke stacks; however emission levels of various pollutants are still on average several times greater than natural gas power plants. In these modern designs, pollution from coal-fired power plants comes from the emission of gases such as carbon dioxide, nitrogen oxides, and sulphur dioxide into the air.

Acid rain is caused by the emission of nitrogen oxides and sulphur dioxide. These gases may be only mildly acidic themselves, yet when they react with the atmosphere, they create acidic compounds such as sulphurous acid, nitric acid and sulphuric acid which fall as rain, hence the term acid rain. In Europe and the U.S.A., stricter emission laws and decline in heavy industries have reduced the environmental hazards associated with this problem, leading to lower emissions after their peak in 1960s.

European Environment Agency (EEA) gives fuel-dependent emission factors based on actual emissions from power plants in EU.

Pollutant	*Hard Coal*	*Brown Coal*	*Fuel Oil*	*Other Oil*	*Gas*
CO_2 (g/GJ)	94600	101000	77400	74100	56100
SO_2 (g/GJ)	765	1361	1350	228	0.68
NO_x (g/GJ)	292	183	195	129	93.3
CO (g/GJ)	89.1	89.1	15.7	15.7	14.5
Non methane organic compounds (g/GJ)	4.92	7.78	3.70	3.24	1.58
Particulate matter (g/GJ)	1203	3254	16	1.91	0.1
Flue gas volume total (m^3/GJ)	360	444	279	276	272

Carbon Dioxide

Electricity generation using carbon based fuels is responsible for a large fraction of carbon dioxide (CO_2) emissions worldwide and for

41% of U.S. man-made carbon dioxide emissions. Of fossil fuels, coal combustion in thermal power stations result in greater amounts of carbon dioxide emissions per unit of electricity generated (2249 lbs/MWh) while oil produces less (1672 lb/(MW·h) or 211 kg/GJ) and natural gas produces the least 1135 lb/(MW·h) (143 kg/GJ).)

The Intergovernmental Panel on Climate Change states that carbon dioxide is a greenhouse gas and that increased quantities within the atmosphere will "very likely" lead to higher average temperatures on a global scale (global warming); concerns regarding the potential for such warming to change the global climate prompted IPCC recommendations calling for large cuts to CO_2 emissions worldwide.

Emissions may be reduced through more efficient and higher combustion temperature and through more efficient production of electricity within the cycle. Carbon capture and storage (CCS) of emissions from coal fired power stations is another alternative but the technology is still being developed and will increase the cost of fossil fuel-based production of electricity. CCS may not be economically viable, unless the price of emitting CO_2 to the atmosphere rises.

Particulate Matter

Another problem related to coal combustion is the emission of particulates that have a serious impact on public health. Power plants remove particulate from the flue gas with the use of a bag house or electrostatic precipitator. Several newer plants that burn coal use a different process, Integrated Gasification Combined Cycle in which synthesis gas is made out of a reaction between coal and water. The synthesis gas is processed to remove most pollutants and then used initially to power gas turbines. Then the hot exhaust gases from the gas turbines are used to generate steam to power a steam turbine. The pollution levels of such plants are drastically lower than those of "classic" coal power plants.

Particulate matter from coal-fired plants can be harmful and have negative health impacts. Studies have shown that exposure to particulate matter is related to an increase of respiratory and cardiac mortality. Particulate matter can irritate small airways in the lungs, which can lead to increased problems with asthma, chronic bronchitis, airway obstruction, and gas exchange.

There are different types of particulate matter, depending on the chemical composition and size. The dominant form of particulate matter from coal-fired plants is coal fly ash, but secondary sulphate and nitrate also comprise a major portion of the particulate matter from coal-fired plants. Coal fly ash is what remains after the coal has been combusted, so it consists of the incombustible materials that are found in the coal.

The size and chemical composition of these particles affects the impacts on human health. Currently coarse (diametre greater than 2.5 ìm) and fine (diametre between 0.1 ìm and 2.5 ìm) particles are regulated, but ultrafine particles (diametre less than 0.1 ìm) are currently unregulated, yet they pose many dangers. Unfortunately much is still unknown as to which kinds of particulate matter pose the most harm, which makes it difficult to come up with adequate legislation for regulating particulate matter. There are several methods of helping to reduce the particulate matter emissions from coal-fired plants. Roughly 80% of the ash falls into an ash hopper, but the rest of the ash then gets carried into the atmosphere to become coal-fly ash. Methods of reducing these emissions of particulate matter include:

1. a baghouse
2. an electrostatic precipitator (ESP)
3. cyclone collector

The baghouse has a fine filter that collects the ash particles, electrostatic precipitators use an electric field to trap ash particles on high-voltage plates, and cyclone collectors use centrifugal force to trap particles to the walls.

Radioactive Trace Elements

Coal is a sedimentary rock formed primarily from accumulated plant matter, and it includes many inorganic minerals and elements which were deposited along with organic material during its formation. As the rest of the Earth's crust, coal also contains low levels of uranium, thorium, and other naturally occurring radioactive isotopes whose release into the environment leads to radioactive contamination. While these substances are present as very small trace impurities, enough coal is burned that significant amounts of these substances are released. A 1,000 MW coal-burning power plant could have an uncontrolled release of as much as 5.2 metric tons per year of uranium (containing 74 pounds (34 kg) of uranium-235) and 12.8 metric tons per year of thorium. In comparison, a 1,000 MW nuclear plant will generate about 30 short tons of high-level radioactive solid packed waste per year. It is estimated that during 1982, US coal burning released 155 times as much uncontrolled radioactivity into the atmosphere as the Three Mile Island incident. The collective radioactivity resulting from all coal burning worldwide between 1937 and 2040 is estimated to be 2,700,000 curies or 0.101 EBq. It should also be noted that during normal operation, the effective dose equivalent from coal plants is 100 times that from nuclear plants. But it is also worth noting that normal operation is a deceiving baseline for comparison: just the Chernobyl nuclear disaster released, in of I-131 alone, an estimated 1,76 EBq of radioactivity, a value one order of magnitude above this value for total emissions from all coal burned within a century. But at the same time, it shall also be understood that the I-131, the major radioactive substance which comes out in accident situations has a half life of just 8 days. Hence, it is not going to cause as much as damage as the Uranium and Thorium which are released from Coal fired power plants which have much higher half lives. Also, I-131 can be easily removed by eating normal Iodine Tablets.

Water and Air Contamination by Coal Ash

A study released in August 2010 that examined state pollution data in the United States by the organisations Environmental Integrity Project, the Sierra Club and Earthjustice found that coal ash produced by coal-fired power plants dumped at sites across 21 U.S. states has contaminated ground water with toxic elements. The contaminants including the poisons arsenic and lead. Arsenic has been shown to cause skin cancer, bladder cancer and lung cancer, and lead damages the nervous system. Coal ash contaminants are also linked to respiratory diseases and other health and developmental problems, and have disrupted local aquatic life. Coal ash also releases a variety of toxic contaminants into nearby air, posing a health threat to those who breath in fugitive coal dust. Currently, the EPA does not regulate the disposal of coal ash; regulation is up to the states and the electric power industry has been lobbying to maintain this status quo. Most states require no monitoring of drinking water near coal ash dump sites. The study found an additional 39 contaminated U.S. sites and concluded that the problem of coal ash-caused water contamination is even more extensive in the United States than has been estimated. The study brought to 137 the number of ground water sites across the United States that are contaminated by power plant-produced coal ash.

Range of Mercury Contamination in Fish

U.S. government scientists tested fish in 291 streams around the country for mercury contamination. They found mercury in every fish tested, according to the study by the U.S. Department of the Interior. They found mercury even in fish of isolated rural waterways. Twenty five percent of the fish tested had mercury levels above the safety levels determined by the U.S. Environmental Protection Agency for people who eat the fish regularly. The largest source of mercury contamination in the United States is coal-fuelled power plant emissions.

Greening of Fossil Fuel Power Plants

Several methods exist to improve the efficiency of fossil fuel power plants. A frequently used and cost-efficient method is to convert a plant to run on a different fuel. This includes conversions for biomass and waste. Conversions to waste-fired power plants have the benefit of reducing landfilling. In addition, waste fired power plants can be equipped with material recovery, which is also beneficial to the environment.

Regardless of the conversion, a truly green fossil fuel power plant implements carbon capture and storage (CCS). CCS means that the exhaust CO_2 is not released into the environment and the fossil fuel power plant becomes an emissionless power plant. An example of a CCS fossil fuel power plant is the Elsam power station near Esbjerg, Denmark.

Low NOx Burners

A common retrofit in fossil fuelled power stations is the replacement of original burners with Low NOx burners. Careful consideration of fluid dynamics and flame thermodynamics has enabled substantial reduction in flame temperature, leading to reduced formation of Nitrous Oxides.

Clean Coal

"Clean coal" is the name attributed to a process whereby coal is chemically washed of minerals and impurities, sometimes gasified, burned and the resulting flue gases treated with steam, with the purpose of removing sulphur dioxide, and reburned so as to make the carbon dioxide in the flue gas economically recoverable. The coal industry uses the term "clean coal" to describe technologies designed to enhance both the efficiency and the environmental acceptability of coal extraction, preparation and use, but has provided no specific quantitative limits on any emissions, particularly carbon dioxide. Whereas contaminates like sulphur or mercury can be removed from coal, carbon cannot be effectively removed while still leaving a usable fuel, and clean coal plants without carbon sequestration and storage do not significantly reduce carbon dioxide emissions.

James Hansen in an open letter to U.S. President Barack Obama has advocated a "moratorium and phase-out of coal plants that do not capture and store CO_2". In his book *Storms of My Grandchildren*, similarly, Hansen discusses his *Declaration of Stewardship* the first principle of which requires "a moratorium on coal-fired power plants that do not capture and sequester carbon dioxide".

Combined Heat and Power

Combined heat and power (CHP), also known as cogeneration, is the use of a power station to provide both electric power and process heat or district heating. While rejecting heat at a higher than normal temperature to enable building heating lowers overall plant electric power efficiency, the extra fuel burnt is more than offset by the reduction in fossil fuel that would otherwise be used for heating buildings. This technology is widely practiced in for example Denmark, other Scandinavian countries and parts of Germany. Calculations show that CHPDH is the cheapest method of carbon emissions reductions.

Alternatives to Fossil Fuel Power Plants

Alternatives to fossil fuel power plants include nuclear power, solar power, geothermal power, wind power, tidal power, hydroelectric power (hydroelectricity) and other renewable energies. Some of these are proven technologies on an industrial scale (i.e. nuclear, wind, tidal and hydroelectric power) others are still in prototype form.

Nuclear power, and geothermal power may be classed as heat pollutants as they add heat energy to the biosphere that would not otherwise be released. The nett quantity of energy conversion within the biosphere due to the utilisation of wind power, solar power, tidal power, hydroelectric power (hydroelectricity) is static and is derived from the effects of sunlight and the movement of the moon and planets. It may be argued that the nett release of geothermal energy into the biosphere through volcanic activity is mitigated slightly over time due to the draw down of heat energy by geothermal power stations. This would entail an extremely complex analysis to resolve for individual installations.

Generally, the cost of electrical energy produced by non fossil fuel burning power plants is greater than that produced by burning fossil fuels. This statement however only includes the cost to produce the electrical energy and does not take into account indirect costs associated with the many pollutants created by burning fossil fuels (e.g. increased hospital admissions due respiratory diseases caused by fine smoke particles).

Relative Cost by Generation Source

When comparing power plant costs, it is customary to start by calculating the cost of power at the generator terminals by considering several main factors. External costs such as connections costs, the effect of each plant on the distribution grid are considered separately as an additional cost to the calculated power cost at the terminals.

Initial factors considered are:

- Capital costs (including waste disposal and decommissioning costs for nuclear energy)
- Operating and maintenance costs
- Fuel costs (for fossil fuel and biomass sources, and which may be negative for wastes)
- Likely annual hours per year run or load factor (may be 30% for wind energy, but 90% for nuclear energy)
- Offset sales of heat (for example in combined heat and power district heating (CHP/DH)).

These costs occur over the 30–50 year life of the fossil fuel power plants, using discounted cash flows. In general large fossil plants are attractive due to their low initial capital costs—typically around £750–£1000 per kilowatt electrical compared to perhaps £1500 per kilowatt for onshore wind.

Power Station

Figure 7: The Susquehanna Steam Electric Station, a boiling water reactor nuclear power plant.

Figure 8: St. Clair Power Plant, a coal-fired plant in Michigan.

Fiogure 9: The Three Gorges Dam, a hydroelectric dam.

A power station (also referred to as a generating station, power plant, or powerhouse) is an industrial facility for the generation of electric energy.

At the centre of nearly all power stations is a generator, a rotating machine that converts mechanical energy into electrical energy by creating relative motion between a magnetic field and a conductor. The energy source harnessed to turn the generator varies widely. It depends chiefly on which fuels are easily available and on the types of technology that the power company has access to.

History

The world's first power station was built by Sigmund Schuckert in the Bavaria town of Ettal and went into operation in 1878. The station consisted of 24 dynamo electric generators which were driven by a steam engine. It was used to illuminate a grotto in the gardens of Linderhof Palace.

The first public power station was the *Edison Electric Light Station*, built in London at 57, Holborn Viaduct, which started operation in January 1882. This was an initiative of Thomas Edison that was organised and managed by his partner, Edward Johnson. A Babcock and Wilcox boiler powered a 125 horsepower steam engine that drove a 27 ton generator called Jumbo, after the celebrated elephant. This supplied electricity to premises in the area that could be reached through the culverts of the viaduct without digging up the road, which was the monopoly of the gas companies. The customers included the City Temple and the Old Bailey. Another important customer was the Telegraph Office of the General Post Office but this could not be reached though the culverts. Johnson arranged for the supply cable to be run overhead, via Holborn Tavern and Newgate.

In September 1882 in New York, the Pearl Street Station was established by Edison to provide electric lighting in the lower Manhattan Island area; the station ran until destroyed by fire in 1890. The station used reciprocating steam engines to turn direct-current generators.

Because of the DC distribution, the service area was small, limited by voltage drop in the feeders. The War of Currents eventually resolved in favour of AC distribution and utilisation, although some DC systems persisted to the end of the 20th century. DC systems with a service radius of a mile (kilometre) or so were necessarily smaller, less efficient of fuel consumption, and more labour intensive to operate than much larger central AC generating stations.

AC systems used a wide range of frequencies depending on the type of load; lighting load using higher frequencies, and traction systems and heavy motor load systems preferring lower frequencies. The economics of central station generation improved greatly when unified light and power systems, operating at a common frequency, were developed.

The same generating plant that fed large industrial loads during the day, could feed commuter railway systems during rush hour and then serve lighting load in the evening, thus improving the system load factor and reducing the cost of electrical energy overall. Many exceptions existed, generating stations were dedicated to power or light by the choice of frequency, and rotating frequency changers and rotating converters were particularly common to feed electric railway systems from the general lighting and power network.

Throughout the first few decades of the 20th century central stations became larger, using higher steam pressures to provide greater efficiency, and relying on interconnections of multiple generating stations to improve reliability and cost. High-voltage AC transmission allowed hydroelectric power to be conveniently moved from distant waterfalls to city markets. The advent of the steam turbine in central station service, around 1906, allowed great expansion of generating capacity. Generators were no longer limited by the power transmission of belts or the relatively slow speed of reciprocating engines, and could grow to enormous sizes.

For example, Sebastian Ziani de Ferranti planned what would have been the largest reciprocating steam engine ever built for a proposed new central station, but scrapped the plans when turbines became available in the necessary size. Building power systems out of central stations required combinations of engineering skill and financial acumen in equal measure. Pioneers of central station generation include George Westinghouse and Samuel Insull in the United States, Ferranti and Charles Hesterman Merz in UK, and many others.

Thermal Power Stations

In thermal power stations, mechanical power is produced by a heat engine that transforms thermal energy, often from combustion of a fuel, into rotational energy. Most thermal power stations produce steam, and these are sometimes called steam power stations. Not all thermal energy can be transformed into mechanical power, according to the second law of thermodynamics. Therefore, there is always heat lost to the environment.

If this loss is employed as useful heat, for industrial processes or district heating, the power plant is referred to as a cogeneration power plant or CHP (combined heat-and-power) plant. In countries where district heating is common, there are dedicated heat plants called heat-only boiler stations. An important class of power stations in the Middle East uses by-product heat for the desalination of water.

The efficiency of a steam turbine is limited by the maximum temperature of the steam produced and is not directly a function of the fuel used. For the same steam conditions, coal, nuclear and gas power plants all have the same theoretical efficiency. Overall, if a system is on constantly (base load) it will be more efficient than one that is used intermittently (peak load).

Besides use of reject heat for process or district heating, one way to improve overall efficiency of a power plant is to combine two different thermodynamic cycles. Most commonly, exhaust gases from a gas turbine are used to generate steam for a boiler and steam turbine. The combination of a "top" cycle and a "bottom" cycle produces higher overall efficiency than either cycle can attain alone.

Classification

Figure 10: CHP plant in Warsaw, Poland.

Figure 11: Geothermal power station in Iceland.

Figure 12: Coal Power Station in Tampa, United States.

Thermal power plants are classified by the type of fuel and the type of prime mover installed.

By Fuel

- Fossil fuelled power plants may also use a steam turbine generator or in the case of natural gas fired plants may use a combustion turbine. A coal-fired power station produces electricity by burning coal to generate steam, and has the side-effect of producing large amounts of sulphur dioxide which pollutes air and water and carbon dioxide, which allegedly contributes to global warming. About 50% of electric generation in the USA is produced by coal fired power plants

- Nuclear power plants use a nuclear reactor's heat to operate a steam turbine generator. About 20% of electric generation in the USA is produced by nuclear power plants.
- Geothermal power plants use steam extracted from hot underground rocks.
- Biomass-fuelled power plants may be fuelled by waste from sugar cane, municipal solid waste, landfill methane, or other forms of biomass.
- In integrated steel mills, blast furnace exhaust gas is a low-cost, although low-energy-density, fuel.
- Waste heat from industrial processes is occasionally concentrated enough to use for power generation, usually in a steam boiler and turbine.
- Solar thermal electric plants use sunlight to boil water, which turns the generator.

By Prime Mover

- Steam turbine plants use the dynamic pressure generated by expanding steam to turn the blades of a turbine. Almost all large non-hydro plants use this system. About 90% of all electric power produced in the world is by use of steam turbines.
- Gas turbine plants use the dynamic pressure from flowing gases (air and combustion products) to directly operate the turbine. Natural-gas fuelled (and oil fuelled) combustion turbine plants can start rapidly and so are used to supply "peak" energy during periods of high demand, though at higher cost than base-loaded plants. These may be comparatively small units, and sometimes completely unmanned, being remotely operated. This type was pioneered by the UK, Princetown being the world's first, commissioned in 1959.
- Combined cycle plants have both a gas turbine fired by natural gas, and a steam boiler and steam turbine which use the hot exhaust gas from the gas turbine to produce electricity. This greatly increases the overall efficiency of the plant, and many

new baseload power plants are combined cycle plants fired by natural gas.

- Internal combustion reciprocating engines are used to provide power for isolated communities and are frequently used for small cogeneration plants. Hospitals, office buildings, industrial plants, and other critical facilities also use them to provide backup power in case of a power outage. These are usually fuelled by diesel oil, heavy oil, natural gas, and landfill gas.
- Microturbines, Stirling engine and internal combustion reciprocating engines are low-cost solutions for using opportunity fuels, such as landfill gas, digester gas from water treatment plants and waste gas from oil production.

Cooling Towers

All thermal power plants produce waste heat energy as a byproduct of the useful electrical energy produced. The amount of waste heat energy equals or exceeds the amount of electrical energy produced. Gas-fired power plants can achieve 50% conversion efficiency while coal and oil plants achieve around 30–49%. The waste heat produces a temperature rise in the atmosphere which is small compared to that of greenhouse-gas emissions from the same power plant. Natural draft wet cooling towers at many nuclear power plants and large fossil fuel fired power plants use large hyperbolic chimney-like structures (as seen in the image at the left) that release the waste heat to the ambient atmosphere by the evapouration of water. However, the mechanical induced-draft or forced-draft wet cooling towers in many large thermal power plants, nuclear power plants, fossil fired power plants, petroleum refineries, petrochemical plants, geothermal, biomass and waste to energy plants use fans to provide air movement upward through downcoming water and are not hyperbolic chimney-like structures. The induced or forced-draft cooling towers are typically rectangular, box-like structures filled with a material that enhances the contacting of the upflowing air and the downflowing water.

In areas with restricted water use a dry cooling tower or radiators, directly air cooled, may be necessary, since the cost or environmental consequences of obtaining make-up water for evapourative cooling would be prohibitive. These have lower efficiency and higher energy consumption in fans than a wet, evapourative cooling tower.

Where economically and environmentally possible, electric companies prefer to use cooling water from the ocean, or a lake or river, or a cooling pond, instead of a cooling tower. This type of cooling can save the cost of a cooling tower and may have lower energy costs for pumping cooling water through the plant's heat exchangers. However, the waste heat can cause the temperature of the water to rise detectably. Power plants using natural bodies of water for cooling must be designed to prevent intake of organisms into the cooling cycle. A further environmental impact would be organisms that adapt to the warmer plant water and may be injured if the plant shuts down in cold weather.

Water consumption by power stations is a developing issue.

In recent years, recycled wastewater, or grey water, has been used in cooling towers. The Calpine Riverside and the Calpine Fox power stations in Wisconsin as well as the Calpine Mankato power station in Minnesota are among these facilities.

Other Sources of Energy

Other power stations use the energy from wave or tidal motion, wind, sunlight or the energy of falling water, hydroelectricity. These types of energy sources are called renewable energy.

Hydroelectricity

Dams built to produce hydroelectricity impound a reservoir of water and release it through one or more water turbines, connected to generators, and generate electricity, from the energy provided by difference in water level upstream and downstream.

Pumped Storage

A pumped-storage hydroelectric power plant is a net consumer of energy but can be used to smooth peaks and troughs in overall electricity demand. Pumped storage plants typically use "spare" electricity during off peak periods to pump water from a lower reservoir or dam to an upper reservoir. Because the electricity is consumed "off peak" it is typically cheaper than power at peak times.

This is because the "base load" power stations, which are typically coal fired, cannot be switched on and off quickly so remain in service even when demand is low. During hours of peak demand, when the electricity price is high, the water pumped to the high reservoir is allowed to flow back to the lower reservoir through a water turbine connected to an electricity generator. Unlike coal power stations, which can take more than 12 hours to start up from cold, the hydroelectric plant can be brought into service in a few minutes, ideal to meet a peak load demand. Two substantial pumped storage schemes are in South Africa, one to the East of Cape Town (Palmiet) and one in the Drakensberg, Natal

Solar

Solar energy can be turned into electricity either directly in solar cells, or in a concentrating solar power plant by focusing the light to run a heat engine.

Figure 13: A solar photovoltaic power plant converts sunlight into direct current electricity using the photoelectric effect. Inverters change the direct current into alternating current for connection to the electrical grid. This type of plant does not use rotating machines for energy conversion.

Solar thermal power plants are another type of solar power plant. They use either parabolic troughs or heliostats to direct sunlight onto a pipe containing a heat transfer fluid, such as oil. The heated oil is then used to boil water into steam, which turns a turbine that drives an electrical generator. The central tower type of solar thermal power plant uses hundreds or thousands of mirrors, depending on size, to direct sunlight onto a receiver on top of a tower. Again, the heat is used to produce steam to turn turbines that drive electrical generators.

Wind

Wind turbines can be used to generate electricity in areas with strong, steady winds, sometimes offshore. Many different designs have been used in the past, but almost all modern turbines being produced today use a three-bladed, upwind design. Grid-connected wind turbines now being built are much larger than the units installed during the 1970s, and so produce power more cheaply and reliably than earlier models. With larger turbines (on the order of one megawatt), the blades move more slowly than older, smaller, units, which makes them less visually distracting and safer for airborne animals.

Typical Power Output

The power generated by a power station is measured in multiples of the watt, typically megawatts (10^6 watts) or gigawatts (10^9 watts). Power stations vary greatly in capacity depending on the type of power plant and on historical, geographical and economic factors. The following examples offer a sense of the scale.

Many of the largest operational onshore wind farms are located in the USA. As of 2011, the Roscoe Wind Farm is the largest onshore wind farm in the world, producing 781.5 MW of power, followed by the Horse Hollow Wind Energy Centre (735.5 MW). As of November 2010, the Thanet Offshore Wind Project in United Kingdom is the largest offshore wind farm in the world at 300 MW, followed by Horns Rev II (209 MW) in Denmark.

As of 2011, the largest photovoltaic (PV) power plants in the world are rated up to 97 megawatts. Larger power stations are under construction, some proposed will have a capacity of 150 MW or more. A planned installation in China will produce 2000 megawatts at peak.

Solar thermal power stations in the U.S. have the following output:

The country's largest solar facility at Kramer Junction has an output of 354 MW

The planned Blythe Solar Power Project will produce an estimated 968 MW

Large coal-fired, nuclear, and hydroelectric power stations can generate hundreds of Megawatts to multiple Gigawatts. Some examples:

The Three Mile Island Nuclear Generating Station in the USA has a rated capacity of 802 megawatts.

The coal-fired Ratcliffe-on-Soar Power Station in the UK has a rated capacity of 2 gigawatts.

The Aswan Dam hydro-electric plant in Egypt has a capacity of 2.1 gigawatts.

The Three Gorges Dam hydro-electric plant in China will have a capacity of 22.5 gigawatts when complete; 18.2 gigawatts capacity is operating as of 2010.

Gas turbine power plants can generate tens to hundreds of megawatts. Some examples:

The Indian Queens simple-cycle peaking power station in Cornwall UK, with a single gas turbine is rated 140 megawatts.

The Medway Power Station, a combined-cycle power station in Kent, UK with two gas turbines and one steam turbine, is rated 700 megawatts.

The rated capacity of a power station is nearly the maximum electrical power that that power station can produce. Some power plants are run at almost exactly their rated capacity all the time, as a non-load-following base load power plant, except at times of scheduled or unscheduled maintenance.

However, many power plants usually produce much less power than their rated capacity.

In some cases a power plant produces much less power than its rated capacity because it uses an intermittent energy source. Operators try to pull maximum available power from such power plants, because their marginal cost is practically zero, but the available power varies widely—in particular, it may be zero during heavy storms at night.

In some cases operators deliberately produce less power for economic reasons. The cost of fuel to run a load following power plant may be relatively high, and the cost of fuel to run a peaking power plant is even higher—they have relatively high marginal costs. Operators keep them turned off ("operational reserve") or running at minimum fuel consumption ("spinning reserve") most of the time. Operators feed more fuel into load following power plants only when the demand rises above what lower-cost plants (i.e., intermittent and base load plants) can produce, and then feed more fuel into peaking power plants only when the demand rises faster than the load following power plants can follow.

Operations

The power station operator has several duties in the electricity-generating facility. Operators are responsible for the safety of the work crews that frequently do repairs on the mechanical and electrical equipment. They maintain the equipment with periodic inspections and log temperatures, pressures and other important information at regular intervals. Operators are responsible for starting and stopping the generators depending on need. They are able to synchronise and adjust the voltage output of the added generation with the running electrical system without upsetting the system. They must know the electrical and mechanical systems in order to troubleshoot problems in the facility and add to the reliability of the facility. Operators must be able to respond to an emergency and know the procedures in place to deal with it.

Geothermal Electricity

Geothermal electricity is electricity generated from geothermal energy. Technologies in use include dry steam power plants, flash steam power plants and binary cycle power plants. Geothermal electricity generation is currently used in 24 countries, while geothermal heating is in use in 70 countries.

Estimates of the electricity generating potential of geothermal energy vary from 35 to 2,000 GW. Current worldwide installed capacity is 10,715 megawatts (MW), with the largest capacity in the United States (3,086 MW), Philippines, and Indonesia.

Geothermal power is considered to be sustainable because the heat extraction is small compared with the Earth's heat content. The emission intensity of existing geothermal electric plants is on average 122 kg of CO_2 per megawatt-hour (MW·h) of electricity, about one-eighth of a conventional coal-fired plant.

History and Development

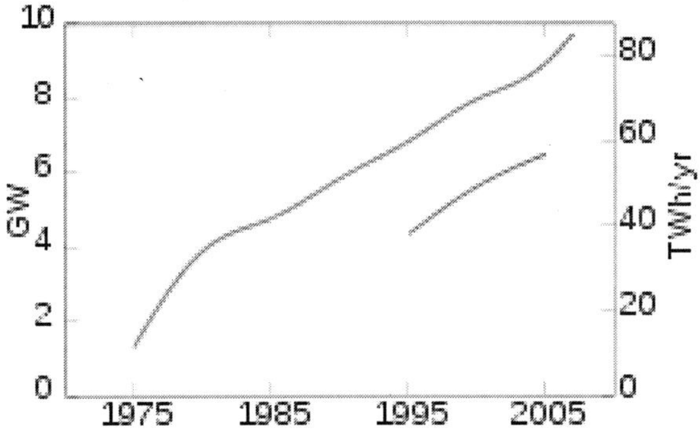

Figure 14: Global geothermal electric capacity. Upper red line is installed capacity; lower green line is realised production.

In the 20th century, demand for electricity led to the consideration of geothermal power as a generating source. Prince Piero Ginori Conti tested the first geothermal power generator on 4 July 1904 in Larderello, Italy. It successfully lit four light bulbs. Later, in 1911, the world's first commercial geothermal power plant was built there. Experimental generators were built in Beppu, Japan and the Geysers, California, in the 1920s, but Italy was the world's only industrial producer of geothermal electricity until 1958.

In 1958, New Zealand became the second major industrial producer of geothermal electricity when its Wairakei station was commissioned. Wairakei was the first plant to use flash steam technology.

In 1960, Pacific Gas and Electric began operation of the first successful geothermal electric power plant in the United States at The Geysers in California. The original turbine lasted for more than 30 years and produced 11 MW net power.

The binary cycle power plant was first demonstrated in 1967 in Russia and later introduced to the USA in 1981, following the 1970s energy crisis and significant changes in regulatory policies.

This technology allows the use of much lower temperature resources than were previously recoverable. In 2006, a binary cycle plant in Chena Hot Springs, Alaska, came on-line, producing electricity from a record low fluid temperature of 57°C (135°F).

Geothermal electric plants have until recently been built exclusively where high temperature geothermal resources are available near the surface.

The development of binary cycle power plants and improvements in drilling and extraction technology may enable enhanced geothermal systems over a much greater geographical range. Demonstration projects are operational in Landau-Pfalz, Germany, and Soultz-sous-Forêts, France, while an earlier effort in Basel, Switzerland was shut down after it triggered earthquakes.

Other demonstration projects are under construction in Australia, the United Kingdom, and the United States of America.

The thermal efficiency of geothermal electric plants is low, around 10-23%, because geothermal fluids are at a low temperature compared with steam from boilers. By the laws of thermodynamics this low temperature limits the efficiency of heat engines in extracting useful energy during the generation of electricity.

Exhaust heat is wasted, unless it can be used directly and locally, for example in greenhouses, timber mills, and district heating. The efficiency of the system does not affect operational costs as it would for a coal or other fossil fuel plant, but it does factor into the viability of the plant.

In order to produce more energy than the pumps consume, electricity generation requires high temperature geothermal fields and specialised heat cycles. Because geothermal power does not rely on variable sources of energy, unlike, for example, wind or solar, its capacity factor can be quite large – up to 96% has been demonstrated. The global average was 73% in 2005.

Resources

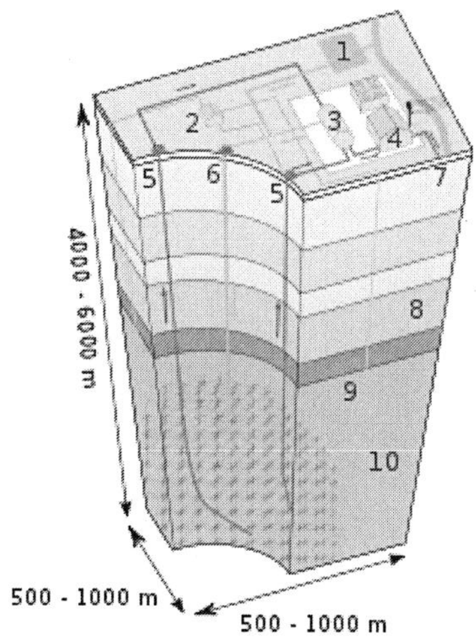

Figure 15: Enhanced geothermal system 1: Reservoir 2: Pump house 3: Heat exchanger 4: Turbine hall 5: Production well 6: Injection well 7: Hot water to district heating 8: Porous sediments 9: Observation well 10: Crystalline bedrock

The earth's heat content is 10^{31} joules. This heat naturally flows to the surface by conduction at a rate of 44.2 terawatts, (TW,) and is replenished by radioactive decay at a rate of 30 TW. These power rates are more than double humanity's current energy consumption from primary sources, but most of this power is too diffuse (approximately 0.1 W/m^2 on average) to be recoverable. The Earth's crust effectively acts as a thick insulating blanket which must be pierced by fluid conduits (of magma, water or other) to release the heat underneath.

Electricity generation requires high temperature resources that can only come from deep underground. The heat must be carried to the surface by fluid circulation, either through magma conduits, hot springs, hydrothermal circulation, oil wells, drilled water wells, or a combination of these. This circulation sometimes exists naturally where the crust is thin: magma conduits bring heat close to the surface, and hot springs bring the heat to the surface. If no hot spring is available, a well must be drilled into a hot aquifer. Away from tectonic plate boundaries the geothermal gradient is 25-30°C per kilometre (km) of depth in most of the world, and wells would have to be several kilometres deep to permit electricity generation. The quantity and quality of recoverable resources improves with drilling depth and proximity to tectonic plate boundaries.

In ground that is hot but dry, or where water pressure is inadequate, injected fluid can stimulate production. Developers bore two holes into a candidate site, and fracture the rock between them with explosives or high pressure water. Then they pump water or liquefied carbon dioxide down one borehole, and it comes up the other borehole as a gas. This approach is called hot dry rock geothermal energy in Europe, or enhanced geothermal systems in North America. Much greater potential may be available from this approach than from conventional tapping of natural aquifers.

Estimates of the electricity generating potential of geothermal energy vary from 35 to 2000 GW depending on the scale of investments. This does not include non-electric heat recovered by co-generation, geothermal heat pumps and other direct use. A 2006 report by the Massachusetts Institute of Technology (MIT), that included the potential of enhanced geothermal systems, estimated that investing 1 billion US dollars in research and development over 15 years would allow the creation of 100 GW of electrical generating capacity by 2050 in the United States alone. The MIT report estimated that over 200 zettajoules (ZJ) would be extractable, with the potential to increase this to over 2,000 ZJ with technology improvements - sufficient to provide all the world's present energy needs for several millennia.

At present, geothermal wells are rarely more than 3 kilometres (2 mi) deep. Upper estimates of geothermal resources assume wells as deep as 10 kilometres (6 mi). Drilling at this depth is now possible in the petroleum industry, although it is an expensive process. The deepest research well in the world, the Kola superdeep borehole, is 12 kilometres (7 mi) deep. This record has recently been imitated by commercial oil wells, such as Exxon's Z-12 well in the Chayvo field, Sakhalin. Wells drilled to depths greater than 4 kilometres (2 mi) generally incur drilling costs in the tens of millions of dollars. The technological challenges are to drill wide bores at low cost and to break larger volumes of rock.

Geothermal power is considered to be sustainable because the heat extraction is small compared to the Earth's heat content, but extraction must still be monitored to avoid local depletion. Although geothermal sites are capable of providing heat for many decades, individual wells may cool down or run out of water. The three oldest sites, at Larderello, Wairakei, and the Geysers have all reduced production from their peaks. It is not clear whether these plants extracted energy faster than it was replenished from greater depths, or whether the aquifers supplying them are being depleted. If production is reduced, and water is reinjected, these wells could theoretically recover their full potential. Such mitigation strategies have already been implemented at some sites. The long-term sustainability of geothermal energy has been demonstrated at the Lardarello field in Italy since 1913, at the Wairakei field in New Zealand since 1958, and at The Geysers field in California since 1960.

Power Station Types

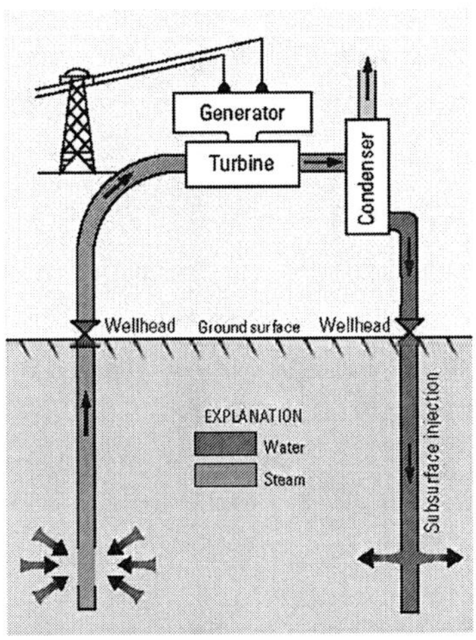

Figure 16: Dry steam plant

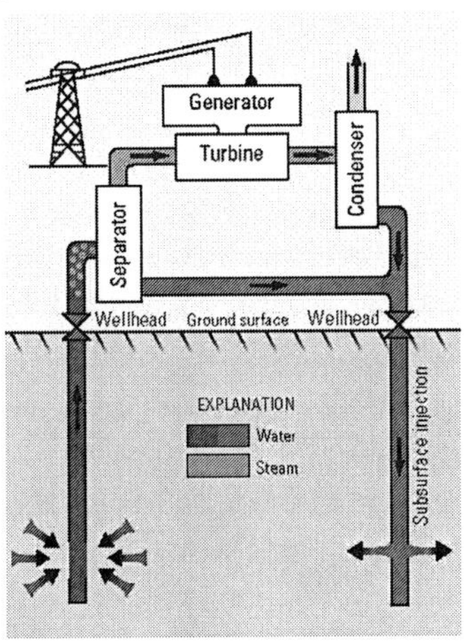

Figure 17: Flash steam plant

Geothermal power stations are not dissimilar to other steam turbine thermal power stations - heat from a fuel source (in geothermal's case, the earth's core) is used to heat water or another working fluid. The working fluid is then used to turn a turbine of a generator, thereby producing electricity. The fluid is then cooled and returned to the heat source.

Dry Steam Power Plants

Dry steam plants are the simplest and oldest design. They directly use geothermal steam of 150°C or greater to turn turbines.

Flash Steam Power Plants

Flash steam plants pull deep, high-pressure hot water into lower-pressure tanks and use the resulting flashed steam to drive turbines. They require fluid temperatures of at least 180°C, usually more. This is the most common type of plant in operation today.

Binary Cycle Power Plants

Binary cycle power plants are the most recent development, and can accept fluid temperatures as low as 57°C. The moderately hot geothermal water is passed by a secondary fluid with a much lower boiling point than water. This causes the secondary fluid to flash vapourise, which then drives the turbines. This is the most common type of geothermal electricity plant being constructed today. Both Organic Rankine and Kalina cycles are used. The thermal efficiency of this type plant is typically about 10-13%.

Worldwide Production

The International Geothermal Association (IGA) has reported that 10,715 megawatts (MW) of geothermal power in 24 countries is online, which is expected to generate 67,246 GWh of electricity in 2010. This represents a 20% increase in geothermal power online capacity since 2005. IGA projects this will grow to 18,500 MW by 2015, due to the large number of projects presently under consideration, often in areas previously assumed to have little exploitable resource.

In 2010, the United States led the world in geothermal electricity production with 3,086 MW of installed capacity from 77 power plants; the largest group of geothermal power plants in the world is located at The Geysers, a geothermal field in California. The Philippines follows the US as the second highest producer of geothermal power in the world, with 1,904 MW of capacity online; geothermal power makes up approximately 18% of the country's electricity generation.

January 2011: Al Gore said in The Climate Project Asia Pacific Summit that Indonesia could become a super power country in electricity production from geothermal energy.

Utility-grade Plants

The largest group of geothermal power plants in the world is located at The Geysers, a geothermal field in California, United States. As of 2004, five countries (El Salvador, Kenya, the Philippines, Iceland, and Costa Rica) generate more than 15% of their electricity from geothermal sources. Naknek Electric Association (NEA) is going to make an exploration well near King Salmon, in Southwest Alaska. It could cut the cost of electricity production by 71 percent and the planned power is 25 megawatts.

During 2005, contracts were placed for an additional 500 MW of electrical capacity in the United States, while there were also plants under construction in 11 other countries. Enhanced geothermal systems that are several kilometres in depth are operational in France and Germany and are being developed or evaluated in at least four other countries.

Table 1: Installed geothermal electric capacity

Country	*Capacity (MW) 2007*	*Capacity (MW) 2010*	*Percentage of National Production*
USA	2687	3086	0.3%
Philippines	1969.7	1904	27%
Indonesia	992	1197	3.7%
Mexico	953	958	3%
Italy	810.5	843	1.5%
New Zealand	471.6	628	10%
Iceland	421.2	575	30%
Japan	535.2	536	0.1%
El Salvador	204.4	204	25%
Kenya	128.8	167	11.2%
Costa Rica	162.5	166	14%
Turkey	38	94	0.3%
Nicaragua	87.4	88	10%
Russia	79	82	
Papua-New Guinea	56	56	
Guatemala	53	52	

Portugal	23	29
China	27.8	24
France	14.7	16
Ethiopia	7.3	7.3
Germany	8.4	6.6
Austria	1.1	1.4
Australia	0.2	1.1
Thailand	0.3	0.3
TOTAL	9,731.9	10,709.7

Environmental Impact

Figure 18: Krafla Geothermal Station in northeast Iceland

Fluids drawn from the deep earth carry a mixture of gases, notably carbon dioxide (CO_2), hydrogen sulfide (H_2S), methane (CH_4), and ammonia (NH_3). These pollutants contribute to global warming, acid rain, and noxious smells if released. Existing geothermal electric plants emit an average of 122 kg of CO_2 per megawatt-hour (MW·h) of electricity, a small fraction of the emission intensity of conventional fossil fuel plants. Plants that experience high levels of acids and volatile chemicals are usually equipped with emission-control systems to reduce the exhaust. Geothermal plants could theoretically inject these gases back into the earth, as a form of carbon capture and storage.

In addition to dissolved gases, hot water from geothermal sources may hold in solution trace amounts of toxic chemicals, such as mercury, arsenic, boron, antimony, and salt. These chemicals come out of solution as the water cools, and can cause environmental damage if released. The modern practice of injecting geothermal fluids back into the

Earth to stimulate production has the side benefit of reducing this environmental risk.

Plant construction can adversely affect land stability. Subsidence has occurred in the Wairakei field in New Zealand. Enhanced geothermal systems can trigger earthquakes as part of hydraulic fracturing. The project in Basel, Switzerland was suspended because more than 10,000 seismic events measuring up to 3.4 on the Richter Scale occurred over the first 6 days of water injection. The risk of geothermal drilling leading to uplift has been experienced in Staufen im Breisgau.

Geothermal has minimal land and freshwater requirements. Geothermal plants use 404 square metres per GWh versus 3,632 and 1,335 square kilometres for coal facilities and wind farms respectively. They use 20 litres of freshwater per MW·h versus over 1000 litres per MW·h for nuclear, coal, or oil.

Economics

Geothermal power requires no fuel, it is therefore immune to fuel cost fluctuations. However, capital costs tend to be high. Drilling accounts for over half the costs, and exploration of deep resources entails significant risks. A typical well doublet in Nevada can support 4.5 megawatt (MW) of electricity generation and costs about $10 million to drill, with a 20% failure rate. In total, electrical plant construction and well drilling cost about 2-5 million € per MW of electrical capacity, while the levelised energy cost is 0.04-0.10 € per kW·h. Enhanced geothermal systems tend to be on the high side of these ranges, with capital costs above $4 million per MW and levelised costs above $0.054 per kW·h in 2007.

Geothermal power is highly scalable: a small power plant can supply a rural village, though capital can be high.

Chevron Corporation is the world's largest private producer of geothermal electricity. The most developed geothermal field is the Geysers in California. In 2008, this field supported 15 plants, all owned by Calpine, with a total generating capacity of 725 MW.

Renewable Energy ower Station

A Renewable energy power station is a broad term relating to power stations that generate electricity in a centralised manner from sustainable sources such as solar, wind, water, biomass or geothermal which is then supplied to a public grid for consumption.

This is opposed to Fossil fuel power stations and decentralised renewable energy generation such as Photovoltaics panels on residential houses and solar water heating. An example of a Renewable energy power station is the PS20 solar power tower in Seville Spain.

Thermal Power Station

Figure 19: Republika Power Plant, a thermal power station in Pernik, Bulgaria

Figure 20: Mohave Generating Station, a 1,580 MW thermal power station near Laughlin, Nevada fuelled by coal

Figure 21: A thermal power station in Richemont, France.

Figure 22: Geothermal power station in Iceland

A thermal power station is a power plant in which the prime mover is steam driven. Water is heated, turns into steam and spins a steam turbine which drives an electrical generator. After it passes through the turbine, the steam is condensed in a condenser and recycled to where it was heated; this is known as a Rankine cycle. The greatest variation in the design of thermal power stations is due to the different fuel sources. Some prefer to use the term *energy centre* because such facilities convert forms of heat energy into electricity. Some thermal power plants also deliver heat energy for industrial purposes, for district heating, or for desalination of water as well as delivering electrical power. A large part of human CO_2 emissions comes from fossil fuelled thermal power plants; efforts to reduce these outputs are various and widespread.

Introductory Overview

Almost all coal, nuclear, geothermal, solar thermal electric, and waste incineration plants, as well as many natural gas power plants are thermal. Natural gas is frequently combusted in gas turbines as well as boilers. The waste heat from a gas turbine can be used to raise steam, in a combined cycle plant that improves overall efficiency. Power plants burning coal, fuel oil, or natural gas are often called *fossil-fuel power plants*. Some biomass-fuelled thermal power plants have appeared also. Non-nuclear thermal power plants, particularly fossil-fuelled plants, which do not use co-generation are sometimes referred to as *conventional power plants*.

Commercial electric utility power stations are usually constructed on a large scale and designed for continuous operation. Electric power plants typically use three-phase electrical generators to produce alternating current (AC) electric power at a frequency of 50 Hz or 60 Hz. Large companies or institutions may have their own power plants to supply heating or electricity to their facilities, especially if steam is created anyway for other purposes. Steam-driven power plants have been used in various large ships, but are now usually used in large naval ships. Shipboard power plants usually directly couple the turbine to the ship's propellers through gearboxes. Power plants in such ships also provide steam to smaller turbines driving electric generators to supply electricity. Shipboard steam power plants can be either fossil fuel or nuclear. Nuclear marine propulsion is, with few exceptions, used only in naval vessels. There have been perhaps about a dozen turbo-electric ships in which a steam-driven turbine drives an electric generator which powers an electric motor for propulsion.

combined heat and power (CH&P) plants, often called *co-generation plants*, produce both electric power and heat for process heat or space heating. Steam and hot water lose energy when piped over substantial distance, so carrying heat energy by steam or hot water is often only worthwhile within a local area, such as a ship, industrial plant, or district heating of nearby buildings.

History

Reciprocating steam engines have been used for mechanical power sources since the 18th Century, with notable improvements being made by James Watt. The very first commercial central electrical generating stations in the Pearl Street Station, New York and the Holborn Viaduct power station, London, in 1882, also used reciprocating steam engines. The development of the steam turbine allowed larger and more efficient central generating stations to be built. By 1892 it was considered as an alternative to reciprocating engines Turbines offered higher speeds, more compact machinery, and stable speed regulation allowing for parallel synchronous operation of generators on a common bus. Turbines entirely replaced reciprocating engines in large central stations after about 1905. The largest reciprocating engine-generator sets ever built were completed in 1901 for the Manhattan Elevated Railway. Each of seventeen units weighed about 500 tons and was rated 6000 kilowatts; a contemporary turbine-set of similar rating would have weighed about 20% as much.

Efficiency

The energy efficiency of a conventional thermal power station, considered as salable energy as a percent of the heating value of the fuel consumed, is typically 33% to 48%. This efficiency is limited as all heat engines are governed by the laws of thermodynamics. The rest of the energy must leave the plant in the form of heat. This waste heat can go through a condenser and be disposed of with cooling water or in cooling towers. If the waste heat is instead utilised for district heating, it is called co-generation. An important class of thermal power station are associated with desalination facilities; these are typically found in desert countries with large supplies of natural gas and in these plants, freshwater production and electricity are equally important co-products.

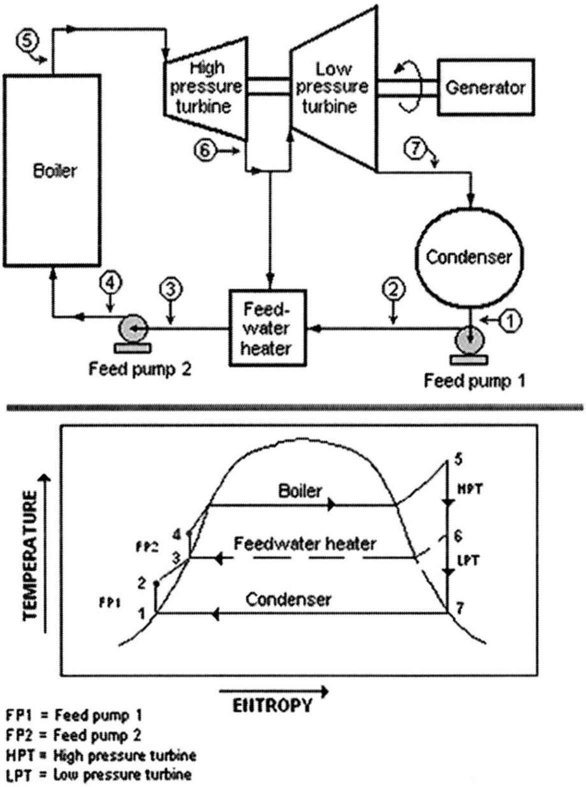

Figure 23: A Rankine cycle with a two-stage steam turbine and a single feed water heater.

Above the critical point for water of 705 °F (374 °C) and 3212 psi (22.06 MPa), there is no phase transition from water to steam, but only a gradual decrease in density. Boiling does not occur and it is not possible to remove impurities via steam separation. In this case a super critical steam plant is required to utilise the increased thermodynamic efficiency by operating at higher temperatures. These plants, also called *once-through* plants because boiler water does not circulate multiple times, require additional water purification steps to ensure that any impurities picked up during the cycle will be removed. This purification takes the form of high pressure ion exchange units called condensate polishers between the steam condenser and the feed water heaters. Sub-critical fossil fuel power plants can achieve 36–40% efficiency. Super critical designs have efficiencies in the low to mid 40% range, with new "ultra critical" designs using pressures of 4400 psi (30.3 MPa) and dual stage reheat reaching about 48% efficiency.

Current nuclear power plants operate below the temperatures and pressures that coal-fired plants do. This limits their thermodynamic efficiency to 30–32%. Some advanced reactor designs being studied, such as the Very high temperature reactor, Advanced gas-cooled reactor and Super critical water reactor, would operate at temperatures and pressures similar to current coal plants, producing comparable thermodynamic efficiency.

Electricity Cost

The direct cost of electric energy produced by a thermal power station is the result of cost of fuel, capital cost for the plant, operator labour, maintenance, and such factors as ash handling and disposal. Indirect, social or environmental costs such as the economic value of environmental impacts, or environmental and health effects of the complete fuel cycle and plant decommissioning, are not usually assigned to generation costs for thermal stations in utility practice, but may form part of an environmental impact assessment.

Diagram of a Typical Coal-fired Thermal Power Station

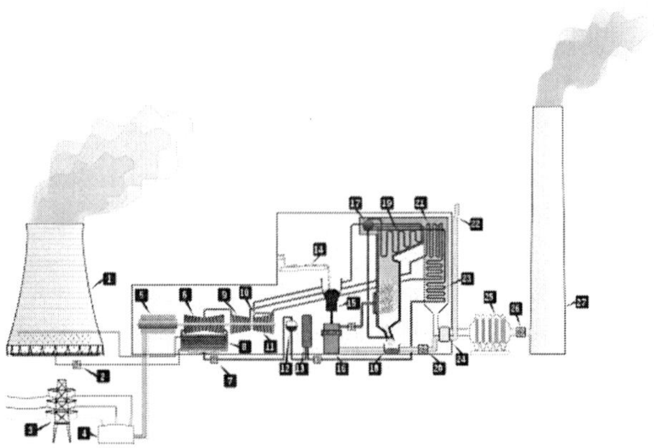

Figure 24.

Typical Diagram of a Coal-fired Thermal Power Station

1. Cooling tower	10. Steam Control valve	19. Superheater
2. Cooling water pump	11. High pressure steam turbine	20. Forced draught (draft) fan
3. transmission line (3-phase)	12. Deaerator	21. Reheater
4. Step-up (3-phase) transformer	13. Feedwater heater	22. Combustion
5. Electrical generator (3-phase)	14. Coal conveyor	23. Economiser
6. Low pressure steam turbine	15. Coal hopper	24. Air preheater
7. Condensate pump	16. Coal pulveriser	25. Precipitator
8. Surface condenser (draft) fan	17. Boiler steam drum	26. Induced draught
9. Intermediate pressure steam turbine	18. Bottom ash hopper	27. Flue gas stack

For units over about 200 MW capacity, redundancy of key components is provided by installing duplicates of the forced and induced draft fans, air preheaters, and fly ash collectors. On some units of about 60 MW, two boilers per unit may instead be provided.

Boiler and Steam Cycle

In fossil-fuelled power plants, *steam generator* refers to a furnace that burns the fossil fuel to boil water to generate steam.

In the nuclear plant field, *steam generator* refers to a specific type of large heat exchanger used in a pressurised water reactor (PWR) to thermally connect the primary (reactor plant) and secondary (steam plant) systems, which generates steam. In a nuclear reactor called a boiling water reactor (BWR), water is boiled to generate steam directly in the reactor itself and there are no units called steam generators.

In some industrial settings, there can also be steam-producing heat exchangers called *heat recovery steam generators* (HRSG) which utilise heat from some industrial process. The steam generating boiler has to produce steam at the high purity, pressure and temperature required for the steam turbine that drives the electrical generator.

Geothermal plants need no boiler since they use naturally occurring steam sources. Heat exchangers may be used where the geothermal steam is very corrosive or contains excessive suspended solids. A fossil fuel steam generator includes an economizer, a steam drum, and the furnace with its steam generating tubes and superheater coils. Necessary safety valves are located at suitable points to avoid excessive boiler pressure. The air and flue gas path equipment include: forced draft (FD) fan, Air Preheater (AP), boiler furnace, induced draft (ID) fan, fly ash collectors (electrostatic precipitator or baghouse) and the flue gas stack.

Feed Water Heating and Deaeration

The feed water used in the steam boiler is a means of transferring heat energy from the burning fuel to the mechanical energy of the spinning steam turbine. The total feed water consists of recirculated *condensate* water and purified *makeup water*. Because the metallic materials it contacts are subject to corrosion at high temperatures and pressures, the makeup water is highly purified before use. A system of water softeners and ion exchange demineralisers produces water so pure that it coincidentally becomes an electrical insulator, with conductivity in the range of 0.3–1.0 microsiemens per centimetre. The makeup water in a 500 MWe plant amounts to perhaps 20 US gallons per minute (1.25 L/s) to offset the small losses from steam leaks in the system. The feed water cycle begins with condensate water being pumped out of the condenser after travelling through the steam turbines. The condensate flow rate at full load in a 500 MW plant is about 6,000 US gallons per minute (400 L/s).

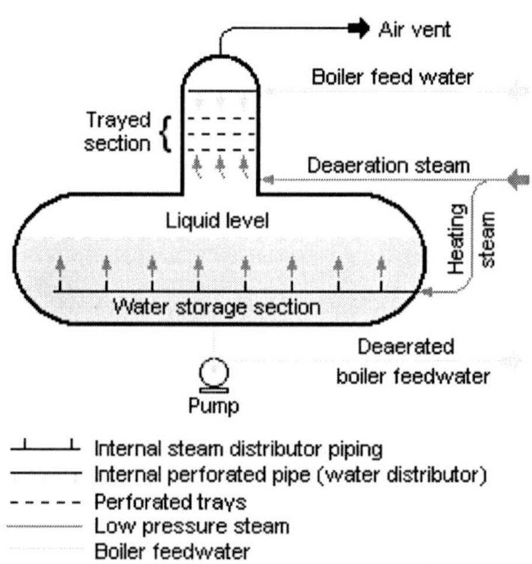

Figure 25: Diagram of boiler feed water deaerator (with vertical, domed aeration section and horizontal water storage section

The water flows through a series of six or seven intermediate feed water heaters, heated up at each point with steam extracted from an appropriate duct on the turbines and gaining temperature at each stage. Typically, the condensate plus the makeup water then flows through a deaerator that removes dissolved air from the water, further purifying and reducing its corrosiveness. The water may be dosed following this point with hydrazine, a chemical that removes the remaining oxygen in the water to below 5 parts per billion (ppb). It is also dosed with pH control agents such as ammonia or morpholine to keep the residual acidity low and thus non-corrosive.

Boiler Operation

The boiler is a rectangular furnace about 50 feet (15 m) on a side and 130 feet (40 m) tall. Its walls are made of a web of high pressure steel tubes about 2.3 inches (58 mm) in diametre.

Pulverised coal is air-blown into the furnace from fuel nozzles at the four corners and it rapidly burns, forming a large fireball at the centre. The thermal radiation of the fireball heats the water that circulates through the boiler tubes near the boiler perimetre. The water circulation rate in the boiler is three to four times the throughput and is typically driven by pumps. As the water in the boiler circulates it absorbs heat and changes into steam at 700 °F (371 °C) and 3,200 psi (Template:Convert/MP). It is separated from the water inside a drum at the top of the furnace. The saturated steam is introduced into superheat pendant tubes that hang in the hottest part of the combustion gases as they exit the furnace. Here the steam is superheated to 1,000 °F (500 °C) to prepare it for the turbine.

Plants designed for lignite (brown coal) are increasingly used in locations as varied as Germany, Victoria, and North Dakota. Lignite is a much younger form of coal than black coal. It has a lower energy density than black coal and requires a much larger furnace for equivalent heat output. Such coals may contain up to 70% water and ash, yielding lower furnace temperatures and requiring larger induced-draft fans. The firing systems also differ from black coal and typically draw hot gas from the furnace-exit level and mix it with the incoming coal in fan-type mills that inject the pulverised coal and hot gas mixture into the boiler. Plants that use gas turbines to heat the water for conversion into steam use boilers known as heat recovery steam generators (HRSG). The exhaust heat from the gas turbines is used to make superheated steam that is then used in a conventional water-steam generation cycle, as described in gas turbine combined-cycle plants section below.

Boiler Furnace and Steam Drum

Once water inside the boiler or steam generator, the process of adding the latent heat of vapourisation or enthalpy is underway. The boiler transfers energy to the water by the chemical reaction of burning some type of fuel.

The water enters the boiler through a section in the convection pass called the economizer. From the economizer it passes to the steam drum. Once the water enters the steam drum it goes down to the lower inlet water wall headers. From the inlet headers the water rises through the water walls and is eventually turned into steam due to the heat being generated by the burners located on the front and rear water walls (typically). As the water is turned into steam/vapour in the water walls, the steam/vapour once again enters the steam drum. The steam/vapour is passed through a series of steam and water separators and then dryers inside the steam drum. The steam separators and dryers remove water droplets from the steam and the cycle through the water walls is repeated. This process is known as natural circulation.

The boiler furnace auxiliary equipment includes coal feed nozzles and igniter guns, soot blowers, water lancing and observation ports (in the furnace walls) for observation of the furnace interior. Furnace explosions due to any accumulation of combustible gases after a trip-out are avoided by flushing out such gases from the combustion zone before igniting the coal.

The steam drum (as well as the super heater coils and headers) have air vents and drains needed for initial start up. The steam drum has internal devices that removes moisture from the wet steam entering the drum from the steam generating tubes. The dry steam then flows into the super heater coils.

Superheater

Fossil fuel power plants can have a superheater and/or re-heater section in the steam generating furnace. In a fossil fuel plant, after the steam is conditioned by the drying equipment inside the steam drum, it is piped from the upper drum area into tubes inside an area of the furnace known as the superheater, which has an elaborate set up of tubing where the steam vapour picks up more energy from hot flue gases outside the tubing and its temperature is now superheated above the saturation temperature. The superheated steam is then piped through the main steam lines to the valves before the high pressure turbine.

Nuclear-powered steam plants do not have such sections but produce steam at essentially saturated conditions. Experimental nuclear plants were equipped with fossil-fired super heaters in an attempt to improve overall plant operating cost.

Steam Condensing

The condenser condenses the steam from the exhaust of the turbine into liquid to allow it to be pumped. If the condenser can be made cooler, the pressure of the exhaust steam is reduced and efficiency of the cycle increases.

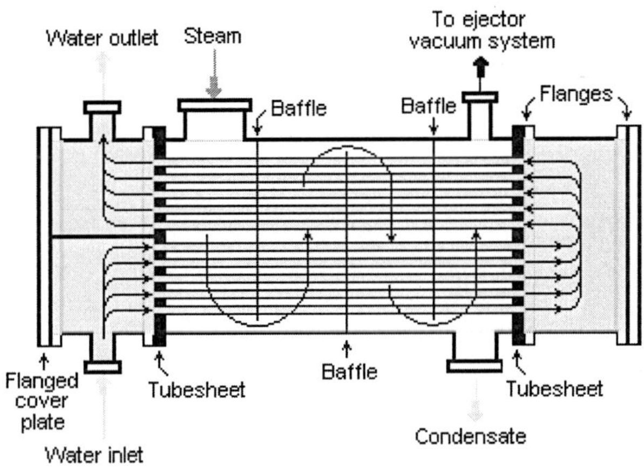

Figure 26: Diagram of a typical water-cooled surface condenser.

The surface condenser is a shell and tube heat exchanger in which cooling water is circulated through the tubes. The exhaust steam from the low pressure turbine enters the shell where it is cooled and converted to condensate (water) by flowing over the tubes as shown in the adjacent diagram. Such condensers use steam ejectors or rotary motor-driven exhausters for continuous removal of air and gases from the steam side to maintain vacuum.

For best efficiency, the temperature in the condenser must be kept as low as practical in order to achieve the lowest possible pressure in the condensing steam. Since the condenser temperature can almost always be kept significantly below 100 °C where the vapour pressure of water is much less than atmospheric pressure, the condenser generally works under vacuum. Thus leaks of non-condensible air into the closed loop must be prevented.

Typically the cooling water causes the steam to condense at a temperature of about 35 °C (95 °F) and that creates an absolute pressure in the condenser of about 2–7 kPa (0.59–2.1 inHg), i.e. a vacuum of about "95 kPa ("28.1 inHg) relative to atmospheric pressure. The large decrease in volume that occurs when water vapour condenses to liquid creates the low vacuum that helps pull steam through and increase the efficiency of the turbines.

The limiting factor is the temperature of the cooling water and that, in turn, is limited by the prevailing average climatic conditions at the power plant's location (it may be possible to lower the temperature beyond the turbine limits during winter, causing excessive condensation in the turbine). Plants operating in hot climates may have to reduce output if their source of condenser cooling water becomes warmer; unfortunately this usually coincides with periods of high electrical demand for air conditioning.

The condenser generally uses either circulating cooling water from a cooling tower to reject waste heat to the atmosphere, or once-through water from a river, lake or ocean.

Figure 27: A Marley mechanical induced draft cooling tower

The heat absorbed by the circulating cooling water in the condenser tubes must also be removed to maintain the ability of the water to cool as it circulates. This is done by pumping the warm water from the condenser through either natural draft, forced draft or induced draft cooling towers (as seen in the image to the right) that reduce the temperature of the water by evapouration, by about 11 to 17 °C (20 to 30 °F)—expelling waste heat to the atmosphere. The circulation flow rate of the cooling water in a 500 MW unit is about 14.2 m³/s (500 ft³/s or 225,000 US gal/min) at full load.

The condenser tubes are made of brass or stainless steel to resist corrosion from either side. Nevertheless they may become internally fouled during operation by bacteria or algae in the cooling water or by mineral scaling, all of which inhibit heat transfer and reduce thermodynamic efficiency. Many plants include an automatic cleaning system that circulates sponge rubber balls through the tubes to scrub them clean without the need to take the system off-line.

The cooling water used to condense the steam in the condenser returns to its source without having been changed other than having been warmed. If the water returns to a local water body (rather than a circulating cooling tower), it is tempered with cool 'raw' water to prevent thermal shock when discharged into that body of water.

Another form of condensing system is the air-cooled condenser. The process is similar to that of a radiator and fan. Exhaust heat from the low pressure section of a steam turbine runs through the condensing tubes, the tubes are usually finned and ambient air is pushed through the fins with the help of a large fan. The steam condenses to water to be reused in the water-steam cycle. Air-cooled condensers typically operate at a higher temperature than water-cooled versions. While saving water, the efficiency of the cycle is reduced (resulting in more carbon dioxide per megawatt of electricity).

From the bottom of the condenser, powerful condensate pumps recycle the condensed steam (water) back to the water/steam cycle.

Reheater

Power plant furnaces may have a reheater section containing tubes heated by hot flue gases outside the tubes. Exhaust steam from the high pressure turbine is passed through these heated tubes to collect more energy before driving the intermediate and then low pressure turbines.

Air Path

External fans are provided to give sufficient air for combustion. The forced draft fan takes air from the atmosphere and, first warming it in the air preheater for better combustion, injects it via the air nozzles on the furnace wall.

The induced draft fan assists the FD fan by drawing out combustible gases from the furnace, maintaining a slightly negative pressure in the furnace to avoid backfiring through any opening.

Steam Turbine Generator

The turbine generator consists of a series of steam turbines interconnected to each other and a generator on a common shaft.

There is a high pressure turbine at one end, followed by an intermediate pressure turbine, two low pressure turbines, and the generator. As steam moves through the system and loses pressure and thermal energy it expands in volume, requiring increasing diametre and longer blades at each succeeding stage to extract the remaining energy. The entire rotating mass may be over 200 metric tons and 100 feet (30 m) long. It is so heavy that it must be kept turning slowly even when shut down (at 3 rpm) so that the shaft will not bow even slightly and become unbalanced. This is so important that it is one of only five functions of blackout emergency power batteries on site. Other functions are emergency lighting, communication, station alarms and turbogenerator lube oil.

Superheated steam from the boiler is delivered through 14–16-inch (360–410 mm) diametre piping to the high pressure turbine where it falls in pressure to 600 psi (4.1 MPa) and to 600 °F (320 °C) in temperature through the stage. It exits via 24–26-inch (610–660 mm) diametre cold reheat lines and passes back into the boiler where the steam is reheated in special reheat pendant tubes back to 1,000 °F (500 °C). The hot reheat steam is conducted to the intermediate pressure turbine where it falls in both temperature and pressure and exits directly to the long-bladed low pressure turbines and finally exits to the condenser.

The generator, 30 feet (9 m) long and 12 feet (3.7 m) in diametre, contains a stationary stator and a spinning rotor, each containing miles of heavy copper conductor—no permanent magnets here. In operation it generates up to 21,000 amperes at 24,000 volts AC (504 MWe) as it spins at either 3,000 or 3,600 rpm, synchronised to the power grid. The rotor spins in a sealed chamber cooled with hydrogen gas, selected because it has the highest known heat transfer coefficient of any gas and for its low viscosity which reduces windage losses. This system requires special handling during startup, with air in the chamber first displaced by carbon dioxide before filling with hydrogen. This ensures that the highly explosive hydrogen–oxygen environment is not created.

The power grid frequency is 60 Hz across North America and 50 Hz in Europe, Oceania, Asia (Korea and parts of Japan are notable exceptions) and parts of Africa.

The electricity flows to a distribution yard where transformers increase the voltage for transmission to its destination. The steam turbine-driven generators have auxiliary systems enabling them to work satisfactorily and safely. The steam turbine generator being rotating equipment generally has a heavy, large diametre shaft. The shaft therefore requires not only supports but also has to be kept in position while running. To minimise the frictional resistance to the rotation, the shaft has a number of bearings. The bearing shells, in which the shaft rotates, are lined with a low friction material like Babbitt metal. Oil lubrication is provided to further reduce the friction between shaft and bearing surface and to limit the heat generated.

Stack Gas Path and Cleanup

As the combustion flue gas exits the boiler it is routed through a rotating flat basket of metal mesh which picks up heat and returns it to incoming fresh air as the basket rotates, This is called the air preheater. The gas exiting the boiler is laden with fly ash, which are tiny spherical ash particles. The flue gas contains nitrogen along with combustion products carbon dioxide, sulphur dioxide, and nitrogen oxides. The fly ash is removed by fabric bag filters or electrostatic precipitators. Once removed, the fly ash byproduct can sometimes be used in the manufacturing of concrete. This cleaning up of flue gases, however, only occurs in plants that are fitted with the appropriate technology. Still, the majority of coal fired power plants in the world do not have these facilities. Legislation in Europe has been efficient to reduce flue gas pollution. Japan has been using flue gas cleaning technology for over 30 years and the US has been doing the same for over 25 years. China is now beginning to grapple with the pollution caused by coal fired power plants.

Where required by law, the sulphur and nitrogen oxide pollutants are removed by stack gas scrubbers which use a pulverised limestone or other alkaline wet slurry to remove those pollutants from the exit stack gas. Other devices use catalysts to remove Nitrous Oxide compounds from the flue gas stream. The gas travelling up the flue gas stack may by this time have dropped to about 50 °C (120 °F). A typical flue gas stack may be 150–180 metres (490–590 ft) tall to disperse the remaining flue gas components in the atmosphere. The tallest flue gas stack in the world is 419.7 metres (1,377 ft) tall at the GRES-2 power plant in Ekibastuz, Kazakhstan.

In the United States and a number of other countries, atmospheric dispersion modelling studies are required to determine the flue gas stack height needed to comply with the local air pollution regulations. The United States also requires the height of a flue gas stack to comply with what is known as the "Good Engineering Practice (GEP)" stack height. In the case of existing flue gas stacks that exceed the GEP stack height, any air pollution dispersion modelling studies for such stacks must use the GEP stack height rather than the actual stack height.

Fly Ash Collection

Fly ash is captured and removed from the flue gas by electrostatic precipitators or fabric bag filters (or sometimes both) located at the outlet of the furnace and before the induced draft fan. The fly ash is periodically removed from the collection hoppers below the precipitators or bag filters. Generally, the fly ash is pneumatically transported to storage silos for subsequent transport by trucks or railroad cars.

Bottom Ash Collection and Disposal

At the bottom of the furnace, there is a hopper for collection of bottom ash. This hopper is always filled with water to quench the ash and clinkers falling down from the furnace. Some arrangement is included to crush the clinkers and for conveying the crushed clinkers and bottom ash to a storage site .

Auxiliary Systems—Boiler Make-up Water Treatment Plant and Storage

Since there is continuous withdrawal of steam and continuous return of condensate to the boiler, losses due to blowdown and leakages have to be made up to maintain a desired water level in the boiler steam drum. For this, continuous make-up water is added to the boiler water system. Impurities in the raw water input to the plant generally consist of calcium and magnesium salts which impart hardness to the water. Hardness in the make-up water to the boiler will form deposits on the tube water surfaces which will lead to overheating and failure of the tubes. Thus, the salts have to be removed from the water, and that is done by a water demineralising treatment plant (DM). A DM plant generally consists of cation, anion, and mixed bed exchangers. Any ions in the final water from this process consist essentially of hydrogen ions and hydroxide ions, which recombine to form pure water. Very pure DM water becomes highly corrosive once it absorbs oxygen from the atmosphere because of its very high affinity for oxygen.

The capacity of the DM plant is dictated by the type and quantity of salts in the raw water input. However, some storage is essential as the DM plant may be down for maintenance. For this purpose, a storage tank is installed from which DM water is continuously withdrawn for boiler make-up. The storage tank for DM water is made from materials not affected by corrosive water, such as PVC. The piping and valves are generally of stainless steel. Sometimes, a steam blanketing arrangement or stainless steel doughnut float is provided on top of the water in the tank to avoid contact with air. DM water make-up is generally added at the steam space of the surface condenser (i.e., the vacuum side). This arrangement not only sprays the water but also DM water gets deaerated, with the dissolved gases being removed by an air ejector attached to the condenser.

Fuel Preparation System

In coal-fired power stations, the raw feed coal from the coal storage area is first crushed into small pieces and then conveyed to the coal feed hoppers at the boilers. The coal is next pulverised into a very fine powder. The pulverisers may be ball mills, rotating drum grinders, or other types of grinders. Some power stations burn fuel oil rather than coal. The oil must kept warm (above its pour point) in the fuel oil storage tanks to prevent the oil from congealing and becoming unpumpable. The oil is usually heated to about 100 °C before being pumped through the furnace fuel oil spray nozzles. Boilers in some power stations use processed natural gas as their main fuel. Other power stations may use processed natural gas as auxiliary fuel in the event that their main fuel supply (coal or oil) is interrupted. In such cases, separate gas burners are provided on the boiler furnaces.

Barring Gear

Barring gear (or "turning gear") is the mechanism provided to rotate the turbine generator shaft at a very low speed after unit stoppages. Once the unit is "tripped" (i.e., the steam inlet valve is closed), the turbine coasts down towards standstill. When it stops completely, there is a tendency for the turbine shaft to deflect or bend if allowed to remain in one position too long. This is because the heat inside the turbine casing tends to concentrate in the top half of the casing, making the top half portion of the shaft hotter than the bottom half. The shaft therefore could warp or bend by millionths of inches.

This small shaft deflection, only detectable by eccentricity metres, would be enough to cause damaging vibrations to the entire steam turbine generator unit when it is restarted. The shaft is therefore automatically turned at low speed (about one percent rated speed) by the barring gear until it has cooled sufficiently to permit a complete stop.

Oil System

An auxiliary oil system pump is used to supply oil at the start-up of the steam turbine generator. It supplies the hydraulic oil system required for steam turbine's main inlet steam stop valve, the governing control valves, the bearing and seal oil systems, the relevant hydraulic relays and other mechanisms. At a preset speed of the turbine during start-ups, a pump driven by the turbine main shaft takes over the functions of the auxiliary system.

Generator Cooling

While small generators may be cooled by air drawn through filters at the inlet, larger units generally require special cooling arrangements. Hydrogen gas cooling, in an oil-sealed casing, is used because it has the highest known heat transfer coefficient of any gas and for its low viscosity which reduces windage losses. This system requires special handling during start-up, with air in the generator enclosure first displaced by carbon dioxide before filling with hydrogen. This ensures that the highly flammable hydrogen does not mix with oxygen in the air. The hydrogen pressure inside the casing is maintained slightly higher than atmospheric pressure to avoid outside air ingress. The hydrogen must be sealed against outward leakage where the shaft emerges from the casing. Mechanical seals around the shaft are installed with a very small annular gap to avoid rubbing between the shaft and the seals. Seal oil is used to prevent the hydrogen gas leakage to atmosphere. The generator also uses water cooling. Since the generator coils are at a potential of about 22 kV, an insulating barrier such as Teflon is used to interconnect the water line and the generator high voltage windings. Demineralised water of low conductivity is used.

Generator High Voltage System

The generator voltage for modern utility-connected generators ranges from 11 kV in smaller units to 22 kV in larger units. The generator high voltage leads are normally large aluminium channels because of their high current as compared to the cables used in smaller machines. They are enclosed in well-grounded aluminium bus ducts and are supported on suitable insulators. The generator high voltage leads are connected to step-up transformers for connecting to a high voltage electrical substation (usually in the range of 115 kV to 765 kV) for further transmission by the local power grid.

The necessary protection and metreing devices are included for the high voltage leads. Thus, the steam turbine generator and the transformer form one unit. Smaller units, may share a common generator step-up transformer with individual circuit breakers to connect the generators to a common bus.

Monitoring and Alarm System

Most of the power plant operational controls are automatic. However, at times, manual intervention may be required. Thus, the plant is provided with monitors and alarm systems that alert the plant operators when certain operating parameters are seriously deviating from their normal range.

Battery Supplied Emergency Lighting and Communication

A central battery system consisting of lead acid cell units is provided to supply emergency electric power, when needed, to essential items such as the power plant's control systems, communication systems, turbine tube oil pumps, and emergency lighting. This is essential for a safe, damage-free shutdown of the units in an emergency situation.

Transport of Coal Fuel to Site and to Storage

Most thermal stations use coal as the main fuel. Raw coal is transported from coal mines to a power station site by trucks, barges, bulk cargo ships or railway cars. Generally, when shipped by railways, the coal cars are sent as a full train of cars. The coal received at site may be of different sizes. The railway cars are unloaded at site by rotary dumpers or side tilt dumpers to tip over onto conveyor belts below. The coal is generally conveyed to crushers which crush the coal to about ¾ inch (6 mm) size. The crushed coal is then sent by belt conveyors to a storage pile. Normally, the crushed coal is compacted by bulldozers, as compacting of highly volatile coal avoids spontaneous ignition. The crushed coal is conveyed from the storage pile to silos or hoppers at the boilers by another belt conveyor system.

Cogeneration

Cogeneration (also combined heat and power, CHP) is the use of a heat engine or a power station to simultaneously generate both electricity and useful heat. All thermal power plants emit a certain amount of heat during electricity generation. This can be released into the natural environment through cooling towers, flue gas, or by other means. By contrast, CHP captures some or all of the by-product heat for heating purposes, either very close to the plant, or—especially in Scandinavia and eastern Europe—as hot water for district heating with temperatures ranging from approximately 80 to 130 °C.

This is also called Combined Heat and Power District Heating or CHPDH. Small CHP plants are an example of decentralised energy.

Cogeneration is common in industries that use both steam and electricity, such as pulp and paper mills, refineries and chemical plants. Steam turbines for cogeneration are designed for *extraction* of steam at lower pressures after it has passed through a number of turbine stages, or they may be designed for final exhaust at *back pressure* (non-condensing), or both. A typical power generation turbine in a paper mill may have extraction pressures of 160 psig (1.103 MPa) and 60 psig (0.41 MPa). A typical back pressure may be 60 psig (0.41 MPa). In practice these pressures are custom designed for each facility. The extracted or exhaust steam is used for process heating, such as drying paper, evapouration, heat for chemical reactions or distillation.

In the United States, Con Edison distributes 66 billion kilograms of 350 °F/180 °C steam each year through its seven cogeneration plants to 100,000 buildings in Manhattan—the biggest steam district in the United States.

The peak delivery is 10 million pounds per hour (corresponding to approx. 2.5 GW) (This steam distribution system is the reason for the steaming manholes often seen in "gritty" New York movies.) Other major cogeneration companies in the United States include Recycled Energy Development and leading advocates include Tom Casten and Amory Lovins. By-product heat at moderate temperatures (212-356°F/ 100-180°C) can also be used in absorption chillers for cooling. A plant producing electricity, heat and cold is sometimes called trigeneration or more generally polygeneration plant. Cogeneration is a thermodynamically efficient use of fuel. In separate production of electricity, some energy must be rejected as waste heat, but in cogeneration this thermal energy is put to good use.

Overview

Figure 28: Masnedø CHP power station in Denmark. This station burns straw as fuel. The adjacent greenhouses are heated by district heating from the plant.

Thermal power plants (including those that use fissile elements or burn coal, petroleum, or natural gas), and heat engines in general, do not convert all of their thermal energy into electricity. In most heat engines, a bit more than half is lost as excess heat. By capturing the excess heat, CHP uses heat that would be wasted in a conventional power plant, potentially reaching an efficiency of up to 80%, for the best conventional plants. This means that less fuel needs to be consumed to produce the same amount of useful energy.

Some tri-cycle plants have used a combined cycle in which several thermodynamic cycles produced electricity, then a heating system was used as a condenser of the power plant's bottoming cycle. For example, the RU-25 MHD generator in Moscow heated a boiler for a conventional steam powerplant, whose condensate was then used for space heat. A more modern system might use a gas turbine powered by natural gas, whose exhaust powers a steam plant, whose condensate provides heat. Tri-cycle plants can have thermal efficiencies above 80%.

The viability of CHP (sometimes termed utilisation factor), especially in smaller CHP installations, depends on a good baseload of operation, both in terms of an on-site (or near site) electrical demand and heat demand. In practice, an exact match between the heat and electricity needs rarely exists. A CHP plant can either meet the need for heat (*heat driven operation*) or be run as a power plant with some use of its waste heat, the latter being less advantageous in terms of its utilisation factor and thus its overall efficiency. The viability can be greatly increased where opportunities for Trigeneration exist. In such cases, the heat from the CHP plant is also used as a primary energy source to deliver cooling by means of an absorption chiller. CHP is most efficient when heat can be used on-site or very close to it. Overall efficiency is reduced when the heat must be transported over longer distances. This requires heavily insulated pipes, which are expensive and inefficient; whereas electricity can be transmitted along a comparatively simple wire, and over much longer distances for the same energy loss.

A car engine becomes a CHP plant in winter when the reject heat is useful for warming the interior of the vehicle. The example illustrates the point that deployment of CHP depends on heat uses in the vicinity of the heat engine. Cogeneration plants are commonly found in district heating systems of cities, hospitals, prisons, oil refineries, paper mills, wastewater treatment plants, thermal enhanced oil recovery wells and industrial plants with large heating needs.

Thermally enhanced oil recovery (TEOR) plants often produce a substantial amount of excess electricity. After generating electricity, these plants pump leftover steam into heavy oil wells so that the oil will flow more easily, increasing production. TEOR cogeneration plants in Kern County, California produce so much electricity that it cannot all be used locally and is transmitted to Los Angeles. CHP is one of the most cost-efficient methods of reducing carbon emissions of heating in cold climates.

Operating Pressures and Boiler Feed Water

Cogeneration facilities typically operate at pressures significantly below that of 100% condensing power plants because of possible contamination of returned condensate from process steam and because much condensate is not recovered, meaning having to treat more make-up water. Boiler feed water must be completely oxygen free and de-mineralised, and the higher the pressure the more critical the level of purity of the feed water.

Types of Plants

Topping cycle plants primarily produce electricity from a steam turbine. The exhausted steam is then condensed and the low temperature heat released from this condensation is utilised for e.g. district heating or water desalination. Bottoming cycle plants produce high temperature heat for industrial processes, then a waste heat recovery boiler feeds an electrical plant. Bottoming cycle plants are only used when the industrial process requires very high temperatures such as furnaces for glass and metal manufacturing, so they are less common. Large cogeneration systems provide heating water and power for an industrial site or an entire town. Common CHP plant types are:

- Gas turbine CHP plants using the waste heat in the flue gas of gas turbines. The fuel used is typically natural gas
- Gas engine CHP plants use a reciprocating gas engine which is generally more competitive than a gas turbine up to about 5 MW. The gaseous fuel used is normally natural gas. These plants are generally manufactured as fully packaged units that can be installed within a plantroom or external plant compound with simple connections to the site's gas supply and electrical distribution and heating systems.
- Biofuel engine CHP plants use an adapted reciprocating gas engine or diesel engine, depending upon which biofuel is being

used, and are otherwise very similar in design to a Gas engine CHP plant. The advantage of using a biofuel is one of reduced hydrocarbon fuel consumption and thus reduced carbon emissions. These plants are generally manufactured as fully packaged units that can be installed within a plantroom or external plant compound with simple connections to the site's electrical distribution and heating systems. Another variant is the wood gasifier CHP plant whereby a wood pellet or wood chip biofuel is gasified in a zero oxygen high temperature environment; the resulting gas is then used to power the gas engine.

- Combined cycle power plants adapted for CHP
- Steam turbine CHP plants that use the heating system as the steam condenser for the steam turbine.
- Molten-carbonate fuel cells and solid oxide fuel cells have a hot exhaust, very suitable for heating.
- Nuclear Power plants can be fitted with steam drains after the high, mid, and/or low pressure turbines to provide heat to a heat system. With a heat system temperature of 95°C it is possible to extract about 10 MW heat for every MW electricity lost. With a temperature of 130°C the gain is slightly smaller, about 7 MW for every MWe lost.

Smaller cogeneration units may use a reciprocating engine or Stirling engine. The heat is removed from the exhaust and radiator. The systems are popular in small sizes because small gas and diesel engines are less expensive than small gas- or oil-fired steam-electric plants. Some cogeneration plants are fired by biomass, or industrial and municipal waste.

Heat Recovery Steam Generators

A heat recovery steam generator (HRSG) is a steam boiler that uses hot exhaust gases from the gas turbines or reciprocating engines in a CHP plant to heat up water and generate steam. The steam, in turn, drives a steam turbine and/or is used in industrial processes that require heat. HRSGs used in the CHP industry are distinguished from conventional steam generators by the following main features:

- The HRSG is designed based upon the specific features of the gas turbine or reciprocating engine that it will be coupled to.
- Since the exhaust gas temperature is relatively low, heat transmission is accomplished mainly through convection.

- The exhaust gas velocity is limited by the need to keep head losses down. Thus, the transmission coefficient is low, which calls for a large heating surface area.
- Since the temperature difference between the hot gases and the fluid to be heated (steam or water) is low, and with the heat transmission coefficient being low as well, the evapourator and economizer are designed with plate fin heat exchangers.

Costs

Typically, for a gas-fired plant the fully installed cost per kW electrical is around £400/kW, which is comparable with large central power stations.

History—Cogeneration in Europe

Figure 29: A cogeneration thermal power plant in Ferrera Erbognone (PV), Italy

Europe has actively incorporated cogeneration into its energy policy via the CHP Directive. In September 2008 at a hearing of the European Parliament's Urban Lodgment Intergroup, Energy Commissioner Andris Piebalgs is quoted as saying, "security of supply really starts with energy efficiency." Energy efficiency and cogeneration are recognised in the opening paragraphs of the European Union's Cogeneration Directive 2004/08/EC. This directive intends to support cogeneration and establish a method for calculating cogeneration abilities per country. The development of cogeneration has been very uneven over the years and has been dominated throughout the last decades by national circumstances. As a whole, the European Union generates 11% of its electricity using cogeneration, saving Europe an estimated 35 Mtoe per annum a day. However, there is large difference between Member States with variations of the energy savings between 2% and 60%. Europe has the three countries with the world's most intensive cogeneration economies: Denmark, the Netherlands and Finland.

Other European countries are also making great efforts to increase efficiency. Germany reported that at present, over 50% of the country's total electricity demand could be provided through cogeneration. So far, Germany has set the target to double its electricity cogeneration from 12.5% of the country's electricity to 25% of the country's electricity by 2020 and has passed supporting legislation accordingly in "Federal Ministry of Economics and Technology, (BMWi), Germany, August 2007. The UK is also actively supporting combined heat and power. In light of UK's goal to achieve a 60% reduction in carbon dioxide emissions by 2050, the government has set the target to source at least 15% of its government electricity use from CHP by 2010. Other UK measures to encourage CHP growth are financial incentives, grant support, a greater regulatory framework, and government leadership and partnership.

According to the IEA 2008 modelling of cogeneration expansion for the G8 countries, the expansion of cogeneration in France, Germany, Italy and the UK alone would effectively double the existing primary fuel savings by 2030. This would increase Europe's savings from today's 155.69 Twh to 465 Twh in 2030. It would also result in a 16% to 29% increase in each country's total cogenerated electricity by 2030. Governments are being assisted in their CHP endeavours by organisations like COGEN Europe who serve as an information hub for the most recent updates within Europe's energy policy. COGEN is Europe's umbrella organisation representing the interests of the cogeneration industry, users of the technology and promoting its benefits in the EU and the wider Europe. The association is supported by the key players in the industry including gas and electricity companies, ESCOs, equipment suppliers, consultancies, national promotion organisations, financial and other service companies.

Cogeneration in the United States

Perhaps the first modern use of energy recycling was done by Thomas Edison. His 1882 Pearl Street Station, the world's first commercial power plant, was a combined heat and power plant, producing both electricity and thermal energy while using waste heat to warm neighbouring buildings. Recycling allowed Edison's plant to achieve approximately 50 percent efficiency. By the early 1900s, regulations emerged to promote rural electrification through the construction of centralised plants managed by regional utilities. These regulations not only promoted electrification throughout the countryside, but they also discouraged decentralised power generation, such as cogeneration. As Recycled Energy Development CEO Sean Casten testified to Congress, they even went so far as to make it illegal for non-utilities to sell power.

By 1978, Congress recognised that efficiency at central power plants had stagnated and sought to encourage improved efficiency with the Public Utility Regulatory Policies Act (PURPA), which encouraged utilities to buy power from other energy producers. The U.S. DOE has an aggressive goal of having CHP comprise of *20% of the US generation capacity by the year 2030.* Eight Clean Energy Application Centres have been established across the nation whose mission is to develop the required technology application knowledge and educational infrastructure necessary to lead "clean energy" (Combined Heat and Power, Waste Heat Recovery and District Energy) technologies as viable energy options and reduce any perceived risks associated with their implementation. The focus of the Application Centres is to provide an outreach and technology deployment program for end users, policy makers, utilities, and industry stakeholders.

Percentage of US Energy Produced by Cogeneration

Cogeneration plants proliferated, soon producing about 8 percent of all energy in the United States. However, the bill left implementation and enforcement up to individual states, resulting in little or nothing being done in many parts of the country. In 2008 Tom Casten, chairman of Recycled Energy Development, said that "*We think we could make about 19 to 20 percent of U.S. electricity with heat that is currently thrown away by industry.*" Outside of the United States, energy recycling is more common. Denmark is probably the most active energy recycler, obtaining about 55% of its energy from cogeneration and waste heat recovery. Other large countries, including Germany, Russia, and India, also obtain a much higher share of their energy from decentralised sources.

MicroCHP

"Micro cogeneration" is a so called distributed energy resource (DER). The installation is usually less than 5 kW_e in a house or small business. Instead of burning fuel to merely heat space or water, some of the energy is converted to electricity in addition to heat. This electricity can be used within the home or business or, if permitted by the grid management, sold back into the electric power grid. This recent development of small scale CHP systems has provided the opportunity for in-house power backup of residential-scale photovoltaic (PV) arrays. The results of a recent study show that a PV+CHP hybrid system not only has the potential to radically reduce energy waste in the status quo electrical and heating systems, but it also enables the share of solar PV to be expanded by about a factor of five. In some regions, in order to reduce waste from excess heat, an absorption chiller has been proposed to utilise the CHP-produced thermal energy for cooling of PV-CHP system. These trigen+PV systems have the potential to save even more energy.

MiniCHP

Mini cogeneration is a so called distributed energy resource (DER). The installation is usually more than 5 kW_e and less than 500 kW_e in a building or medium sized business. In this size range the viability or utilisation factor of the CHP plant is very important to consider since it will greatly affect the efficiency and cost effectiveness (payback) of the CHP plant. The utilisation factor is essentially the calculated hours of operation of the CHP plant expressed as a percentage of the total number of hours in a year. If less than 40% then the application of CHP is considered to be unavailable. To be viable a good baseload for electrical demand and heat demand must exist. Such baseloads arise where building occupation or process activities are extended or continuous in operation. This typically includes for hospitals, prisons, manufacturing processes, swimming pools, airports, hotels, apartment blocks, etc.

Current (2007) Micro- and MiniCHP installations use five different technologies: microturbines, internal combustion engines, stirling engines, closed cycle steam engines and fuel cells. One author indicates that MicroCHP based on Stirling engines is the most cost effective of the so called microgeneration technologies in abating carbon emissions; however, advances in reciprocation engine technology are adding efficiency to CHP plant, particularly in the biogas field. MiniCHP has a large role to play in the field of CO_2 reduction from buildings where more than 14% of emissions can be saved by 2010 using CHP in buildings according to the author.

Chapter 9

Cooling Tower

Cooling towers are heat removal devices used to transfer process waste heat to the atmosphere. Cooling towers may either use the evapouration of water to remove process heat and cool the working fluid to near the wet-bulb air temperature or in the case of *closed circuit dry cooling towers* rely solely on air to cool the working fluid to near the dry-bulb air temperature. Common applications include cooling the circulating water used in oil refineries, chemical plants, power stations and building cooling. The towers vary in size from small roof-top units to very large hyperboloid structures (as in Image 1) that can be up to 200 metres tall and 100 metres in diametre, or rectangular structures (as in Image 2) that can be over 40 metres tall and 80 metres long. Smaller towers are normally factory-built, while larger ones are constructed on site. They are often associated with nuclear power plants in popular culture, although cooling towers are constructed on many types of buildings.

A hyperboloid cooling tower was patented by Frederik van Iterson and Gerard Kuypers in 1918. The first hyperboloid cooling towers were built in the last 1920s in Liverpool, England to cool water used at a electrical power station that used coal.

Classification by Use

HVAC: An HVAC (heating, ventilating, and air conditioning) cooling tower is a subcategory rejecting heat from a chiller. Water-cooled chillers are normally more energy efficient than air-cooled chillers due to heat rejection to tower water at or near wet-bulb temperatures. Air-cooled chillers must reject heat at the dry-bulb temperature, and thus have a lower average reverse-Carnot cycle effectiveness. Large office buildings, hospitals, and schools typically use one or more cooling towers as part of their air conditioning systems. Generally, industrial cooling towers are much larger than HVAC towers.

HVAC use of a cooling tower pairs the cooling tower with a water-cooled chiller or water-cooled condenser. A *ton* of air-conditioning is the removal of 12,000 Btu/hour (3500 W). The *equivalent ton* on the cooling tower side actually rejects about 15,000 Btu/hour (4400 W) due to the heat-equivalent of the energy needed to drive the chiller's compressor. This *equivalent ton* is defined as the heat rejection in cooling 3 U.S. gallons/minute (1,500 pound/hour) of water 10 °F (6 °C), which amounts to 15,000 Btu/hour, or a chiller coefficient of performance (COP) of 4.0. This COP is equivalent to an energy efficiency ratio (EER) of 14.

Cooling towers are also used in HVAC systems that have multiple water source heat pumps that share a common piping *water loop*. In this type of system, the water circulating inside the water loop removes heat from the condenser of the heat pumps whenever the heat pumps are working in the cooling mode, then the cooling tower is used to remove heat from the water loop and reject it to the atmosphere. When the heat pumps are working in heating mode, the condensers draw heat out of the loop water and reject it into the space to be heated.

Industrial Cooling Towers

Figure 1: Industrial cooling towers for a power plant

Industrial cooling towers can be used to remove heat from various sources such as machinery or heated process material. The primary use of large, industrial cooling towers is to remove the heat absorbed in the circulating cooling water systems used in power plants, petroleum refineries, petrochemical plants, natural gas processing plants, food processing plants, semi-conductor plants, and for other industrial facilities such as in condensers of distillation columns, for cooling liquid in crystallisation, etc. The circulation rate of cooling water in a typical 700 MW coal-fired power plant with a cooling tower amounts to about 71,600 cubic metres an hour (315,000 U.S. gallons per minute) and the circulating water requires a supply water make-up rate of perhaps 5 percent (i.e., 3,600 cubic metres an hour).

If that same plant had no cooling tower and used once-through cooling water, it would require about 100,000 cubic metres an hour and that amount of water would have to be continuously returned to the ocean, lake or river from which it was obtained and continuously re-supplied to the plant. Furthermore, discharging large amounts of hot water may raise the temperature of the receiving river or lake to an unacceptable level for the local ecosystem. Elevated water temperatures can kill fish and other aquatic organisms. A cooling tower serves to dissipate the heat into the atmosphere instead and wind and air diffusion spreads the heat over a much larger area than hot water can distribute heat in a body of water. Some coal-fired and nuclear power plants located in coastal areas do make use of once-through ocean water. But even there, the offshore discharge water outlet requires very careful design to avoid environmental problems.

Petroleum refineries also have very large cooling tower systems. A typical large refinery processing 40,000 metric tonnes of crude oil per day (300,000 barrels (48,000 m^3) per day) circulates about 80,000 cubic metres of water per hour through its cooling tower system.

The world's tallest cooling tower is the 200 metres tall cooling tower of Niederaussem Power Station.

Classification by Build—Package Type

Figure 2: Package type cooling towers

This type of cooling towers are preassembled and can be simply transported on trucks as they are compact machines. The capacity of package type towers is limited and for that reason, they are usually preferred by facilities with low heat rejection requirements such as food processing plants, textile plants, buildings like hospitals, hotels, malls, chemical processing plants, automotive factories etc.

Due to intensive use in domestic areas, sound level control is a relatively more important issue for package type cooling towers.

Field Erected Type

Figure 3: Field erected cooling towers

Such facilities as power plants, steel processing plants, petroleum refineries, petrochemical plants prefer field erected type cooling towers due to their large requirements of heat rejection. These towers are a lot larger in size compared to the package type cooling towers. A typical field erected cooling tower has pultruded FRP structure, FRP cladding, a mechanical unit for air draft, drift eliminator and fill.

Heat Transfer Methods

Figure 4: Mechanical draft crossflow cooling tower used in an HVAC application

With respect to the heat transfer mechanism employed, the main types are:

- *Wet cooling towers* or simply *open circuit cooling towers* operate on the principle of evapouration. The working fluid and the evapourated fluid (usually water) are one and the same.
- *Dry cooling towers* operate by heat transfer through a surface that separates the working fluid from ambient air, such as in a tube to air heat exchanger, utilising convective heat transfer. They do not use evapouration.
- *Fluid coolers* or *closed circuit cooling towers* are hybrids that pass the working fluid through a tube bundle, upon which clean water is sprayed and a fan-induced draft applied. The resulting heat transfer performance is much closer to that of a wet cooling tower, with the advantage provided by a dry cooler of protecting the working fluid from environmental exposure and contamination.

In a wet cooling tower (or open circuit cooling tower), the warm water can be cooled to a temperature lower than the ambient air dry-bulb temperature, if the air is relatively dry.

As ambient air is drawn past a flow of water, a small portion of the water evapourate, the energy required by that portion of the water to evapourate is taken from the remaining mass of water reducing his temperature (approximately by 970 BTU for each pound of evapourated water).

Evapouration results in saturated air conditions, lowering the temperature of the water process by the tower to a value close to wet bulb air temperature, which is lower than the ambient dry bulb air temperature, the difference determined by the humidity of the ambient air.

To achieve better performance (more cooling), a medium called *fill* is used to increase the surface area and the time of contact between the air and water flows.

Splash fill consists of material placed to interrupt the water flow causing splashing. *Film fill* is composed of thin sheets of material (usually PVC) upon which the water flows. Both methods create increased surface area and time of contact between the fluid (water) and the gas (air).

Air Flow Generation Methods

Figure 5: A forced draft cooling tower

Figure 6: Forced draft counter flow package type cooling towers

With respect to drawing air through the tower, there are three types of cooling towers:

- *Natural draft*, which utilises buoyancy via a tall chimney. Warm, moist air *naturally* rises due to the density differential to the dry, cooler outside air. Warm moist air is less dense than drier air at the same pressure. This moist air buoyancy produces a current of air through the tower.
- *Mechanical draft*, which uses power driven fan motors to force or draw air through the tower.
  - o *Induced draft*: A mechanical draft tower with a fan at the discharge which pulls air through tower. The fan *induces*

hot moist air out the discharge. This produces low entering and high exiting air velocities, reducing the possibility of *recirculation* in which discharged air flows back into the air intake. This fan/fin arrangement is also known as *draw-through*.

- o *Forced draft*: A mechanical draft tower with a blower type fan at the intake. The fan *forces* air into the tower, creating high entering and low exiting air velocities. The low exiting velocity is much more susceptible to recirculation. With the fan on the air intake, the fan is more susceptible to complications due to freezing conditions. Another disadvantage is that a forced draft design typically requires more motor horsepower than an equivalent induced draft design. The forced draft benefit is its ability to work with high static pressure. They can be installed in more confined spaces and even in some indoor situations. This fan/fill geometry is also known as *blow-through*.

- Fan assisted natural draft. A hybrid type that appears like a natural draft though airflow is assisted by a fan.

Hyperboloid (sometimes incorrectly known as hyperbolic) cooling towers (Image 1) have become the design standard for all natural-draft cooling towers because of their structural strength and minimum usage of material. The hyperboloid shape also aids in accelerating the upward convective air flow, improving cooling efficiency. They are popularly associated with nuclear power plants.

However, this association is misleading, as the same kind of cooling towers are often used at large coal-fired power plants as well. Similarly, not all nuclear power plants have cooling towers, instead cooling their heat exchangers with lake, river or ocean water.

Categorisation by Air-to-Water Flow

Crossflow: Crossflow is a design in which the air flow is directed perpendicular to the water flow. Air flow enters one or more vertical faces of the cooling tower to meet the fill material. Water flows (perpendicular to the air) through the fill by gravity.

The air continues through the fill and thus past the water flow into an open plenum area. A *distribution* or *hot water basin* consisting of a deep pan with holes or *nozzles* in the bottom is utilised in a crossflow tower. Gravity distributes the water through the nozzles uniformly across the fill material.

Counterflow

In a counterflow design the air flow is directly opposite to the water flow. Air flow first enters an open area beneath the fill media and is then drawn up vertically. The water is sprayed through pressurised nozzles and flows downward through the fill, opposite to the air flow.

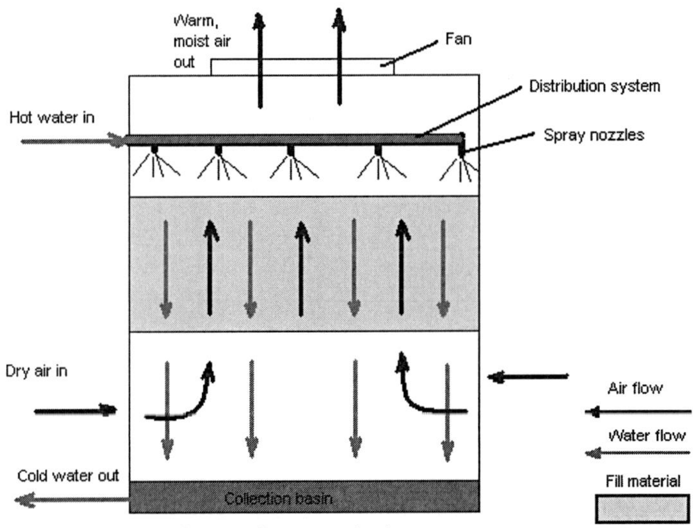

Counterflow type design

Figure 7.

Common to both designs:

- The interaction of the air and water flow allow a partial equalisation and evapouration of water.
- The air, now saturated with water vapour, is discharged from the cooling tower.
- A *collection* or *cold water basin* is used to contain the water after its interaction with the air flow.

Both crossflow and counterflow designs can be used in natural draft and mechanical draft cooling towers.

Cooling Tower as a Flue Gas Stack

At some modern power stations, equipped with flue gas purification like the Power Station Staudinger Grosskrotzenburg and the Power Station Rostock, the cooling tower is also used as a flue gas stack (industrial chimney). At plants without flue gas purification, problems with corrosion may occur.

Wet Cooling Tower Material Balance

Quantitatively, the material balance around a wet, evapourative cooling tower system is governed by the operational variables of makeup flow rate, evapouration and windage losses, draw-off rate, and the concentration cycles:

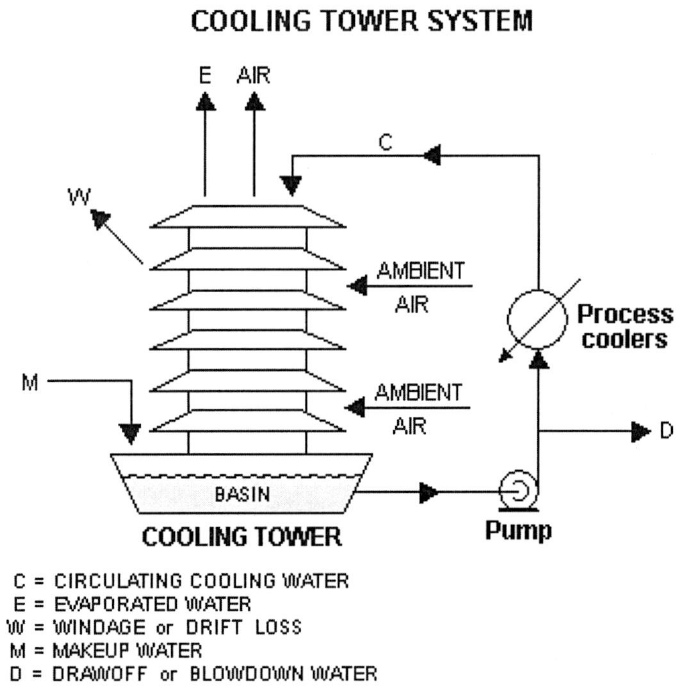

Figure 8.

M = Make-up water in m^3/h

C = Circulating water in m^3/h

D = Draw-off water in m^3/h

E = Evapourated water in m^3/h

W = Windage loss of water in m^3/h

X = Concentration in ppmw (of any completely soluble salts ... usually chlorides)

X_M = Concentration of chlorides in make-up water (M), in ppmw

X_C = Concentration of chlorides in circulating water (C), in ppmw

Cycles= Cycles of concentration = X_C / X_M (dimensionless)

ppmw = parts per million by weight

In the above sketch, water pumped from the tower basin is the cooling water routed through the process coolers and condensers in an industrial facility. The cool water absorbs heat from the hot process streams which need to be cooled or condensed, and the absorbed heat warms the circulating water (C). The warm water returns to the top of the cooling tower and trickles downward over the fill material inside the tower. As it trickles down, it contacts ambient air rising up through the tower either by natural draft or by forced draft using large fans in the tower.

That contact causes a small amount of the water to be lost as windage (W) and some of the water (E) to evapourate. The heat required to evapourate the water is derived from the water itself, which cools the water back to the original basin water temperature and the water is then ready to recirculate. The evapourated water leaves its dissolved salts behind in the bulk of the water which has not been evapourated, thus raising the salt concentration in the circulating cooling water. To prevent the salt concentration of the water from becoming too high, a portion of the water is drawn off (D) for disposal. Fresh water makeup (M) is supplied to the tower basin to compensate for the loss of evapourated water, the windage loss water and the draw-off water.

A water balance around the entire system is:

$$M = E + D + W$$

Since the evapourated water (E) has no salts, a chloride balance around the system is:

$$M (X_M) = D (X_C) + W (X_C) = X_C (D + W)$$

and, therefore:

$$X_C / X_M = \text{Cycles of concentration}$$

$$= M \div (D + W) = M \div (M - E) = 1 + [E \div (D + W)]$$

From a simplified heat balance around the cooling tower:

$$E = C \cdot \Delta T \cdot c_p \div H_V$$

where:

H_V = latent heat of vapourisation of water = ca. 2260 kJ / kg

ΔT = water temperature difference from tower top to tower bottom, in °C

c_p = specific heat of water = ca. 4.184 kJ / (kg°C)

Windage (or drift) losses (W) from large-scale industrial cooling towers, in the absence of manufacturer's data, may be assumed to be:

W = 0.3 to 1.0 percent of C for a natural draft cooling tower without windage drift eliminators

W = 0.1 to 0.3 percent of C for an induced draft cooling tower without windage drift eliminators

W = about 0.005 percent of C (or less) if the cooling tower has windage drift eliminators

Cycles of concentration represents the accumulation of dissolved minerals in the recirculating cooling water. Draw-off (or blowdown) is used principally to control the buildup of these minerals.

The chemistry of the makeup water including the amount of dissolved minerals can vary widely. Makeup waters low in dissolved minerals such as those from surface water supplies (lakes, rivers etc.) tend to be aggressive to metals (corrosive). Makeup waters from ground water supplies (wells) are usually higher in minerals and tend to be scaling (deposit minerals). Increasing the amount of minerals present in the water by cycling can make water less aggressive to piping however excessive levels of minerals can cause scaling problems.

As the cycles of concentration increase the water may not be able to hold the minerals in solution. When the solubility of these minerals have been exceeded they can precipitate out as mineral solids and cause fouling and heat exchange problems in the cooling tower or the heat exchangers. The temperatures of the recirculating water, piping and heat exchange surfaces determine if and where minerals will precipitate from the recirculating water. Often a professional water treatment consultant will evaluate the makeup water and the operating conditions of the cooling tower and recommend an appropriate range for the cycles of concentration. The use of water treatment chemicals, pretreatment such as water softening, pH adjustment, and other techniques can affect the acceptable range of cycles of concentration.

Concentration cycles in the majority of cooling towers usually range from 3 to 7. In the United States the majority of water supplies are well waters and have significant levels of dissolved solids. On the other hand one of the largest water supplies, New York City, has a surface supply quite low in minerals and cooling towers in that city are often allowed to concentrate to 7 or more cycles of concentration.

Besides treating the circulating cooling water in large industrial cooling tower systems to minimise scaling and fouling, the water should be filtered and also be dosed with biocides and algaecides to prevent growths that could interfere with the continuous flow of the water. For closed loop evapourative towers, corrosion inhibitors may be used, but caution should be taken to meet local environmental regulations as some inhibitors use chromates.

Ambient conditions dictate the efficiency of any given tower due to the amount of water vapour the air is able to absorb and hold, as can be determined on a psychrometric chart.

Cooling Towers and Legionnaires' Disease

Figure 9: Cooling tower and water discharge of a nuclear power plant

Another very important reason for using biocides in cooling towers is to prevent the growth of *Legionella*, including species that cause legionellosis or *Legionnaires' disease*, most notably *L. pneumophila*, or Mycobacterium avium. The various *Legionella* species are the cause of *Legionnaires' disease* in humans and transmission is via exposure to aerosols—the inhalation of mist droplets containing the bacteria. Common sources of *Legionella* include cooling towers used in open recirculating evapourative cooling water systems, domestic hot water systems, fountains, and similar disseminators that tap into a public water supply. Natural sources include freshwater ponds and creeks.

French researchers found that *Legionella* bacteria travelled up to 6 kilometres through the air from a large contaminated cooling tower at a petrochemical plant in Pas-de-Calais, France. That outbreak killed 21 of the 86 people who had a laboratory-confirmed infection.

Drift (or windage) is the term for water droplets of the process flow allowed to escape in the cooling tower discharge. Drift eliminators are used in order to hold drift rates typically to 0.001%-0.005% of the circulating flow rate. A typical drift eliminator provides multiple directional changes of airflow while preventing the escape of water droplets. A well-designed and well-fitted drift eliminator can greatly reduce water loss and potential for Legionella or other chemical exposure.

Cooling Tower Fog

Under certain ambient conditions, plumes of water vapour (fog) can be seen rising out of the discharge from a cooling tower, and can be mistaken as smoke from a fire. If the outdoor air is at or near saturation, and the tower adds more water to the air, saturated air with liquid water droplets can be discharged—which is seen as fog. This phenomenon typically occurs on cool, humid days, but is rare in many climates.

This phenomenon can be prevented by decreasing the relative humidity of the saturated discharge air. For that purpose, in hybrid towers, saturated discharge air is mixed with heated low relative humidity air. Some air enters the tower above drift eliminator level, passing through heat exchangers. The relative humidity of the dry air is even more decreased instantly as being heated while entering the tower. The discharged mixture has a relatively lower relative humidity and the fog is invisible.

Cooling Tower Operation in Freezing Weather

Cooling towers with malfunctions can freeze during very cold weather. Typically, freezing starts at the corners of a cooling tower with a reduced or absent heat load. Increased freezing conditions can create growing volumes of ice, resulting in increased structural loads. During the winter, some sites continuously operate cooling towers with 40 °F (4 °C) water leaving the tower. Basin heaters, tower draindown, and other freeze protection methods are often employed in cold climates.

- Do not operate the tower unattended.
- Do not operate the tower without a heat load. This can include basin heaters and heat trace. Basin heaters maintain the temperature of the water in the tower pan at an acceptable level. Heat trace is a resistive element that runs along water pipes located in cold climates to prevent freezing.
- Maintain design water flow rate over the fill.
- Manipulate airflow to maintain water temperature above freezing point.

Some Commonly Used Terms in the Cooling Tower Industry

- Drift - Water droplets that are carried out of the cooling tower with the exhaust air. Drift droplets have the same concentration of impurities as the water entering the tower. The drift rate is typically reduced by employing baffle-like devices, called drift eliminators, through which the air must travel after leaving the fill and spray zones of the tower. Drift can also be reduced by using warmer entering cooling tower temperatures.
- Blow-out - Water droplets blown out of the cooling tower by wind, generally at the air inlet openings. Water may also be lost, in the absence of wind, through splashing or misting. Devices such as wind screens, louvers, splash deflectors and water diverters are used to limit these losses.
- Plume - The stream of saturated exhaust air leaving the cooling tower. The plume is visible when water vapour it contains condenses in contact with cooler ambient air, like the saturated air in one's breath fogs on a cold day. Under certain conditions, a cooling tower plume may present fogging or icing hazards to its surroundings. Note that the water evapourated in the cooling process is "pure" water, in contrast to the very small percentage of drift droplets or water blown out of the air inlets.
- Blow-down - The portion of the circulating water flow that is removed in order to maintain the amount of dissolved solids and other impurities at an acceptable level. It may be noted that higher TDS (total dissolved solids) concentration in solution results in greater potential cooling tower efficiency. However the higher the TDS concentration, the greater the risk of scale, biological growth and corrosion. The amount of blow-down is primarily designated by the electrical conductivity of the circulating water. Biological growth, scaling and corrosion can be prevented by chemicals (Biocid, Sulphuric Acid, corrosion inhibitor). On the other hand, the only way to decrease the electrical conductivity is increasing the amount of blow-down and subsequently increasing the amount of make up.
- Make up - The water that is needed to be added to the circulating water system in order to compensate water losses such as, evapouration, drift loss, blow-out, etc.
- Leaching - The loss of wood preservative chemicals by the washing action of the water flowing through a wood structure cooling tower.

- Noise - Sound energy emitted by a cooling tower and heard (recorded) at a given distance and direction. The sound is generated by the impact of falling water, by the movement of air by fans, the fan blades moving in the structure, and the motors, gearboxes or drive belts.
- Approach - The approach is the difference in temperature between the cooled-water temperature and the entering-air wet bulb temperature (twb). Since the cooling towers are based on the principles of evapourative cooling, the maximum cooling tower efficiency depends on the wet bulb temperature of the air. The wet-bulb temperature is a type of temperature measurement that reflects the physical properties of a system with a mixture of a gas and a vapour, usually air and water vapour
- Range - The range is the temperature difference between the water inlet and water exit.
- Fill - Inside the tower, fills are added to increase contact surface as well as contact time between air and water. Thus they provide better heat transfer. The efficiency of the tower also depends on them. There are two types of fills that may be used:
    - o Film type fill (causes water to spread into a thin film)
    - o Splash type fill (breaks up water and interrupts its vertical progress)
- Full-Flow Filtration- Full-flow filtration continuously strains the entire system flow. For example, in a 100-ton system, the flow rate would be roughly 300 gal/min. A filter would be selected to accommodate the entire 300 gal/min flow rate. In this case, the filter typically is installed after the cooling tower on the discharge side of the pump. While this is the preferred method of filtration, for higher flow systems, it may be cost prohibitive.
- Side-Stream Filtration- Side-stream filtration, although popular, does not provide complete protection, but it can be effective. With side-stream filtration, a portion of the water is filtered continuously. This method works on the principle that continuous particle removal will keep the system clean. Manufacturers typically package side-stream filters on a skid,

complete with a pump and controls. For high flow systems, this method is cost-effective.

Properly sizing a side-stream filtration system is critical to obtain satisfactory filter performance. There is some debate over how to properly size the side-stream system.

Many engineers size the system to continuously filter the cooling tower basin water at a rate equivalent to 10% of the total circulation flow rate. For example, if the total flow of a system is 1,200 gal/min (a 400-ton system), a 120 gal/min side-stream system is specified.

- Cycle of concentration - Maximum allowed multiplier for the amount of miscellaneous substances in circulating water by the amount of those substances in make-up water.
- Treated timber - A structural material for cooling towers which was largely abandoned about 10 years ago. It is still used due to its low costs. The life of treated timber varies a lot, depending on the operating conditions of tower, such as frequency of stops, treatment of the circulating water, etc. Considering proper working conditions, estimated life of treated timber structural member is about 10 years.
- Pultruded FRP - A common structure material for cooling towers, known for its high corrosion resistance capabilities. Pultuded FRP is produced using pultrusion technology and became the most common structure material for cooling towers as it offers lower costs and requires less maintenance compared to reinforced concrete.

Fire Hazards

Cooling towers which are constructed in whole or in part of combustible materials can support propagating internal fires. The resulting damage can be sufficiently severe to require the replacement of the entire cell or tower structure.

For this reason, some codes and standards recommend combustible cooling towers be provided with an automatic fire sprinkler system.

Fires can propagate internally within the tower structure during maintenance when the cell is not in operation (such as for maintenance or construction), and even when the tower is in operation, especially those of the induced-draft type because of the existence of relatively dry areas within the towers.

Stability

Figure 10: Ferrybridge power station. Two hyperboloid cooling towers on Kharkov Power Station #5

Being very large structures, they are susceptible to wind damage, and several spectacular failures have occurred in the past. At Ferrybridge power station on 1 November 1965, the station was the site of a major structural failure, when three of the cooling towers collapsed owing to vibrations in 85 mph (137 km/h) winds. Although the structures had been built to withstand higher wind speeds, the shape of the cooling towers meant that westerly winds were funnelled into the towers themselves, creating a vortex.

Three out of the original eight cooling towers were destroyed and the remaining five were severely damaged. The towers were rebuilt and all eight cooling towers were strengthened to tolerate adverse weather conditions. Building codes were changed to include improved structural support, and wind tunnel tests introduced to check tower structures and configuration.

European Market

The Cooling Tower market is composed of one third of “close wet” units and two thirds of “open wet” units, and reaches in 2010 a total of 190,24M€.

Table 1: The market by country is splitted in 2010 as following:

Countries	*Sales Volume in M□*	*Share*
Benelux	18.70	9.8%
France	22.48	11.8%
Germany	41.40	21.8%
Greece	1.82	1.0%

Italy	20.48	10.8%
Poland	6.31	3.3%
Portugal	2.72	1.4%
Russia, Ukraine and CIS countries	16.59	8.7%
Scandinavia	4.64	2.4%
Spain	9.47	5.0%
Switzerland	5.33	2.8%
Turkey	8.50	4.5%
UK and Ireland	13.66	7.2%
Eastern Europe	18.16	9.5%

Architectural Engineering

Architectural engineering, also known as building engineering, is the application of engineering principles and technology to building design and construction. Definitions of an architectural engineer may refer to:

- An engineer in the structural, mechanical, electrical, construction or other engineering fields of building design and construction.
- A licensed engineering professional in parts of the United States.
- In informal contexts, and formally in some places, a professional synonymous with or similar to an architect. In some languages, "architect" is literally translated as "architectural engineer."

Engineering for Building—Structural Engineering

Structural engineering involves the analysis and design of physical objects such as buildings, bridges, equipment supports, towers and walls. Those concentrating on buildings are responsible for the structural performance of a large part of the built environment and are, sometimes, informally referred to as "building engineers". Structural engineers require expertise in strength of materials and in the seismic design of structures covered by earthquake engineering. Architectural Engineers sometimes practice structural as one aspect of their designs; the structural discipline when practiced as a speciality works closely with architects and other engineering specialists.

Mechanical, Electrical, and Plumbing (MEP)

Mechanical engineering | Electrical engineering | engineers are specialists, commonly referred to as "MEP" (mechanical, electrical, and plumbing) when engaged in the building design fields. Also known as "building services engineering" in the United Kingdom, Canada, and Australia. Mechanical engineers design and oversee the heating ventilation and air conditioning (HVAC), plumbing, and rain gutter systems. Plumbing designers often include design specifications for simple active fire protection systems, but for more complicated projects, fire protection engineers are often separately retained. Electrical engineers are responsible for the building's power distribution, telecommunication, fire alarm, signalisation, lightning protection and control systems, as well as lighting systems.

The Architectural Engineer (PE) in the United States

In many jurisdictions of the United States, the architectural engineer is a licensed engineering professional, usually a graduate of an architectural engineering university program preparing students to perform whole-building design in competition with architect-engineer teams; or for practice in one of structural, mechanical or electrical fields of building design, but with an appreciation of integrated architectural requirements.

Formal architectural engineering education, following the engineering model of earlier disciplines, developed in the late 19th century, and became widespread in the United States by the mid-20th century. With the establishment of a specific "architectural engineering" NCEES Professional Engineering registration examination in the 1990s, and first offering in April 2003, architectural engineering became recognised as a distinct engineering discipline in the United States. Architectural engineers are not entitled to practice architecture unless they are also licensed as architects.

The Architect as Architectural Engineer

In some countries architecture, as a profession providing architectural services, is sometimes referred to as "architectural engineering". In others, such as in Japan, the terms "architecture" and "building engineering" are used synonymously. The practice of architecture includes the planning, designing and overseeing the building's construction.

In some languages, such as Korean and Arabic, "architect" is literally translated as "architectural engineer". In some countries, an "architectural engineer" (such as the *ingegnere edile* in Italy) is entitled to practice architecture and is often referred to as an architect. These individuals are often also structural engineers. In other countries, such as Germany, Austria, Pakistan and most of the Arabic countries, architecture graduates receive an engineering degree (*Dipl.-Ing. - Diplom-Ingenieur*).

In Brazil, architects and engineers currently share the same accreditation process (CREA - Regional Council of Engineers and Architects). Besides traditional architecture design training, Brazilian architecture courses also offer complementary training in engineering disciplines such as structural, electrical, hydraulic and mechanical engineering.

After graduation, architects can be fully responsible for most engineering design and construction, except highly specialised tasks such as road design and high voltage electrical.

Education

The architectural, structural, mechanical and electrical engineering branches each have well established educational requirements that are usually fulfilled by completion of a university program.

Architectural Engineering as a Single Integrated Field of Study

What differentiates architectural engineering as a separate and single, integrated field of study, compared to other engineering disciplines, is its multi-disciplined engineering approach. Through training in and appreciation of architecture, the field seeks integration of building systems within its overall building design. Architectural engineering includes the design of building systems including Heating, ventilation and air conditioning (HVAC), plumbing, fire protection, electrical, lighting, architectural acoustics, and structural systems. In some university programs, students are required to concentrate on one of the systems; in others, they can receive a generalist architectural or building engineering degree.

Grid-tied Electrical System

A grid-tied electrical system, also called *tied to grid* or *grid tie system*, is a semi-autonomous electrical generation or grid energy storage system which links to the mains to feed excess capacity back to the local mains electrical grid. When insufficient electricity is generated, or the batteries are not fully charged, electricity drawn from the mains grid can make up the shortfall.

When batteries are used to store, the system is called battery-to-grid (B2G), which includes vehicle-to-grid (V2G).

How it Works

Direct Current electricity is passed to an inverter which is grid tied. The inverter monitors the alternating current mains supply waveforms and generates electricity that is phase matched. The DC electrical system is attached to the mains via this mechanism. Because the local system is phase matched, electricity can be drawn from the mains in the situation of a generating shortfall or discharge in batteries.

Quality of Service

A key concept of this system is the possibility of creating an electrical micro-system that is not dependent on the grid-tie to provide the high level of service quality which is often delivered by a traditional mains supply. Conversely if the mains supply of the region is unreliable, the local generation system can be used to power important equipment.

Environmentally Friendly

Local generation can be from an environmentally friendly source such as photovoltaic. Individuals can choose to install their own system if an environmentally friendly mains provider is not available in their location. B2G also spare the use of fossil fuel power plants and electrical substations to supply energy in peak loads of the public electric grid.

Small Scale Start

A micro generation facility can be started with a very small system such as a home wind power generation or photovoltaic (solar cells) generation system.

Sell to and Buy from Mains

- Excess electricity can be sold to mains.
- Electrical shortfall can be bought from mains.

List of Countries or Regions that Legally Allow Grid-tied Electrical Systems

- India
- Australia
- Belgium
- Bosnia and Herzegovina
- Canada
- Greece
- Hungary

- Israel
- Japan
- Mexico
- The Netherlands
- New Zealand
- Portugal
- Sri Lanka
- Singapore
- Spain
- USA
- Dominican Republic

Electric Power Transmission

Electric-power transmission is the bulk transfer of electrical energy, from generating power plants to Electrical substations located near demand centres. This is distinct from the local wiring between high-voltage substations and customers, which is typically referred to as electric power distribution. Transmission lines, when interconnected with each other, become transmission networks. In the US, these are typically referred to as "power grids" or just "the grid", while in the UK the network is known as the "national grid." North America has three major grids: The Western Interconnection; The Eastern Interconnection and the Electric Reliability Council of Texas (or ERCOT) grid. Historically, transmission and distribution lines were owned by the same company, but since the turn of the 21st century many countries have liberalised the electricity market in ways that have led to the separation of the electricity transmission business from the distribution business.

Transmission lines mostly use high-voltage three-phase alternating current (AC), although single phase AC is sometimes used in railway electrification systems. High-voltage direct-current (HVDC) technology is for greater efficiency in very long distances (typically greater than 400 miles, or 600 km); submarine power cables (typically longer than 30 miles, or 50 km). HVDC links are increasingly used as isolating features to stabilise against control problems in large power distribution networks where sudden new loads or blackouts in one part of a network can otherwise result in synchronisation problems that may bring the rest of the network down.

Electricity is transmitted at high voltages (110 kV or above) to reduce the energy lost in long-distance transmission. Power is usually transmitted through overhead power lines. Underground power transmission has a significantly higher cost and greater operational limitations but is sometimes used in urban areas or sensitive locations.

A key limitation in the distribution of electricity is that, with minor exceptions, electrical energy cannot be stored, and therefore must be generated as needed. A sophisticated system of control is therefore required to ensure electric generation very closely matches the demand. If supply and demand are not in balance, generation plants and transmission equipment can shut down which, in the worst cases, can lead to a major regional blackout, such as occurred in the US Northeast blackouts of 1965, 1977, 2003, and in the west, one in 1996 and 2011. To reduce the risk of such failures, electric transmission networks are interconnected into regional, national or continental wide networks thereby providing multiple redundant alternate routes for power to flow should (weather or equipment) failures occur. Much analysis is done by transmission companies to determine the maximum reliable capacity of each line (ordinarily less than its physical or thermal limit) to ensure spare capacity is available should there be any such failure in another part of the network.

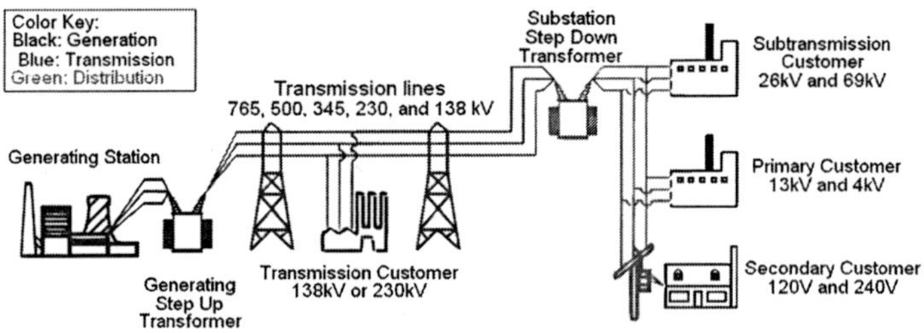

Figure 11: Diagram of an electrical system.

Overhead Transmission

Figure 12: Contiguous United States power transmission grid consists of 300,000 km of lines operated by 500 companies.

High-voltage overhead conductors are not covered by insulation. The conductor material is nearly always an aluminium alloy, made into several strands and possibly reinforced with steel strands.

Copper was sometimes used for overhead transmission but aluminium is lower in weight for only marginally reduced performance, and much lower in cost. Overhead conductors are a commodity supplied by several companies worldwide.

Improved conductor material and shapes are regularly used to allow increased capacity and modernise transmission circuits. Conductor sizes range from 12 mm^2 (#6 American wire gauge) to 750 mm^2 (1,590,000 circular mils area), with varying resistance and current-carrying capacity.

Thicker wires would lead to a relatively small increase in capacity due to the skin effect, that causes most of the current to flow close to the surface of the wire. Because of this current limitation, multiple parallel cables (called bundle conductors) are used when higher capacity is needed. Bundle conductors are also used at high voltages to reduce corona discharge energy loss.

Today, transmission-level voltages are usually considered to be 110 kV and above. Lower voltages such as 66 kV and 33 kV are usually considered subtransmission voltages but are occasionally used on long lines with light loads.

Voltages less than 33 kV are usually used for distribution. Voltages above 230 kV are considered extra high voltage and require different designs compared to equipment used at lower voltages.

Since overhead transmission wires depend on air for insulation, design of these lines requires minimum clearances to be observed to maintain safety.

Adverse weather conditions of high wind and low temperatures can lead to power outages: wind speeds as low as 23 knots (43 km/h) can permit conductors to encroach operating clearances, resulting in a flashover and loss of supply.

Oscillatory motion of the physical line can be termed gallop or flutter depending on the frequency and amplitude of oscillation.

Underground Transmission

Electric power can also be transmitted by underground power cables instead of overhead power lines.

History

Figure 13: New York City streets in 1890. Besides telegraph lines, multiple electric lines were required for each class of device requiring different voltages.

In the early days of commercial electric power, transmission of electric power at the same voltage as used by lighting and mechanical loads restricted the distance between generating plant and consumers. In 1882, generation was with direct current (DC), which could not easily be increased in voltage for long-distance transmission. Different classes of loads (for example, lighting, fixed motors, and traction/ railway systems) required different voltages, and so used different generators and circuits.

Due to this specialisation of lines and because transmission was so inefficient that generators needed to be near their loads, it seemed at the time that the industry would develop into what is now known as a distributed generation system with large numbers of small generators located nearby their loads.

In 1886 in Great Barrington, Massachusetts, a 1 kV alternating current (AC) distribution system was installed. That same year, AC power at 2 kV, transmitted 30 km, was installed at Cerchi, Italy. At an AIEE meeting on May 16, 1888, Nikola Tesla delivered a lecture entitled *A New System of Alternating Current Motors and Transformers*, describing the equipment which allowed efficient generation and use of polyphase alternating currents. The transformer, and Tesla's polyphase and single-phase induction motors, were essential for a combined AC distribution system for both lighting and machinery. Ownership of the rights to the Tesla patents was a key advantage to the Westinghouse Company in offering a complete alternating current power system for both lighting and power.

Figure 14: Nikola Tesla's Alternating current polyphase generators on display at the 1893 World's Fair in Chicago. Tesla's polyphase innovations revolutionised transmission.

Regarded as one of the most influential electrical innovations, the *universal system* used transformers to step-up voltage from generators to high-voltage transmission lines, and then to step-down voltage to local distribution circuits or industrial customers. By a suitable choice of utility frequency, both lighting and motor loads could be served. Rotary converters and later mercury-arc valves and other rectifier equipment allowed DC to be provided where needed. Generating stations and loads using different frequencies could be interconnected using rotary converters. By using common generating plants for every type of load, important economies of scale were achieved, lower overall capital investment was required, load factor on each plant was increased allowing for higher efficiency, a lower cost for the consumer and increased overall use of electric power.

By allowing multiple generating plants to be interconnected over a wide area, electricity production cost was reduced. The most efficient available plants could be used to supply the varying loads during the day. Reliability was improved and capital investment cost was reduced, since stand-by generating capacity could be shared over many more customers and a wider geographic area. Remote and low-cost sources of energy, such as hydroelectric power or mine-mouth coal, could be exploited to lower energy production cost.

The first transmission of three-phase alternating current using high voltage took place in 1891 during the international electricity exhibition in Frankfurt. A 25 kV transmission line, approximately 175 km long, connected Lauffen on the Neckar and Frankfurt.

Voltages used for electric power transmission increased throughout the 20th century. By 1914, fifty-five transmission systems each operating at more than 70 kV were in service. The highest voltage then used was 150 kV.

The rapid industrialisation in the 20th century made electrical transmission lines and grids a critical part of the infrastructure in most industrialised nations. Interconnection of local generation plants and small distribution networks was greatly spurred by the requirements of World War I, where large electrical generating plants were built by governments to provide power to munitions factories. Later these plants were connected to supply civil loads through long-distance transmission.

Bulk Power Transmission

Engineers design transmission networks to transport the energy as efficiently as feasible, while at the same time taking into account economic factors, network safety and redundancy. These networks use components such as power lines, cables, circuit breakers, switches and transformers. The transmission network is usually administered on a regional basis by an entity such as a regional transmission organisation or transmission system operator.

Transmission efficiency is hugely improved by devices that increase the voltage, and proportionately reduce the current in the conductors, thus keeping the power transmitted nearly equal to the power input. The reduced current flowing through the line reduces the losses in the conductors. According to Joule's Law, energy losses are directly proportional to the square of the current. Thus, reducing the current by a factor of 2 will lower the energy lost to conductor resistance by a factor of 4.

This change in voltage is usually achieved in AC circuits using a *step-up transformer*. DC systems require relatively costly conversion equipment which may be economically justified for particular projects, but are less common currently.

A transmission grid is a network of power stations, transmission circuits, and substations. Energy is usually transmitted within a grid with three-phase AC. Single phase AC is used only for distribution to end users since it is not usable for large polyphase induction motors. In the 19th century, two-phase transmission was used but required either three wires with unequal currents or four wires.

Higher order phase systems require more than three wires, but deliver marginal benefits.

The capital cost of electric power stations is so high, and electric demand is so variable, that it is often cheaper to import some portion of the needed power than to generate it locally. Because nearby loads are often correlated (hot weather in the Southwest portion of the US might cause many people to use air conditioners), electricity often comes from distant sources. Because of the economics of load balancing, wide area transmission grids now span across countries and even large portions of continents. The web of interconnections between power producers and consumers ensures that power can flow, even if a few links are inoperative.

The unvarying (or slowly varying over many hours) portion of the electric demand is known as the *base load* and is generally served best by large facilities (which are therefore efficient due to economies of scale) with low variable costs for fuel and operations. Such facilities might be nuclear or coal-fired power stations, or hydroelectric, while other renewable energy sources such as concentrated solar thermal and geothermal power have the potential to provide base load power. Renewable energy sources such as solar photovoltaics, wind, wave, and tidal are, due to their intermittency, not considered "base load" but can still add power to the grid. The remaining power demand, if any, is supplied by peaking power plants, which are typically smaller, faster-responding, and higher cost sources, such as combined cycle or combustion turbine plants fuelled by natural gas.

Long-distance transmission of electricity (thousands of kilometres) is cheap and efficient, with costs of US$0.005–0.02/kWh (compared to annual averaged large producer costs of US$0.01–0.025/kWh, retail rates upwards of US$0.10/kWh, and multiples of retail for instantaneous suppliers at unpredicted highest demand moments). Thus distant suppliers can be cheaper than local sources (e.g., New York City buys a lot of electricity from Canada). Multiple local sources (even if more expensive and infrequently used) can make the transmission grid more fault tolerant to weather and other disasters that can disconnect distant suppliers.

Long distance transmission allows remote renewable energy resources to be used to displace fossil fuel consumption. Hydro and wind sources cannot be moved closer to populous cities, and solar costs are lowest in remote areas where local power needs are minimal. Connection costs alone can determine whether any particular renewable alternative is economically sensible. Costs can be prohibitive for transmission lines, but various proposals for massive infrastructure investment in high capacity, very long distance super grid transmission networks could be recovered with modest usage fees.

Grid Input

At the generating plants the energy is produced at a relatively low voltage between about 2.3 kV and 30 kV, depending on the size of the unit. The generator terminal voltage is then stepped up by the power station transformer to a higher voltage (115 kV to 765 kV AC, varying by country) for transmission over long distances.

Losses

Transmitting electricity at high voltage reduces the fraction of energy lost to resistance, which averages around 7%. For a given amount of power, a higher voltage reduces the current and thus the resistive losses in the conductor. For example, raising the voltage by a factor of 10 reduces the current by a corresponding factor of 10 and therefore the I^2R losses by a factor of 100, provided the same sized conductors are used in both cases. Even if the conductor size (cross-sectional area) is reduced 10-fold to match the lower current the I^2R losses are still reduced 10-fold. Long distance transmission is typically done with overhead lines at voltages of 115 to 1,200 kV. At extremely high voltages, more than 2 MV between conductor and ground, corona discharge losses are so large that they can offset the lower resistance loss in the line conductors. Measures to reduce corona losses include conductors having large diametre; often hollow to save weight, or bundles of two or more conductors.

Transmission and distribution losses in the USA were estimated at 6.6% in 1997 and 6.5% in 2007. In general, losses are estimated from the discrepancy between energy produced (as reported by power plants) and energy sold to end customers; the difference between what is produced and what is consumed constitute transmission and distribution losses.

As of 1980, the longest cost-effective distance for DC electricity was determined to be 7,000 km (4,300 mi). For AC it was 4,000 km (2,500 mi), though all transmission lines in use today are substantially shorter. In an alternating current circuit, the inductance and capacitance of the phase conductors can be significant. The currents that flow in these components of the circuit impedance constitute reactive power, which transmits no energy to the load. Reactive current causes extra losses in the transmission circuit. The ratio of real power (transmitted to the load) to apparent power is the power factor. As reactive current increases, the reactive power increases and the power factor decreases. For systems with low power factors, losses are higher than for systems with high power factors. Utilities add capacitor banks and other components (such as phase-shifting transformers; static VAR compensators; physical transposition of the phase conductors; and flexible AC transmission systems, FACTS) throughout the system to control reactive power flow for reduction of losses and stabilisation of system voltage.

Subtransmission

Subtransmission is part of an electric power transmission system that runs at relatively lower voltages. It is uneconomical to connect all distribution substations to the high main transmission voltage, because the equipment is larger and more expensive. Typically, only larger substations connect with this high voltage. It is stepped down and sent to smaller substations in towns and neighbourhoods. Subtransmission circuits are usually arranged in loops so that a single line failure does not cut off service to a large number of customers for more than a short time. While subtransmission circuits are usually carried on overhead lines, in urban areas buried cable may be used.

There is no fixed cutoff between subtransmission and transmission, or subtransmission and distribution. The voltage ranges overlap somewhat. Voltages of 69 kV, 115 kV and 138 kV are often used for subtransmission in North America. As power systems evolved, voltages formerly used for transmission were used for subtransmission, and subtransmission voltages became distribution voltages. Like transmission, subtransmission moves relatively large amounts of power, and like distribution, subtransmission covers an area instead of just point to point.

Transmission Grid Exit

At the substations, transformers reduce the voltage to a lower level for distribution to commercial and residential users. This distribution is accomplished with a combination of sub-transmission (33 kV to 132 kV) and distribution (3.3 to 25 kV). Finally, at the point of use, the energy is transformed to low voltage.

High-voltage Direct Current

High voltage direct current (HVDC) is used to transmit large amounts of power over long distances or for interconnections between asynchronous grids. When electrical energy is required to be transmitted over very long distances, it is more economical to transmit using direct current instead of alternating current. For a long transmission line, the lower losses and reduced construction cost of a DC line can offset the additional cost of converter stations at each end. Also, at high AC voltages, significant (although economically acceptable) amounts of energy are lost due to corona discharge, the capacitance between phases or, in the case of buried cables, between phases and the soil or water in which the cable is buried.

HVDC is also used for long submarine cables because over about 30 km length AC can no longer be applied. In that case special high voltage cables for DC are built. Many submarine cable connections – up to 600 km length – are in use nowadays.

HVDC links are sometimes used to stabilise against control problems with the AC electricity flow. The power transmitted by an AC line increases as the phase angle between source end voltage and destination ends increases, but too great a phase angle will allow the generators at either end of the line to fall out of step. Since the power flow in a DC link is controlled independently of the phases of the AC networks at either end of the link, this stability limit does not apply to a DC line, and it can transfer its full thermal rating. A DC link stabilises the AC grids at either end, since power flow and phase angle can be controlled independently.

In other words, to transmit AC power as AC when needed in either direction between Seattle and Boston would require the (highly challenging) continuous real-time adjustment of the relative phase of the two electrical grids. With HVDC instead the interconnection would: (1) Convert AC in Seattle into HVDC. (2) Use HVDC for the three thousand miles of cross country transmission. Then (3) convert the HVDC to locally synchronised AC in Boston, and optionally in other cooperating cities along the transmission route. One prominent example of such a transmission line is the Pacific DC Intertie located in the Western United States.

Limitations

The amount of power that can be sent over a transmission line is limited. The origins of the limits vary depending on the length of the line. For a short line, the heating of conductors due to line losses sets a thermal limit. If too much current is drawn, conductors may sag too close to the ground, or conductors and equipment may be damaged by overheating. For intermediate-length lines on the order of 100 km (62 mi), the limit is set by the voltage drop in the line. For longer AC lines, system stability sets the limit to the power that can be transferred. Approximately, the power flowing over an AC line is proportional to the sine of the phase angle of the voltage at the receiving and transmitting ends. Since this angle varies depending on system loading and generation, it is undesirable for the angle to approach 90 degrees. Very approximately, the allowable product of line length and maximum load is proportional to the square of the system voltage. Series capacitors or phase-shifting transformers are used on long lines to improve stability. High-voltage direct current lines are restricted only by thermal and voltage drop limits, since the phase angle is not material to their operation.

Up to now, it has been almost impossible to foresee the temperature distribution along the cable route, so that the maximum applicable current load was usually set as a compromise between understanding of operation conditions and risk minimisation. The availability of industrial Distributed Temperature Sensing (DTS) systems that measure in real time temperatures all along the cable is a first step in monitoring the transmission system capacity. This monitoring solution is based on using passive optical fibres as temperature sensors, either integrated directly inside a high voltage cable or mounted externally on the cable insulation. A solution for overhead lines is also available. In this case the optical fibre is integrated into the core of a phase wire of overhead transmission lines (OPPC). The integrated Dynamic Cable Rating (DCR) or also called Real Time Thermal Rating (RTTR) solution enables not only to continuously monitor the temperature of a high voltage cable circuit in real time, but to safely utilise the existing network capacity to its maximum. Furthermore it provides the ability to the operator to predict the behaviour of the transmission system upon major changes made to its initial operating conditions.

Control

To ensure safe and predictable operation the components of the transmission system are controlled with generators, switches, circuit breakers and loads. The voltage, power, frequency, load factor, and reliability capabilities of the transmission system are designed to provide cost effective performance for the customers.

Load Balancing

The transmission system provides for base load and peak load capability, with safety and fault tolerance margins. The peak load times vary by region largely due to the industry mix. In very hot and very cold climates home air conditioning and heating loads have an effect on the overall load. They are typically highest in the late afternoon in the hottest part of the year and in mid-mornings and mid-evenings in the coldest part of the year. This makes the power requirements vary by the season and the time of day. Distribution system designs always take the base load and the peak load into consideration.

The transmission system usually does not have a large buffering capability to match the loads with the generation. Thus generation has to be kept matched to the load, to prevent overloading failures of the generation equipment.

Multiple sources and loads can be connected to the transmission system and they must be controlled to provide orderly transfer of power. In centralised power generation, only local control of generation is necessary, and it involves synchronisation of the generation units, to prevent large transients and overload conditions.

In distributed power generation the generators are geographically distributed and the process to bring them online and offline must be carefully controlled. The load control signals can either be sent on separate lines or on the power lines themselves. To load balance the voltage and frequency can be used as a signalling mechanism.

In voltage signalling, the variation of voltage is used to increase generation. The power added by any system increases as the line voltage decreases. This arrangement is stable in principle. Voltage based regulation is complex to use in mesh networks, since the individual components and setpoints would need to be reconfigured every time a new generator is added to the mesh.

In frequency signalling, the generating units match the frequency of the power transmission system. In droop speed control, if the frequency decreases, the power is increased. (The drop in line frequency is an indication that the increased load is causing the generators to slow down.)

Wind turbines, v2g and other distributed storage and generation systems can be connected to the power grid, and interact with it to improve system operation.

Failure Protection

Under excess load conditions, the system can be designed to fail gracefully rather than all at once. Brownouts occur when the supply power drops below the demand. Blackouts occur when the supply fails completely. Rolling blackouts (also called load shedding) are intentionally engineered electrical power outages, used to distribute insufficient power when the demand for electricity exceeds the supply.

Communications

Operators of long transmission lines require reliable communications for control of the power grid and, often, associated generation and distribution facilities. Fault-sensing protective relays at each end of the line must communicate to monitor the flow of power into and out of the protected line section so that faulted conductors or equipment can be quickly de-energised and the balance of the system restored. Protection of the transmission line from short circuits and other faults is usually so critical that common carrier telecommunications are insufficiently reliable, and in remote areas a common carrier may not be available. Communication systems associated with a transmission project may use:

- Microwaves
- Power line communication
- Optical fibres

Rarely, and for short distances, a utility will use pilot-wires strung along the transmission line path. Leased circuits from common carriers are not preferred since availability is not under control of the electric power transmission organisation.

Transmission lines can also be used to carry data: this is called power-line carrier, or PLC. PLC signals can be easily received with a radio for the long wave range. Optical fibres can be included in the stranded conductors of a transmission line, in the overhead shield wires. These cables are known as optical ground wire (*OPGW*). Sometimes a standalone cable is used, all-dielectric self-supporting (*ADSS*) cable, attached to the transmission line cross arms.

Some jurisdictions, such as Minnesota, prohibit energy transmission companies from selling surplus communication bandwidth or acting as a telecommunications common carrier. Where the regulatory structure permits, the utility can sell capacity in extra dark fibres to a common carrier, providing another revenue stream.

Electricity Market Reform

Some regulators regard electric transmission to be a natural monopoly and there are moves in many countries to separately regulate transmission. Spain was the first country to establish a regional transmission organisation. In that country transmission operations and market operations are controlled by separate companies. The transmission system operator is Red Electrica de España (REE) and the wholesale electricity market operator is Operador del Mercado Iberico de Energía – Polo Español, S.A. (OMEL). Spain's transmission system is interconnected with those of France, Portugal, and Morocco.

In the United States and parts of Canada, electrical transmission companies operate independently of generation and distribution companies.

Cost of Electric Power Transmission

The cost of high voltage electricity transmission (as opposed to the costs of electricity distribution) is comparatively low, compared to all other costs arising in a consumer's electricity bill. In the UK transmission costs are about 0.2p/kWh compared to a delivered domestic price of around 10 p/kWh.

Research evaluates the level of capital expenditure in the electric power T&D equipment market will be worth $128.9bn in 2011.

Merchant Transmission

Merchant transmission is an arrangement where a third party constructs and operates electric transmission lines through the franchise area of an unrelated utility. Advocates of merchant transmission claim that this will create competition to construct the most efficient and lowest cost additions to the transmission grid. Merchant transmission projects typically involve DC lines because it is easier to limit flows to paying customers.

Operating merchant transmission projects in the United States include the Cross Sound Cable from Long Island, New York to New Haven, Connecticut, Neptune RTS Transmission Line from Sayreville, N.J., to Newbridge, N.Y, ITC Holdings, Inc. transmission system in the midwest, and Path 15 in California. Additional projects are in development or have been proposed throughout the United States.

There is only one unregulated or market interconnector in Australia: Basslink between Tasmania and Victoria. Two DC links originally implemented as market interconnectors Directlink and Murraylink have been converted to regulated interconnectors. NEMMCO

A major barrier to wider adoption of merchant transmission is the difficulty in identifying who benefits from the facility so that the beneficiaries will pay the toll. Also, it is difficult for a merchant transmission line to compete when the alternative transmission lines are subsidized by other utility businesses.

Health Concerns

Some large studies, including a large United States study, have failed to find any link between living near power lines and developing any sickness or diseases such as cancer. One old study from 1997 found that it did not matter how close you were to a power line or a sub-station, there was no increased risk of cancer or illness.

The mainstream scientific evidence suggests that low-power, low-frequency, electromagnetic radiation associated with household currents and high transmission power lines does not constitute a short or long term health hazard. Some studies, however, have found statistical correlations between various diseases and living or working near power lines. No adverse health effects have been substantiated for people not living close to powerlines.

There are established biological effects for acute *high* level exposure to magnetic fields well above 100 μT (1000 mG). In a residential setting, there is "limited evidence of carcinogenicity in humans and less than sufficient evidence for carcinogenicity in experimental animals", in particular, childhood leukaemia, *associated with* average exposure to residential power-frequency magnetic field above 0.3 (3 mG) to 0.4 μT (4 mG). These levels exceed average residential power-frequency magnetic fields in homes which are about 0.07 μT (0.7 mG) in Europe and 0.11 μT (1.1 mG) in North America.

Government Policy

Historically, local governments have exercised authority over the grid and have significant disincentives to take action that would benefit states other than their own. Localities with cheap electricity have a disincentive to making interstate commerce in electricity trading easier, since other regions will be able to compete for local energy and drive up rates. Some regulators in Maine for example do not wish to address congestion problems because the congestion serves to keep Maine rates low. Further, vocal local constituencies can block or slow permitting by pointing to visual impact, environmental, and perceived health concerns. In the US, generation is growing 4 times faster than transmission, but big transmission upgrades require the coordination of multiple states, a multitude of interlocking permits, and cooperation between a significant portion of the 500 companies that own the grid.

From a policy perspective, the control of the grid is balkanized, and even former energy secretary Bill Richardson refers to it as a *third world grid.* There have been efforts in the EU and US to confront the problem. The US national security interest in significantly growing transmission capacity drove passage of the 2005 energy act giving the Department of Energy the authority to approve transmission if states refuse to act. However, soon after using its power to designate two National Interest Electric Transmission Corridors, 14 senators signed a letter stating the DOE was being too aggressive.

Special Transmission

Grids for Railways: In some countries where electric trains run on low frequency AC (e.g., 16.7 Hz and 25 Hz) power, there are separate single phase traction power networks operated by the railways. These grids are fed by separate generators in some traction powerstations or by traction current converter plants from the public three phase AC network.

Superconducting Cables

High-temperature superconductors promise to revolutionise power distribution by providing lossless transmission of electrical power. The development of superconductors with transition temperatures higher than the boiling point of liquid nitrogen has made the concept of superconducting power lines commercially feasible, at least for high-load applications. It has been estimated that the waste would be halved using this method, since the necessary refrigeration equipment would consume about half the power saved by the elimination of the majority of resistive losses. Some companies such as Consolidated Edison and American Superconductor have already begun commercial production of such systems. In one hypothetical future system called a SuperGrid, the cost of cooling would be eliminated by coupling the transmission line with a liquid hydrogen pipeline.

Superconducting cables are particularly suited to high load density areas such as the business district of large cities, where purchase of an easement for cables would be very costly.

Single Wire Earth Return

Single-wire earth return (SWER) or single wire ground return is a single-wire transmission line for supplying single-phase electrical power for an electrical grid to remote areas at low cost. It is principally used for rural electrification, but also finds use for larger isolated loads such as water pumps, and light rail. Single wire earth return is also used for HVDC over submarine power cables.

Wireless Power Transmission

Both Nikola Tesla and Hidetsugu Yagi attempted to devise systems for large scale wireless power transmission, with no commercial success.

In November 2009, LaserMotive won the NASA 2009 Power Beaming Challenge by powering a cable climber 1 km vertically using a ground-based laser transmitter.

The system produced up to 1 kW of power at the receiver end. In August 2010, NASA contracted with private companies to pursue the design of laser power beaming systems to power low earth orbit satellites and to launch rockets using laser power beams.

Wireless power transmission has been studied for transmission of power from solar power satellites to the earth. A high power array of microwave or laser transmitters would beam power to a rectenna. Major engineering and economic challenges face any solar power satellite project.

Security of Control Systems

The Federal government of the United States admits that the power grid is susceptible to cyber-warfare. The United States Department of Homeland Security works with industry to identify vulnerabilities and to help industry enhance the security of control system networks, the federal government is also working to ensure that security is built in as the U.S. develops the next generation of 'smart grid' networks.

Records

- *Highest capacity system:* 6.3 GW HVDC Itaipu (Brazil) (±600 kV DC)
- *Highest transmission voltage (AC):* 1.15 MV on Powerline Ekibastuz-Kokshetau (Kazakhstan)
- Largest double-circuit transmission, Kita-Iwaki Powerline.
- *Highest pylons:* Yangtze River Crossing (height: 345 m/1,132 ft)
- *Longest power line:* Inga-Shaba (length: 1,700 kilometres / 1,056 miles)
- *Longest span of power line:* 5,376 m (17,638 ft) at Ameralik Span

- Longest submarine cables:
  - NorNed, North Sea – (length of submarine cable: 580 kilometres / 360 miles)
  - Basslink, Bass Strait – (length of submarine cable: 290 kilometres / 180 miles, total length: 370.1 kilometres / 230 miles)
  - Baltic-Cable, Baltic Sea – (length of submarine cable: 238 kilometres / 148 miles, HVDC length: 250 kilometres / 155 miles, total length: 262 kilometres / 163 miles)
- Longest underground cables:
  - Murraylink, Riverland/Sunraysia – (length of underground cable: 180 kilometres / 112 miles)

Plant Efficiency

The efficiency of a plant is the percentage of the total energy content of a power plant's fuel that is converted into electricity. The remaining energy is usually lost to the environment as heat unless it is used for district heating.

This is complicated by the fact that there are two different ways to measure the fuel energy input - LCV = Lower Calorific Value, which is the same as NCV = Net Calorific Value, or, HCV = Higher Calorific Value which is the same as GCV, Gross Calorific Value.

Depending on which convention is used, a differences of 10 % in the apparent efficiency of a gas fired plant can arise, so it is very important to know which convention, HCV or LCV (NCV or GCV) is being used.

Boiler

Figure 15: A portable boiler (preserved, Poland)

Figure 16: A stationary boiler (United States)

A boiler is a closed vessel in which water or other fluid is heated. The heated or vapourised fluid exits the boiler for use in various processes or heating applications.

Overview

Materials: The pressure vessel in a boiler is usually made of steel (or alloy steel), or historically of wrought iron. Stainless steel is virtually prohibited (by the ASME Boiler Code) for use in wetted parts of modern boilers, but is used often in superheater sections that will not be exposed to liquid boiler water. In live steam models, copper or brass is often used because it is more easily fabricated in smaller size boilers. Historically, copper was often used for fireboxes (particularly for steam locomotives), because of its better formability and higher thermal conductivity; however, in more recent times, the high price of copper often makes this an uneconomic choice and cheaper substitutes (such as steel) are used instead.

For much of the Victorian "age of steam", the only material used for boilermaking was the highest grade of wrought iron, with assembly by rivetting. This iron was often obtained from specialist ironworks, such as at Cleator Moor (UK), noted for the high quality of their rolled plate and its suitability for high-reliability use in critical applications, such as high-pressure boilers. In the 20th century, design practice instead moved towards the use of steel, which is stronger and cheaper, with welded construction, which is quicker and requires less labour.

Cast iron may be used for the heating vessel of domestic water heaters. Although such heaters are usually termed "boilers" in some countries, their purpose is usually to produce hot water, not steam, and so they run at low pressure and try to avoid actual boiling. The brittleness of cast iron makes it impractical for high pressure steam boilers.

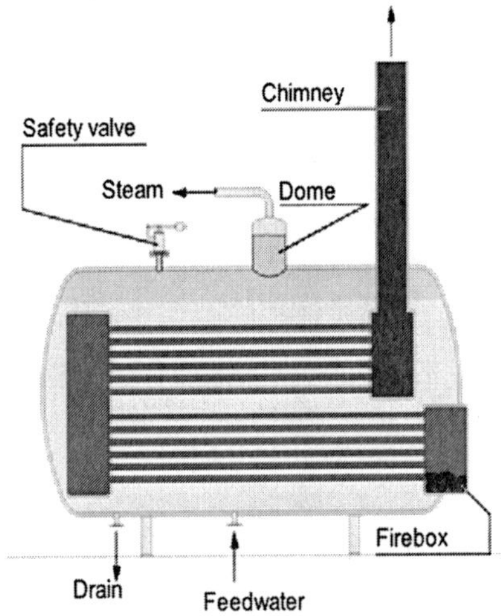

Figure 17: Diagram of a fire-tube boiler

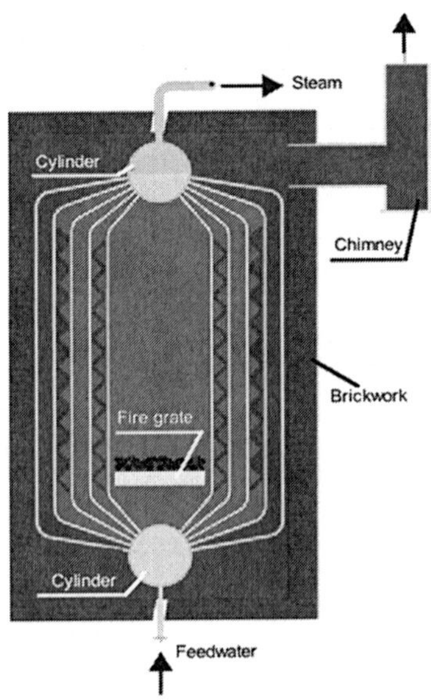

Figure 18: Diagram of a water-tube boiler.

Fuel

The source of heat for a boiler is combustion of any of several fuels, such as wood, coal, oil, or natural gas. Electric steam boilers use resistance- or immersion-type heating elements. Nuclear fission is also used as a heat source for generating steam. Heat recovery steam generators (HRSGs) use the heat rejected from other processes such as gas turbines.

Configurations

Boilers can be classified into the following configurations:

- *"Pot boiler"* or *"Haycock boiler"*: a primitive "kettle" where a fire heats a partially-filled water container from below. 18th century Haycock boilers generally produced and stored large volumes of very low-pressure steam, often hardly above that of the atmosphere. These could burn wood or most often, coal. Efficiency was very low.
- Fire-tube boiler. Here, water partially fills a boiler barrel with a small volume left above to accommodate the steam (*steam space*). This is the type of boiler used in nearly all steam locomotives. The heat source is inside a furnace or *firebox* that has to be kept permanently surrounded by the water in order to maintain the temperature of the *heating surface* just below boiling point. The furnace can be situated at one end of a fire-tube which lengthens the path of the hot gases, thus augmenting the heating surface which can be further increased by making the gases reverse direction through a second parallel tube or a bundle of multiple tubes (two-pass or return flue boiler); alternatively the gases may be taken along the sides and then beneath the boiler through flues (3-pass boiler). In the case of a locomotive-type boiler, a boiler barrel extends from the firebox and the hot gases pass through a bundle of fire tubes inside the barrel which greatly increase the heating surface compared to a single tube and further improve heat transfer. Fire-tube boilers usually have a comparatively low rate of steam production, but high steam storage capacity. Fire-tube boilers mostly burn solid fuels, but are readily adaptable to those of the liquid or gas variety.
- Water-tube boiler. In this type, the water tubes are arranged inside a furnace in a number of possible configurations: often the water tubes connect large drums, the lower ones containing water and the upper ones, steam and water; in other cases, such as a monotube boiler, water is circulated by a pump through a succession of coils. This type generally gives high steam production rates, but less storage capacity than the above. Water tube boilers can be designed to exploit any heat source and are generally preferred in high pressure applications since the high pressure water/steam is contained within small

diametre pipes which can withstand the pressure with a thinner wall.

- Flash boiler. A specialised type of water-tube boiler.

Figure 19: 1950s design steam locomotive boiler, from a Victorian Railways J class

- Fire-tube boiler with Water-tube firebox. Sometimes the two above types have been combined in the following manner: the firebox contains an assembly of water tubes, called thermic siphons. The gases then pass through a conventional firetube boiler. Water-tube fireboxes were installed in many Hungarian locomotives, but have met with little success in other countries.
- Sectional boiler. In a cast iron sectional boiler, sometimes called a "pork chop boiler" the water is contained inside cast iron sections. These sections are assembled on site to create the finished boiler.

Safety

Historically, boilers were a source of many serious injuries and property destruction due to poorly understood engineering principles. Thin and brittle metal shells can rupture, while poorly welded or riveted seams could open up, leading to a violent eruption of the pressurised steam. When water is converted to steam it expands in volume over 1,000 times and travels a down a steam pipes at over 100 kilometres/hr. Because of this steam is a great way of moving energy and heat around a site from a central boiler house to where it is needed, but without the right boiler feed water treatment, a steam-raising plant will suffer from scale formation and corrosion. At best, this increases energy costs and can lead to poor quality steam, reduced efficiency, shorter plant life and an operation which is unreliable. At worst, it can lead to catastrophic failure and loss of life. Collapsed or dislodged boiler tubes could also spray scalding-hot steam and smoke out of the air intake and firing chute, injuring the firemen who loaded coal into the fire chamber. Extremely large boilers providing hundreds of horsepower to operate factories could demolish entire buildings.

A boiler that has a loss of feed water and is permitted to boil dry can be extremely dangerous. If feed water is then sent into the empty boiler, the small cascade of incoming water instantly boils on contact with the superheated metal shell and leads to a violent explosion that cannot be controlled even by safety steam valves. Draining of the boiler could also occur if a leak occurred in the steam supply lines that was larger than the make-up water supply could replace. The *Hartford Loop* was invented in 1919 by the Hartford Steam Boiler and Insurance Company as a method to help prevent this condition from occurring, and thereby reduce their insurance claims.

Superheated Steam Boilers

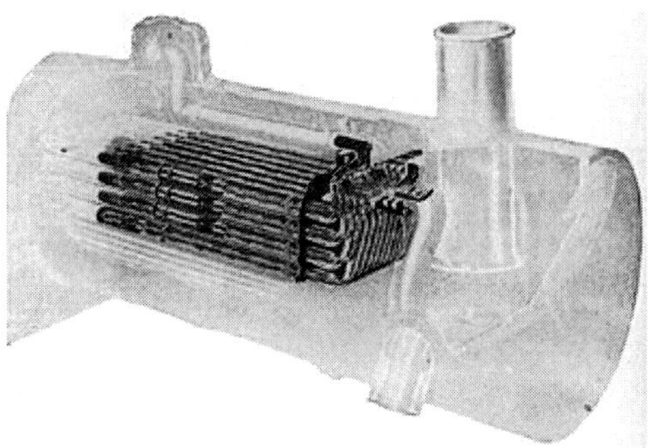

Figure 20: A superheated boiler on a steam locomotive.

Most boilers produce steam to be used at saturation temperature; that is, saturated steam. Superheated steam boilers vapourise the water and then further heat the steam in a *superheater*. This provides steam at much higher temperature, but can decrease the overall thermal efficiency of the steam generating plant because the higher steam temperature requires a higher flue gas exhaust temperature. There are several ways to circumvent this problem, typically by providing an *economizer* that heats the feed water, a combustion air heater in the hot flue gas exhaust path, or both. There are advantages to superheated steam that may, and often will, increase overall efficiency of both steam generation and its utilisation: gains in input temperature to a turbine should outweigh any cost in additional boiler complication and expense. There may also be practical limitations in using *wet* steam, as entrained condensation droplets will damage turbine blades.

Superheated steam presents unique safety concerns because, if any system component fails and allows steam to escape, the high pressure and temperature can cause serious, instantaneous harm to anyone in its path. Since the escaping steam will initially be completely superheated vapour, detection can be difficult, although the intense heat and sound from such a leak clearly indicates its presence.

Superheater operation is similar to that of the coils on an air conditioning unit, although for a different purpose. The steam piping is directed through the flue gas path in the boiler furnace. The temperature in this area is typically between 1300–1600 degrees Celsius (2372–2912 °F). Some superheaters are radiant type; that is, they absorb heat by radiation. Others are convection type, absorbing heat from a fluid. Some are a combination of the two types. Through either method, the extreme heat in the flue gas path will also heat the superheater steam piping and the steam within. While the temperature of the steam in the superheater rises, the pressure of the steam does not: the turbine or moving pistons offer a *continuously expanding space* and the pressure remains the same as that of the boiler. Almost all steam superheater system designs remove droplets entrained in the steam to prevent damage to the turbine blading and associated piping.

Supercritical Steam Generator

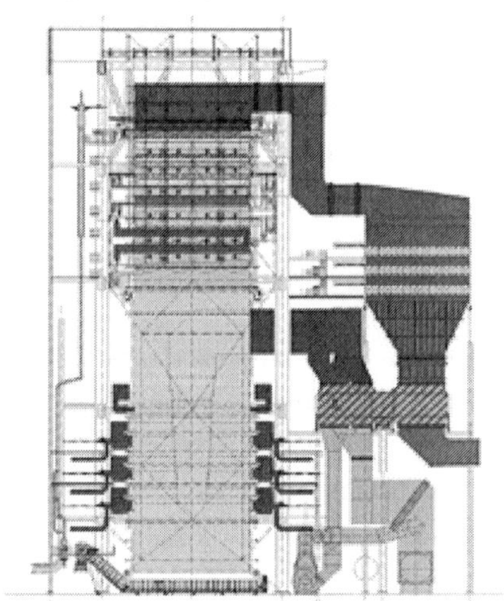

Figure 21: Steam generation power plant.

Supercritical steam generators are frequently used for the production of electric power. They operate at supercritical pressure. In contrast to a "subcritical boiler", a supercritical steam generator operates at such a high pressure that actual boiling ceases to occur, the boiler has no liquid water - steam separation. There is no generation of steam bubbles within the water, because the pressure is above the critical pressure at which steam bubbles can form. It passes below the critical point as it does work in a high pressure turbine and enters the generator's condenser. This results in slightly less fuel use and therefore less greenhouse gas production. The term "boiler" should not be used for a supercritical pressure steam generator, as no "boiling" actually occurs in this device.

Hydronic Boilers

Hydronic boilers are used in generating heat for residential and industrial purposes. They are the typical power plant for central heating systems fitted to houses in northern Europe (where they are commonly combined with domestic water heating), as opposed to the forced-air furnaces or wood burning stoves more common in North America. The hydronic boiler operates by way of heating water/fluid to a preset temperature (or sometimes in the case of single pipe systems, until it boils and turns to steam) and circulating that fluid throughout the home typically by way of radiators, baseboard heaters or through the floors. The fluid can be heated by any means...gas, wood, fuel oil, etc., but in built-up areas where piped gas is available, natural gas is currently the most economical and therefore the usual choice. The fluid is in an enclosed system and circulated throughout by means of a motorized pump.

The name "boiler" can be a misnomer in that, except for systems using steam radiators, the water in a properly functioning hydronic boiler never actually boils. Some new systems are fitted with condensing boilers for greater efficiency. These boilers are referred to as condensing boilers because the boilers are designed to operate at lower temperatures than typical hydronic boilers. These lower temperatures would typically damage a boiler because it would cause condensation in the flue gases which ultimately form carbonic acid which corrodes the flue and fireside boiler heating surfaces. Condensing boilers solve this problem by routing the carbonic acid down the drain and by making the flue out of stainless steel or PVC. Although condensing boilers are becoming more popular, they are still less common than other types of hydronic boilers because they are more expensive.

Hydronic systems are being used more and more in new construction in North America for several reasons. Among the reasons are:

- They are more efficient and more economical than forced-air systems (although initial installation can be more expensive, because of the cost of the copper and aluminium).
- The baseboard copper pipes and aluminium fins take up less room and use less metal than the bulky steel ductwork required for forced-air systems.
- They provide more even, less fluctuating temperatures than forced-air systems. The copper baseboard pipes hold and release heat over a longer period of time than air does, so the furnace does not have to switch off and on as much. (Copper heats mostly through conduction and radiation, whereas forced-air heats mostly through forced convection. Air has much lower thermal conductivity and volumetric heat capacity than copper, so the conditioned space warms up and cools down more quickly than with hydronic.)
- They tend to not dry out the interior air as much as forced air systems, but this is not always true. When forced air duct systems are air-sealed properly, and have return-air paths back to the furnace (thus reducing pressure differentials and therefore air movement between inside and outside the house), this is not an issue.
- They do not introduce any dust, allergens, mould, or (in the case of a faulty heat exchanger) combustion byproducts into the living space.

Forced-air heating does have some advantages, however.

Accessories

Boiler Fittings and Accessories

- Safety valve: It is used to relieve pressure and prevent possible explosion of a boiler.
- Water level indicators: They show the operator the level of fluid in the boiler, also known as a sight glass, water gauge or water column is provided.
- Bottom blowdown valves: They provide a means for removing solid particulates that condense and lie on the bottom of a

boiler. As the name implies, this valve is usually located directly on the bottom of the boiler, and is occasionally opened to use the pressure in the boiler to push these particulates out.

- Continuous blowdown valve: This allows a small quantity of water to escape continuously. Its purpose is to prevent the water in the boiler becoming saturated with dissolved salts. Saturation would lead to foaming and cause water droplets to be carried over with the steam - a condition known as priming. Blowdown is also often used to monitor the chemistry of the boiler water.
- Flash Tank: High pressure blowdown enters this vessel where the steam can 'flash' safely and be used in a low-pressure system or be vented to atmosphere while the ambient pressure blowdown flows to drain.
- Automatic Blowdown/Continuous Heat Recovery System: This system allows the boiler to blowdown only when makeup water is flowing to the boiler, thereby transferring the maximum amount of heat possible from the blowdown to the makeup water. No flash tank is generally needed as the blowdown discharged is close to the temperature of the makeup water.
- Hand holes: They are steel plates installed in openings in "header" to allow for inspections & installation of tubes and inspection of internal surfaces.
- Steam drum internals, A series of screen, scrubber & cans (cyclone separators).
- Low- water cutoff: It is a mechanical means (usually a float switch) that is used to turn off the burner or shut off fuel to the boiler to prevent it from running once the water goes below a certain point. If a boiler is "dry-fired" (burned without water in it) it can cause rupture or catastrophic failure.
- Surface blowdown line: It provides a means for removing foam or other lightweight non-condensible substances that tend to float on top of the water inside the boiler.
- Circulating pump: It is designed to circulate water back to the boiler after it has expelled some of its heat.
- Feedwater check valve or clack valve: A non-return stop valve in the feedwater line. This may be fitted to the side of the boiler, just below the water level, or to the top of the boiler.

- Top feed: A check valve (clack valve) in the feedwater line, mounted on top of the boiler. It is intended to reduce the nuisance of limescale. It does not prevent limescale formation but causes the limescale to be precipitated in a powdery form which is easily washed out of the boiler.
- Desuperheater tubes or bundles: A series of tubes or bundles of tubes in the water drum or the steam drum designed to cool superheated steam. Thus is to supply auxiliary equipment that does not need, or may be damaged by, dry steam.
- Chemical injection line: A connection to add chemicals for controlling feedwater pH.

Steam Accessories

- Main steam stop valve:
- Steam traps:
- Main steam stop/Check valve: It is used on multiple boiler installations.

Combustion Accessories

- Fuel oil system:
- Gas system:
- Coal system:
- Soot blower

Other Essential Items

- Pressure gauges:
- Feed pumps:
- Fusible plug:
- Inspectors test pressure gauge attachment:
- Name plate:
- Registration plate:

Controlling Draught

Most boilers now depend on mechanical draught equipment rather than natural draught. This is because natural draught is subject to outside air conditions and temperature of flue gases leaving the furnace, as well as the chimney height. All these factors make proper draught hard to attain and therefore make mechanical draught equipment much more economical.

There are three types of mechanical draught:

Induced Draught: This is obtained one of three ways, the first being the "stack effect" of a heated chimney, in which the flue gas is less dense than the ambient air surrounding the boiler. The denser column of ambient air forces combustion air into and through the boiler. The second method is through use of a steam jet. The steam jet oriented in the direction of flue gas flow induces flue gasses into the stack and allows for a greater flue gas velocity increasing the overall draught in the furnace. This method was common on steam driven locomotives which could not have tall chimneys. The third method is by simply using an induced draught fan (ID fan) which removes flue gases from the furnace and forces the exhaust gas up the stack. Almost all induced draught furnaces operate with a slightly negative pressure.

Forced Draught: Draught is obtained by forcing air into the furnace by means of a fan (FD fan) and ductwork. Air is often passed through an air heater; which, as the name suggests, heats the air going into the furnace in order to increase the overall efficiency of the boiler. Dampers are used to control the quantity of air admitted to the furnace. Forced draught furnaces usually have a positive pressure.

Balanced Draught: Balanced draught is obtained through use of both induced and forced draught. This is more common with larger boilers where the flue gases have to travel a long distance through many boiler passes. The induced draught fan works in conjunction with the forced draught fan allowing the furnace pressure to be maintained slightly below atmospheric.

Chapter 10

Photovoltaic System

Figure 1: "Solar park" or "PV farm"

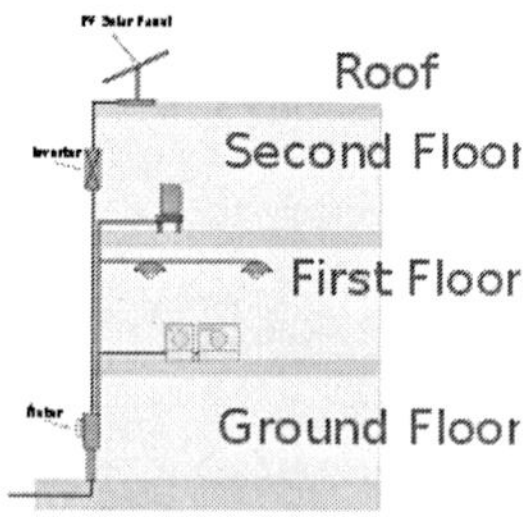

Figure 2: A grid-tied residential PV system

Figure 3: PV array on an old house

A photovoltaic system (or PV system) is a system which uses one or more solar panels to convert sunlight into electricity. It consists of multiple components, including the photovoltaic modules, mechanical and electrical connections and mountings and means of regulating and/or modifying the electrical output.

Photovoltaic Modules

Due to the low voltage of an individual solar cell (typically ca. 0.5V), several cells are wired in series in the manufacture of a "laminate". The laminate is assembled into a protective weatherproof enclosure, thus making a photovoltaic module or solar panel. Modules may then be strung together into a photovoltaic array. The electricity generated can be either stored, used directly (island/standalone plant) or fed into a large electricity grid powered by central generation plants (grid-connected/grid-tied plant) or combined with one or many domestic electricity generators to feed into a small grid (hybrid plant). Depending on the type of application, the rest of the system ("balance of system" or "BOS") consists of different components. The BOS depends on the load profile and the system type. Systems are generally designed in order to ensure the highest energy yield for a given investment.

Photovoltaic Arrays

Figure 4: A photovoltaic array is a linked assembly of PV modules.

Figure 5: The solar panels on this small yacht at sea can charge the 12 volt batteries at up to 9 amperes in full, direct sunlight.

A *photovoltaic array* (or *solar array*) is a linked collection of solar panels.

The power that one module can produce is seldom enough to meet requirements of a home or a business, so the modules are linked together to form an *array*. Most PV arrays use an inverter to convert the DC power produced by the modules into alternating current that can power lights, motors, and other loads. The modules in a PV array are usually first connected in series to obtain the desired voltage; the individual strings are then connected in parallel to allow the system to produce more current. Solar arrays are typically measured under STC (standard test conditions) or PTC (PVUSA test conditions), in watts, kilowatts, or even megawatts.

Costs of production have been reduced in recent years for more widespread use through production and technological advances. One source claims the cost in February 2006 ranged \$3–10/watt while a similar size is said to have cost \$8–10/watt in February 1996, depending on type. For example, crystal silicon solar cells have largely been replaced by less expensive multicrystalline silicon solar cells, and thin film silicon solar cells have also been developed recently at lower costs of production. Although they are reduced in energy conversion efficiency from single crystalline "siwafers", they are also much easier to produce at comparably lower costs.

Applications—Standalone Systems

Figure 6: Solar powered parking metre.

A standalone system does not have a connection to the electricity "mains" (aka "grid"). Standalone systems vary widely in size and application from wristwatches or calculators to remote buildings or spacecraft. If the load is to be supplied independently of solar insolation, the generated power is stored and buffered with a battery. In non-portable applications where weight is not an issue, such as in buildings, lead acid batteries are most commonly used for their low cost. A charge controller may be incorporated in the system to: a) avoid battery damage by excessive charging or discharging and, b) optimising the production of the cells or modules by maximum power point tracking (MPPT).

However, in simple PV systems where the PV module voltage is matched to the battery voltage, the use of MPPT electronics is generally considered unnecessary, since the battery voltage is stable enough to provide near-maximum power collection from the PV module. In small devices (e.g. calculators, parking metres) only direct current (DC) is consumed. In larger systems (e.g. buildings, remote water pumps) AC is usually required. To convert the DC from the modules or batteries into AC, an inverter is used.

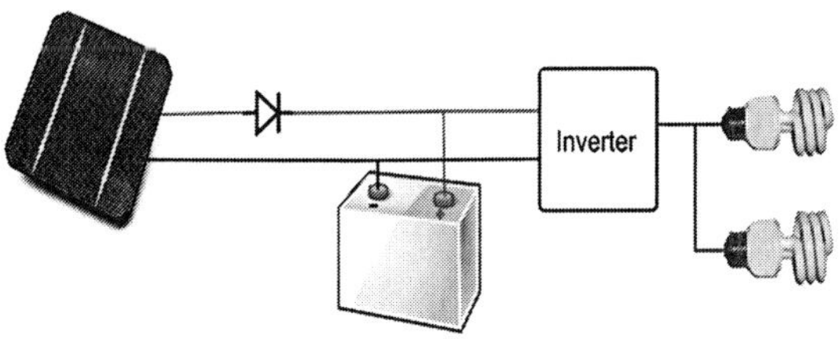

Figure 7: A schematic of a bare-bones off-grid system, consisting (from left to right) of photovoltaic module, a blocking-diode to prevent battery drain during low-insolation, a battery, an inverter, and an AC load such as a fluorescent lamp

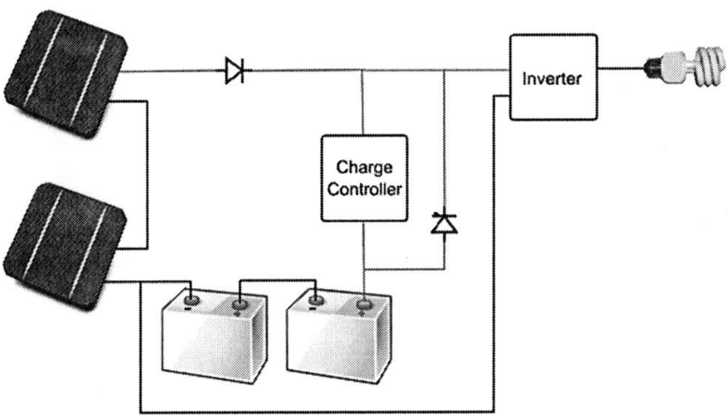

Figure 8: off-grid PV system with battery charger

Solar Vehicles

Ground, water, air or space vehicles may obtain some or all of the energy required for their operation from the sun. Surface vehicles generally require higher power levels than can be sustained by a practically-sized solar array, so a battery is used to meet peak power demand, and the solar array recharges it. Space vehicles have successfully used solar photovoltaic systems for years of operation, eliminating the weight of fuel or primary batteries.

Small Scale DIY Solar Systems

With a growing DIY-community and an increasing interest in environmentally friendly "green energy", some hobbyists have endeavoured to build their own PV solar systems from kits or partly diy. Usually, the DIY-community uses inexpensive and/or high efficiency systems (such as those with solar tracking) to generate their own power. As a result, the DIY-systems often end up cheaper than their commercial counterparts. Often, the system is also hooked up into the regular power grid to repay part of the investment via net metreing.

These systems usually generate power amount of ~2 kW or less. Through the internet, the community is now able to obtain plans to construct the system (at least partly DIY) and there is a growing trend toward building them for domestic requirements. The DIY-PV solar systems are now also being used both in developed countries and in developing countries, to power residences and small businesses.

Grid-connected System

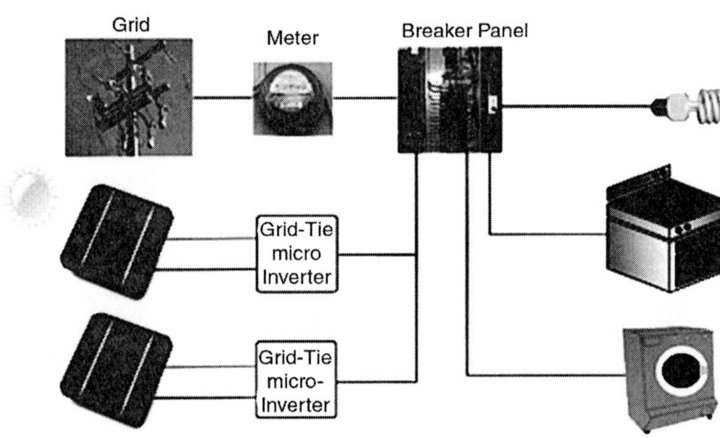

Figure 9: Diagram of a residential grid-connected PV system

A grid connected system is connected to a large independent grid (typically the public electricity grid) and feeds power into the grid. Grid connected systems vary in size from residential (2-10kWp) to solar power stations (up to 10s of GWp). This is a form of decentralised electricity generation.

In the case of residential or building mounted grid connected PV systems, the electricity demand of the building is met by the PV system. Only the excess is fed into the grid when there is an excess. The feeding of electricity into the grid requires the transformation of DC into AC by a special, grid-controlled solar inverter.

In kW sized installations the DC side system voltage is as high as permitted (typically 1000V except US residential 600V) to limit ohmic losses. Most modules (72 crystalline silicon cells) generate about 160W at 36 volts. It is sometimes necessary or desirable to connect the modules partially in parallel rather than all in series. One set of modules connected in series is known as a 'string'.

Building Systems

In urban and suburban areas, photovoltaic arrays are commonly used on rooftops to supplement power use; often the building will have a connection to the power grid, in which case the energy produced by the PV array can be sold back to the utility in some sort of net metreing agreement. Solar trees are arrays that, as the name implies, mimic the look of trees, provide shade, and at night can function as street lights. In agricultural settings, the array may be used to directly power DC pumps, without the need for an inverter. In remote settings such as mountainous areas, islands, or other places where a power grid is unavailable, solar arrays can be used as the sole source of electricity, usually by charging a storage battery.

There is financial support available for people wishing to install PV arrays. In the UK, households are paid a 'Feedback Fee' to buy excess electricity at a flat rate per kWh. This is up to 44.3p/kWh which can allow a home to earn double their usual annual domestic electricity bill. The current UK feed-in tariff system is due for review on 31 March 2012, after which the current scheme may no longer be available.

Power Plants

A *photovoltaic power station* is a power station using photovoltaic modules and inverters for utility scale electricity generation, connected to an electricity transmission grid. Some large photovoltaic power stations like Waldpolenz Solar Park cover a significant area and have a maximum power output of 40-60 MW.

System Performance

Insolation and Energy: At high noon on a cloudless day at the equator, the power of the sun is about 1 kW/m^2, on the Earth's surface, to a plane that is perpendicular to the sun's rays. As such, PV arrays can track the sun through each day to greatly enhance energy collection. However, tracking devices add cost, and require maintenance, so it is more common for PV arrays to have fixed mounts that tilt the array and face due South in the Northern Hemisphere (in the Southern Hemisphere, they should point due North). The tilt angle, from horizontal, can be varied for season, but if fixed, should be set to give optimal array output during the peak electrical demand portion of a typical year.

For the weather and latitudes of the United States and Europe, typical insolation ranges from 4 kWh/m^2/day in northern climes to 6.5 kWh/m^2/day in the sunniest regions. Typical solar panels have an average efficiency of 12%, with the best commercially available panels at 20%. Thus, a photovoltaic installation in the southern latitudes of Europe or the United States may expect to produce 1 kWh/m^2/day. A typical "150 watt" solar panel is about a square metre in size. Such a panel may be expected to produce 1 kWh every day, on average, after taking into account the weather and the latitude. In the Sahara desert, with less cloud cover and a better solar angle, one could ideally obtain closer to 8.3 kWh/m^2/day provided the nearly ever present wind would not blow sand on the units. The unpopulated area of the Sahara desert is over 9 million km^2, which if covered with solar panels would provide 630 terawatts total power. The Earth's current energy consumption rate is around 13.5 TW at any given moment (including oil, gas, coal, nuclear, and hydroelectric).

Tracking the Sun

Trackers and sensors to optimise the performance are often seen as optional, but tracking systems can increase viable output by up to 100%. PV arrays that approach or exceed one megawatt often use solar trackers. Accounting for clouds, and the fact that most of the world is not on the equator, and that the sun sets in the evening, the correct measure of solar power is insolation – the average number of kilowatt-hours per square metre per day. For the weather and latitudes of the United States and Europe, typical insolation ranges from 4kWh/m^2/day in northern climes to 6.5 kWh/m^2/day in the sunniest regions.

For large systems, the energy gained by using tracking systems outweighs the added complexity (trackers can increase efficiency by 30% or more).

Shading and Dirt

Photovoltaic cell electrical output is extremely sensitive to shading. When even a small portion of a cell, module, or array is shaded, while the remainder is in sunlight, the output falls dramatically due to internal 'short-circuiting' (the electrons reversing course through the shaded portion of the p-n junction).

If the current drawn from the series string of cells is no greater than the current that can be produced by the shaded cell, the current (and so power) developed by the string is limited. If enough voltage is available from the rest of the cells in a string, current will be forced through the cell by breaking down the junction in the shaded portion. This breakdown voltage in common cells is between 10 and 30 volts. Instead of adding to the power produced by the panel, the shaded cell absorbs power, turning it into heat. Since the reverse voltage of a shaded cell is much greater than the forward voltage of an illuminated cell, one shaded cell can absorb the power of many other cells in the string, disproportionately affecting panel output. For example, a shaded cell may drop 8 volts, instead of adding 0.5 volts, at a particular current level, thereby absorbing the power produced by 16 other cells. Therefore it is extremely important that a PV installation is not shaded at all by trees, architectural features, flag poles, or other obstructions.

Most modules have bypass diodes between each cell or string of cells that minimise the effects of shading and only lose the power of the shaded portion of the array (The main job of the bypass diode is to eliminate hot spots that form on cells that can cause further damage to the array, and cause fires.). Sunlight can be absorbed by dust, snow, or other impurities at the surface of the module. This can cut down the amount of light that actually strikes the cells by as much as half. Maintaining a clean module surface will increase output performance over the life of the module.

Temperature

Module output and life are also degraded by increased temperature. Allowing ambient air to flow over, and if possible behind, PV modules reduces this problem.

Module Efficiency

In 2010, solar panels available for consumers can have a yield of up to 19%, while commercially available panels can go as far as 27%. Thus, a photovoltaic installation in the southern latitudes of

Europe or the United States may expect to produce 1 kWh/m²/day. A typical "150 watt" solar panel is about a square metre in size. Such a panel may be expected to produce 1 kWh every day, on average, after taking into account the weather and the latitude.

Monitoring

Photovoltaic systems need to be monitored to detect breakdown and optimise their operation. Several photovoltaic monitoring strategies depending on the output of the installation and its nature. Monitoring can be performed on site or remotely. It can measure production only, retrieve all the data from the inverter or retrieve all of the data from the communicating equipment (probes, metres, etc.). Monitoring tools can be dedicated to supervision only or offer additional functions. Individual inverters may include monitoring using manufacturer specific protocols and software.

Energy metreing of an inverter may be of limited accuracy and not suitable for revenue metreing purposes. A third-party data acquistion system can monitor multiple inverters, using the inverter manufacturer's protocols, and also acquire weather-related information. Independent smart metres may measure the total energy production of a PV array system. Separate measures such as satellite image analaysis or a solar radiation metre (a pyranometre) can be used to estimate total insolation.

Data collected from a monitoring system can be displayed remotely over the World Wide Web. Some companies offer analysis software to analyse system performance. Small residential systems may have minimal data analysis requirements other than perhaps total energy production; larger grid-connected power plants can benefit from more detailed investigations of performance.

Performance Factors

Uncertainties in revenue over time relate mostly to the evaluation of the solar resource and to the performance of the system itself. In the best of cases, uncertainties are typically 4% for year-to-year climate variability, 5% for solar resource estimation (in a horizontal plane), 3% for estimation of irradiation in the plane of the array, 3% for power rating of modules, 2% for losses due to dirt and soiling, 1.5% for losses due to snow, and 5% for other sources of error. Identifying and reacting to manageable losses is critical for revenue and O&M efficiency. Monitoring of array performance may be part of contractual agreements between the array owner, the builder, and the utility purchasing the energy produced.

Access to the Internet has allowed a further improvement in energy monitoring and communication. Dedicated systems are available from a number of vendors. For solar PV system that use microInverters (panel-level DC to AC conversion), module power data is automatically provided. These solutions provide data for the system owner and the installer. Installers are able to remotely monitor multiple installations, and see at-a-glance the status of their entire installed base.

Components

Trackers: A solar tracker tilts a solar panel throughout the day. Depending on the type of tracking system, the panel is either aimed directly at the sun or the brightest area of a partly clouded sky. Trackers greatly enhance early morning and late afternoon performance, substantially increasing the total amount of power produced by a system. Trackers are effective in regions that receive a large portion of sunlight directly. In diffuse light (i.e. under cloud or fog), tracking has little or no value. Because most concentrated photovoltaics systems are very sensitive to the sunlight's angle, tracking systems allow them to produce useful power for more than a brief period each day. Tracking systems improve performance for two main reasons. First, when a solar panel is perpendicular to the sunlight, the light it receives is more intense than it would be if angled. Second, direct light is used more efficiently than angled light. Special Anti-reflective coatings can improve solar panel efficiency for direct and angled light, somewhat reducing the benefit of tracking.

Inverters

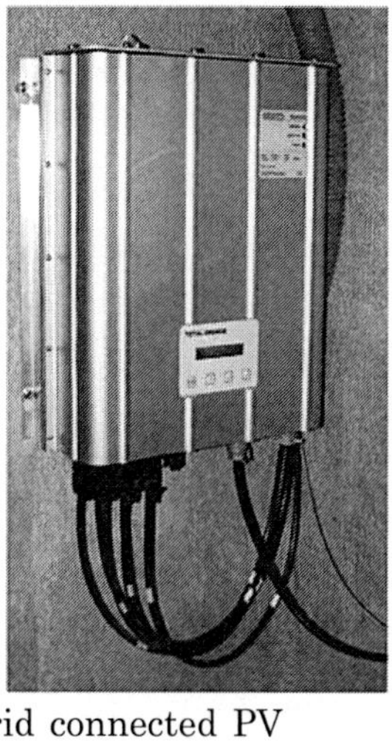

Figure 10: Inverter for grid connected PV

On the AC side, these inverters must supply electricity in sinusoidal form, synchronised to the grid frequency, limit feed in voltage to no higher than the grid voltage including disconnecting from the grid if the grid voltage is turned off.

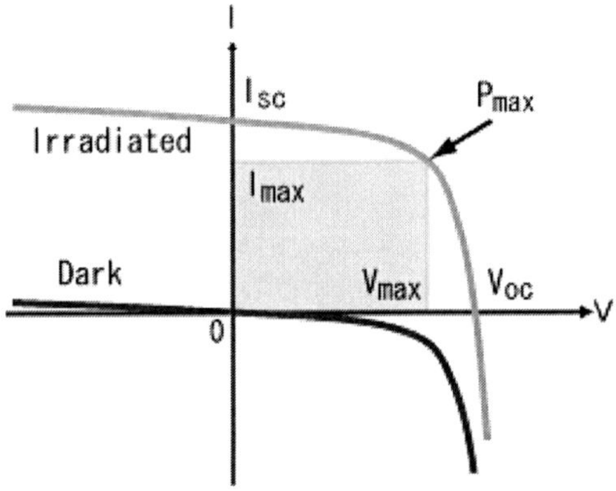

Figure 11.

On the DC side, the power output of a module varies as a function of the voltage in a way that power generation can be optimised by varying the system voltage to find the 'maximum power point'. Most inverters therefore incorporate 'maximum power point tracking'.

A solar inverter may connect to a string of solar panels. In small installations a solar micro-inverter is connected at each solar panel.

For safety reasons a circuit breaker is provided both on the AC and DC side to enable maintenance. AC output may be connected through an electricity metre into the public grid.

The metre must be able to run in both directions.

In some countries, for installations over 30kWp a frequency and a voltage monitor with disconnection of all phases is required.

Mounting Systems

Modules are assembled into arrays on some kind of mounting system. For solar parks a large rack is mounted on the ground, and the modules mounted on the rack.

For buildings, many different racks have been devised for pitched roofs. For flat roofs, racks, bins and building integrated solutions are used.

Connection to a DC Grid

DC grids are only to be found in electric powered transport: railways trams and trolleybuses. A few pilot plants for such applications have been built, such as the tram depots in Hannover Leinhausen and Geneva (Bachet de Pesay). The 150 kW_p Geneva site feeds 600V DC directly into the tram/trolleybus electricity network provided about 15% of the electricity at its opening in 1999.

Hybrid Systems

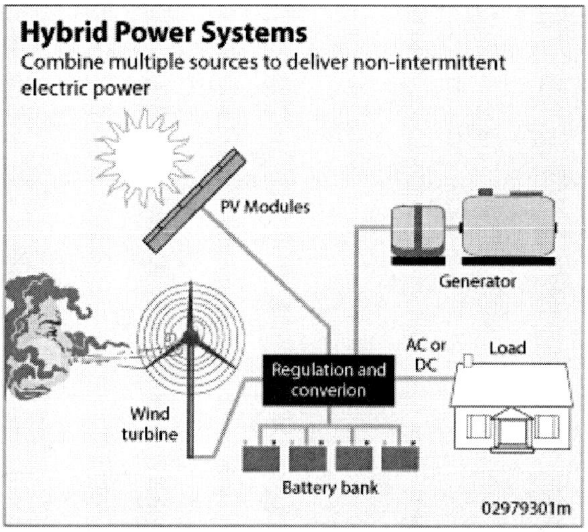

Figure 12: A hybrid system combines PV with other forms of generation, usually a diesel generator. Biogas is also used. The other form of generation may be a type able to modulate power output as a function of demand. However more than one renewable form of energy may be used e.g. wind. The photovoltaic power generation serves to reduce the consumption of non renewable fuel. Hybrid systems are most often found on islands. Pellworm island in Germany and Kythnos island in Greece are notable examples (both are combined with wind). The Kythnos plant has diocane diesel consumption by 11.2%

There has also been recent work showing that the PV penetration limit can be increased by deploying a distributed network of PV+CHP hybrid systems in the U.S. The temporal distribution of solar flux, electrical and heating requirements for representative U.S. single family residences were analysed and the results clearly show that hybridizing CHP with PV can enable additional PV deployment above what is possible with a conventional centralised electric generation system. This theory was reconfirmed with numerical simulations using per second solar flux data to determine that the necessary battery backup to provide for such a hybrid system is possible with relatively small and inexpensive battery systems. In addition, large PV+CHP systems are possible for institutional buildings, which again provide back up for intermittent PV and reduce CHP runtime.

Standardisation

Increasing use of photovoltaic systems and integration of photovoltaic power into existing structures and techniques of supply and distribution increases the value of general standards and definitions for photovoltaic components and systems. The standards are compiled at the International Electrotechnical Commission (IEC) and apply to efficiency, durability and safety of cells, modules, simulation programs, plug connectors and cables, mounting systems, overall efficiency of inverters etc.

Legality of Photovoltaic Systems

The State of California prohibits Homeowners' associations from restricting solar devices.

Many localities require a license to install a photovoltaic system. A grid-tied system normally requires a licensed electrician to make the connection between the system and the grid-connected wiring of the building.

List of Photovoltaic Power Stations

Figure 13: Nellis Solar Power Plant in the United States.

Figure 14: 80 MW Ohotnikovo Solar Park in Ukraine.

The following page lists photovoltaic power stations that are larger than 25 MW in current net capacity. Solar photovoltaic cells convert sunlight into electricity and many solar photovoltaic power stations have been built, mainly in Europe. As of October 2011, the largest photovoltaic (PV) power plants in the world are the Sarnia Photovoltaic Power Plant (Canada, 97 MW), Montalto di Castro Photovoltaic Power Station (Italy, 84.2 MW), Finsterwalde Solar Park (Germany, 80.7 MW), Ohotnikovo Solar Park (Ukraine, 80 MW), Lieberose Photovoltaic Park (Germany, 71.8 MW), Rovigo Photovoltaic Power Plant (Italy, 70 MW), Olmedilla Photovoltaic Park (Spain, 60 MW), and the Strasskirchen Solar Park (Germany, 54 MW).

The Blythe Solar Power Project is a 500 MW photovoltaic station under construction in Riverside County, California. The Agua Caliente Solar Project is a 290 megawatt photovoltaic solar generating facility being built in Yuma County, Arizona. The California Valley Solar Ranch (CVSR) is a 250 megawatt (MW) solar photovoltaic power plant, which is being built by SunPower in the Carrizo Plain, northeast of California Valley. The 230 MW Antelope Valley Solar Ranch is a First Solar photovoltaic project which is under construction in the Antelope Valley area of the Western Mojave Desert, and due to be completed in 2013. Many of these plants are integrated with agriculture and some use innovative tracking systems that follow the sun's daily path across the sky to generate more electricity than conventional fixed-mounted systems. There are no fuel costs or emissions during operation of the power stations.

Powder River Basin

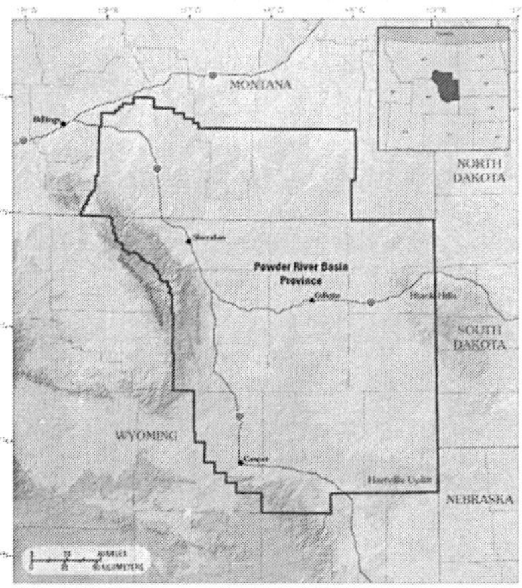

Figure 18: The Powder River Basin

The Powder River Basin is a geologic region in southeast Montana and northeast Wyoming, about 120 miles (190 km) east to west and 200 miles (320 km) north to south, known for its coal deposits. The region supplies about 40 percent of coal in the United States.

It is both a topographic drainage and geologic structural basin. The basin is so named because it is drained by the Powder River, although it is also drained in part by the Cheyenne River, Tongue River, Bighorn River, Little Missouri River, Platte River, and their tributaries.

While the Powder River Basin is unified geologically, residents of areas distant from the Powder River proper do not think of themselves as living in a single geographic region.

For example, residents of the Crow Reservation in the Big Horn River watershed, or of Sheridan in the Tongue River watershed, would locate the Powder River Basin as the region east of the Big Horn Mountains, using a definition based on watershed and topography.

Major cities in the area include Gillette and Sheridan, Wyoming and Miles City, Montana. Outside of these main towns, the area is very sparsely populated and is known for its rolling grasslands and semiarid climate.

It is the single largest source of coal mined in the United States, and contains one of the largest deposits of coal in the world. Most of the active coal mining in the Powder River Basin actually takes place in drainages of the Cheyenne River.

Because of the Powder River Basin, Wyoming has been the top coal-producing state in the United States since 1988. In 2007, the Powder River Basin alone produced 436 million short tons (396 million tonnes) of coal, more than twice the production of second-place West Virginia, and more than the entire Appalachian region.

The Black Thunder Coal Mine is the most productive coal mine in the United States; in 2006 this single mine produced 84 million metric tons of coal, more than any state except Wyoming, West Virginia, and Kentucky.

The region is also a major producer of natural gas through a process called coal-bed methane extraction. Underground coal seams are dewatered through pumping, which reduces pressure and releases trapped gases up to the well head.

Geologic History

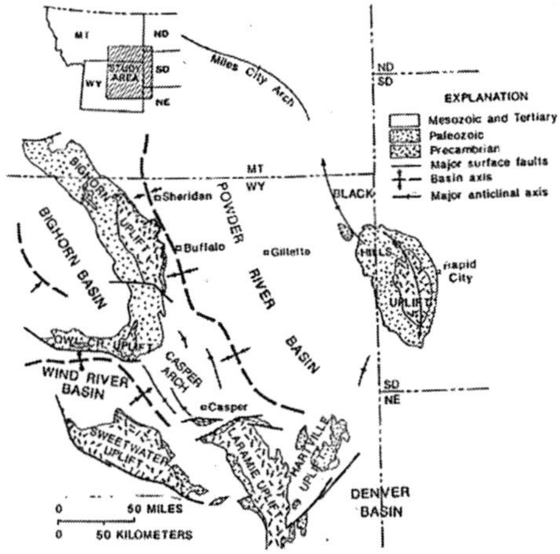

Figure 19: Powder River Basin and surrounding uplifts (US Geological Survey)

The Powder River Basin contains a section of Phanerozoic rocks up to 17,000 feet (5,200 m) thick, from Cambrian to Holocene.

Cretaceous

The thickest section of the Powder River Basin is composed of Cretaceous rocks, an overall regressive sequence of mostly marine shales and sandstones deposited in the Western Interior Seaway.

Tertiary

The coal beds of the region began to form about 60 million years ago when the land began rising from a shallow sea. The rise of the Black Hills uplift on the east and the Hartville uplift on the southeast side of the basin created the present outline of the Powder River Basin.

When the coal beds were forming, the climate in the area was subtropical, averaging about 120 inches (3,048.00 mm) of rainfall a year. For some 25 million years, the basin floor was covered with lakes and swamps. Because of the large area of the swamps, the organic material accumulated into peat bogs instead of being washed to the sea. Periodically the layers of peat were covered with sediments washed in from nearby mountains. Eventually the climate became drier and cooler. The basin filled with sediment and buried the peat under thousands of feet, compressing the layers of peat and forming coal. Over the last several million years, much of the overlying sediment has eroded away, leaving the coal seams near the surface.

Coal

Powder River Basin coal is classified as "sub-bituminous" and contains an average of approximately 8,500 btu/lb, with low SO_2. Contrast this with eastern, Appalachian bituminous coal containing an average of 12,500 btu/lb and high SO_2. PRB coal was essentially worthless until air pollution emissions from power plants became a concern.

A coal-fired plant designed to burn Appalachian coal must be modified to remove SO_2 at a cost estimated in 1999 to be around $113 per ton of SO_2 removal if it switches to burning PRB coal, and $322 per ton of SO_2 removal by installing scrubbers.

The Powder River Basin is the largest coal mining region in the United States, but most of the coal is buried too deeply to be economically accessible. The Powder River Basin coal beds are shaped like elongated bowls and as mines expand from east to west in the Powder River Basin, they will be going "down the sides of the bowl." This means that the overburden (rock lying over the coal) will increase as will the stripping ratio (the ratio of rock that needs to be moved to get to a ton of coal).

The United States Geological Survey (USGS) has conducted a series of studies on the economic accessibility of coal in the major coal producing regions of the country. The studies have typically found that only a small fraction of the coal will be economically accessible at the current price of $10.47/ton. In August 2008, the USGS issued an updated assessment of coal in the Powder River Basin. After considering stripping ratios and production costs, the USGS concluded that at the time of the economic evaluation, only 6 percent of the original resource, or 10.1 billion short tons of coal, was currently economically recoverable. At a price of $60/ton, roughly half (48%) of the coal is economic to produce.

Presently the approximately 15 mines in the Powder River Basin are working in areas where the stripping ratio is between 1:1 (i.e. one ton of rock for one ton of coal) and 3:1. As the mines expand the stripping ratio will increase. As more rock needs to be moved (using large electrically powered draglines and diesel and electric mining trucks) the production cost will also increase.

The United States uses about 1 billion tons of coal a year, with about 40 percent of the coal currently coming from the Powder River

Basin. The amount of coal coming from the Powder River Basin has been increasing over the last 20 years.

Increasing the price paid for coal can increase the amount of economically recoverable coal, but increasing the price of coal will also increase the production cost for the coal. In addition, because coal is a solid, not a liquid, it cannot be produced from many scattered wells as oil and gas can be. Rather, coal has to be produced from mines that expand slowly by moving massive quantities of overburden.

The mines in the Powder River Basin typically have less than 20 years of life remaining. Almost all of the coal in the Powder River Basin is federally owned and further mine expansions will require a series of federal and state approvals, as well as large investments in additional mine equipment to begin the excavations.

The majority of the coal mined in the Powder River Basin is part of the Fort Union Formation (Paleocene), with the low sulphur and ash content of the coal in the region making it very desirable. Coal supplies about half of the United States' electricity supplies, with the Powder River Basin mines supplying around 40 percent of the coal that fuels those stations, mainly to the east of the Rocky Mountains, for generating electricity.

Coal Mining Companies Currently Operating in the Powder River Basin

- Peabody Energy/Powder River Coal Company (North Antelope Rochelle Mine, Caballo Mine, Rawhide Mine)
- Arch Coal (Black Thunder Mine, Coal Creek Mine)
- Cloud Peak Energy(Cordero Rojo Mine Complex, Antelope Mine)
- Alpha Natural Resources (Belle Ayr Mine, Eagle Butte Mine)
- Kiewit Corporation (Buckskin Mine)
- Black Hills Corporation/Wyodak Resources Development (Wyodak Mine)
- Western Fuels Association (Dry Fork Mine)

Power Plants FuelLed from Powder River Basin Coal

- GRDA - Vinita, OK - (Chouteau, OK)
- State Line Generating Station - Dominion Resources - (Hammond, IN)

- Eckert Power Plant - Lansing Board of Water and Light - (Lansing, MI)
- Erickson Power Plant - Lansing Board of Water and Light - (Lansing, MI)
- Fayette Power Project - (La Grange, Texas)
- Monroe Power Plant - Detroit Edison (Monroe, Michigan)
- Pleasant Prairie Power Plant - Wisconsin Energy Corporation - (Pleasant Prairie, Wisconsin)
- Presque Isle Power Plant - Wisconsin Energy Corporation - (Marquette, Michigan)
- St. Clair Power Plant - Detroit Edison - (East China, Michigan)
- Gerald Gentleman Station - Nebraska Public Power District
- Omaha Public Power District - (Omaha, NE) (Nebraska City, NE)
- Arapahoe Station - Xcel Energy (Colorado)
- Pawnee Station - Xcel Energy (Colorado)
- Harrington Station - Xcel Energy (Texas)
- Newton Power Plant - Ameren, Newton, Illinois
- Flint Creek Power Plant - American Electric Power (Arkansas)
- J. Robert Welsh Power Plant - American Electric Power (Texas)
- James H. Miller Generating Station
- Jeffrey Energy Centre (Kansas)
- Comanche Station - Xcel Energy (Colorado)
- Allen S. King Plant - Xcel Energy (Minnesota)
- Robert W Scherer Power Plant - Georgia Power (Georgia)
- Sherburne County (Sherco) Plant - Xcel Energy (Minnesota)
- Laramie River Station (Basin electric Power Cooperative) - (Wheatland, Wyoming)
- Tolk Station - Xcel Energy (Texas)
- Big Cajun II - NRG Energy (New Roads, Louisiana)
- W. A. Parish Station - NRG Energy (Texas)
- Limestone Station - NRG Energy (Texas)
- Warrick Power Plant (Newburgh, IN)
- Centralia Power Plant (Centralia, WA)

- Nanticoke Generating Station - Ontario Power Generation (Nanticoke, Ontario Canada)
- Thunder Bay Generating Station - Ontario Power Generation (Thunder Bay, Ontario)
- Brandon Generating Station - Manitoba Hydro (Brandon, Manitoba)
- Sikeston Power Plant (Sikeston, MO)
- CP Crain Plant (Baltimore, Maryland)
- Drax Power Station, Selby, United Kingdom. Trials are ongoing with a UK sourced Bituminous Coal, blended to increase CV.

Petroleum

The Powder River Basin also contains major deposits of petroleum. The oil and gas are produced from rocks ranging from Pennsylvanian to Tertiary, but most comes from sandstones in the thick section of Cretaceous rocks.

Coalbed Methane

Recent controversy surrounds the extensive coalbed methane extraction in the region. In the last decade, nearly 7000 such wells have been drilled. An extensive network of gas pipeplines connecting these wells has been built, along with a series of pressurisation plants, as well as power lines to provide electricity to operate the system. In addition, thousands of miles of new access roads have been constructed.

Extracting the gas requires that water be pumped to the surface to release gas trapped in the coal seam. While some of the water is successfully utilised in agriculture production such as livestock water and crop irrigation, some waters are naturally high in salinity and sodium adsorption ratio. There has been controversy on how to best manage these saline waters. In 2007, Powder River Basin coalbed field produced 442 billion cubic feet of gas, making the field the 3rd largest source of natural gas in the United States.

Uranium

The region also contains major deposits of uranium, contained in sandstones. The Wasatch Formation (Eocene) contains the uranium ore "roll front" type deposits found in the Pumpkin Buttes District. Cameco Corporation subsidiary Power Resources Inc. operates uranium mines in the basin.

Transportation

In Spring 2005, coal extracted from the mines would retail at the mines for around $5 a ton. However, power stations and plants in the eastern United States were paying over $30 a ton - the difference caused by the cost of transportation. (In October 2008, the mine-mouth price of Powder River Basin coal was closer to $15 per ton.) To transport coal from the basin, there is a joint railway line owned by the BNSF Railway and the Union Pacific Railroad running the length of the southern section of the Powder River Basin, including the climb over Crawford Hill, Nebraska.

A third railroad, the Dakota, Minnesota and Eastern Railroad, has faced strong resistance from many parties for its attempts to extend its rail line into the coal mining area - more so since its purchase by the Canadian Pacific Railway.

History

Originally a single track Burlington Northern Railway line built in stages from 1972 to 1979, the rail line ran south from Donkey Junction in the north 13 miles (21 km) to Caballo, Wyoming; and then for 103 miles (166 km) to Shawnee in Converse County. The Chicago and North Western Railway ran close to the northern section, as did the Union Pacific at Caballo.

Having already formed the Western Railroad to distribute the coal, in 1982 C&NW and the UP formed Western Railroad Properties, Inc. (WRPI), to acquire half interest in the Burlington Northern coal line from Shawnee Junction to Coal Creek Junction. On December 15, 1986 WPRI purchased 11 miles (18 km) more of BN line from Coal Creek Junction to East Caballo Junction. Beginning June 27, 1983 WPRI constructed six miles (10 km) of new railroad from Shawnee Junction. to Shawnee, rebuilt 45 miles (72 km) of C&NW line from Shawnee to Crandall and 56 miles (90 km) of new railroad from Crandall to Joyce, Nebraska. The first commercial train ran on August 16, 1984.

By 1985, the line was single track for almost its entire length, and it was handling 19 million tons of coal. The implementation of the second stage of the Clean Air Act (1990) caused demand for clean coal to rise quickly. The C&NW struggled to upgrade capacity to dual track, resulting in numerous failures on the line in 1994, and eventually Union Pacific's purchase of C&NW in 1995.

The UP spent 855 million dollars over the next five years expanding capacity over its entire network to handle coal shipments from the PRB. By 2005 the Joint Line capacity had grown to handle an all-time record 325 million tons, and was either dual or three track capacity for its entire length.

Due to various trackage and locomotive failures on the Joint Line in late 2004 and early 2005, the line failed to deliver the amount of contracted coal supplies, and electricity rates increased by 15 percent. Coal customers threatened to evaluate alternate sources of energy and transportation, including the Arkansas Electric Cooperative Corporation. As a result, the 280-mile (450 km) expansion of the Dakota, Minnesota and Eastern Railroad line was approved by the Surface Transportation Board. In 2006 UP and BNSF announced a $100 million investment to provide three track capacity for the entire length of the Joint Line plus a fourth track added over the steepest sections, including Logan Hill. These improvements will enable the Joint Line to handle over 400 million tons of coal.

Presently more than eighty train loads of coal, which vary in size from 125 to 150 cars, are shipped from southern PRB mines each day. In 2006, Union Pacific set a record by hauling 194 million tons of coal — an 8% increase compared with 2005 tonnage. The company achieved this by increasing train size, with trains averaging more than 15,000 tons, a 200-ton weight increase compared with fourth-quarter 2005's average. According to the BNSF 2008 Annual Report, the quadruple track project was completed. In 2009, the America's Power Factuality Tour stopped at the Powder River Basin to report on its role in generating electricity in the US.

Inertial Fusion Power Plant

An inertial fusion power plant is intended to produce electric power by use of inertial confinement fusion techniques on an industrial scale. This type of power plant is still in a research phase.

Two established options for possible medium-term implementation of fusion energy production are magnetic confinement - being used in the ITER international project - and laser-based inertial confinement, as used in the United States' NIF. Inertial confinement fusion (ICF), including heavy ion inertial fusion (HIF) has been proposed as a possible additional means of implementing a fusion power plant.

Overall Principles of an Inertial Fusion Energy (IFE) Reactor

For an easier understanding, it is worth using the analogy of operation between an IFE reactor and a gasoline engine. By applying such an analogy, the process may be seen as a four strokes cycle:

- intake of the fusion fuel (microcapsule) into the reactor chamber;
- compression of the microcapsule in order to initiate the fusion reactions;
- explosion of the plasma created during the compression stroke, leading to the release of fusion energy;
- exhaust of the reaction residue, which will be treated afterwards to extract all the reusable elements, mainly tritium.

To allow such an operation, an inertial fusion reactor is made of several subsets:

- the injection system, which delivers to the reaction chamber the fusion fuel capsules, and at the same time the possible devices necessary to initiate fusion:
  - o the container (*hohlraum*), intended to take the fuel capsule to a uniform very high temperature, mainly for laser and ion beam confinement techniques;
  - o the "wires array" and its power transmission line, for z-pinch confinement technique;
- the "driver" used to compress the fusion fuel capsules; depending on the technique, it can be:
  - o lasers;
  - o an ion beam accelerator;
  - o a z-pinch device;
- the reaction chamber, built upon:
  - o an external wall made of metal;
  - o an internal blanket intended to protect the external wall from the fusion shockwave and radiation, to get the emitted energy, and to produce the tritium fuel;
- the system intended to process reaction products and debris.

Further information: An example of a planned IFE plant can be seen in the Z machine article

IFE Projects

Several projects of inertial fusion power plants have been proposed, notably power production plans based on the following experimental devices, either in operation or under construction:

- in Spain there exists the 4D-Pinch called Pulsotron;
- in the United States, the National Ignition Facility (laser

confinement) and Z machine (z-pinch confinement) experiments;

- in France, the Megajoule laser experiment;
- in Japan (Osaka University), the KONGOH experiment (laser confinment).

As may be noted, only first two of these projects is based on z-pinch confinement, all others being based on laser confinement techniques.

The various phases of such a project are the following :

- burning demonstration: reproducible achievement of energy release.
- high gain demonstration: experimental demonstration of the feasibility of a reactor with a sufficient energy gain.
- industrial demonstration: validation of the various technical options, and of the whole data needed to define a commercial reactor.
- commercial demonstration: demonstration of the reactor ability to work over a long period, while respecting all the requirements for safety, liability and cost.

At the moment, according to the available data, inertial confinement fusion experiments have not gone beyond the first phase, as well for laser (although it is strongly expected to reach the objectives of the second phase around 2010, when NIF and Megajoule are complete) as for z-pinch (Z machine); these techniques should now demonstrate their ability to obtain a high fusion energy gain, as well as their capability for repetitive working.

Space-based Solar Power

Figure 20: Part of the solar energy is lost on its way through the atmosphere by the effects of reflection and absorption.

Right: Space-based solar power systems convert sunlight to microwaves outside the atmosphere, avoiding these losses, and the downtime (and cosine losses, for fixed flat-plate collectors) due to the Earth's rotation.

Space-based solar power (SBSP) is the concept of collecting solar power in space for use on Earth. It has been in research since the early 1970s.

SBSP would differ from current solar collection methods in that the means used to collect energy would reside on an orbiting satellite instead of on Earth's surface. Some projected benefits of such a system are:

- *Higher collection rate:* In space, transmission of solar energy is unaffected by the filtering effects of atmospheric gasses. Consequently, collection in orbit is approximately 144% of the maximum attainable on Earth's surface.
- *Longer collection period:* Orbiting satellites can be exposed to a consistently high degree of solar radiation, generally for 24 hours per day, whereas surface panels can collect for 12 hours per day at most.
- Elimination of weather concerns, since the collecting satellite would reside well outside of any atmospheric gasses, cloud cover, wind, and other weather events.
- Elimination of plant and wildlife interference.
- *Redirectable power transmission:* A collecting satellite could possibly direct power on demand to different surface locations based on geographical baseload or peak load power needs.

SBSP also introduces several new hurdles, primarily the problem of transmitting energy from orbit to Earth's surface for use.

Since wires extending from Earth's surface to an orbiting satellite are neither practical nor feasible with current technology, SBSP designs generally include the use of some manner of wireless power transmission.

The collecting satellite would convert solar energy into electrical energy on-board, powering a microwave transmitter or laser emitter, and focus its beam toward a collector (rectenna) on the Earth's surface. Radiation and micrometeoroid damage could also become concerns for SBSP.

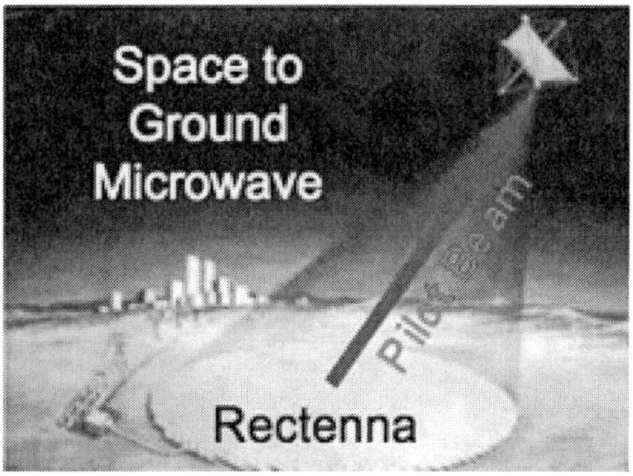

Figure 21: A laser pilot beam guide the microwave power transmission to a rectenna.

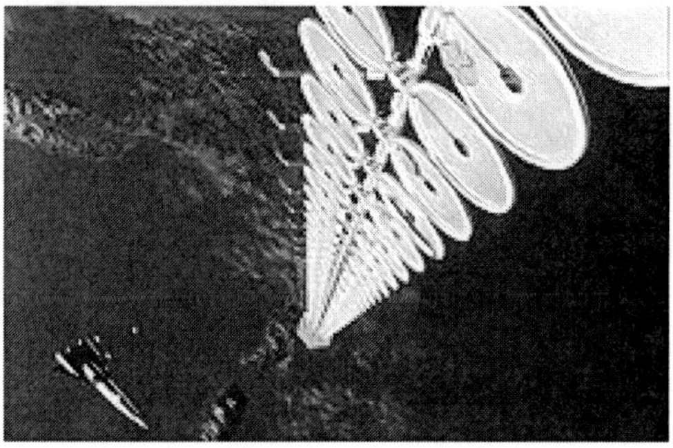

Figure 22: Nasa Suntower concept.

History

The SBSP concept, originally known as Satellite Solar Power System (SSPS), was first described in November 1968. In 1973 Peter Glaser was granted U.S. patent number 3,781,647 for his method of transmitting power over long distances (e.g., from an SPS to Earth's surface) using microwaves from a very large antenna (up to one square kilometre) on the satellite to a much larger one, now known as a rectenna, on the ground.

Glaser then was a vice president at Arthur D. Little, Inc. NASA signed a contract with ADL to lead four other companies in a broader study in 1974. They found that, while the concept had several major problems—chiefly the expense of putting the required materials in orbit and the lack of experience on projects of this scale in space, it showed enough promise to merit further investigation and research.

Between 1978 and 1981, the Congress authorised the Department of Energy and NASA to jointly investigate the concept. They organised the Satellite Power System Concept Development and Evaluation

Program. The study remains the most extensive performed to date (budget 50 millions $). Several reports were published investigating the engineering feasibility of such an engineering project. They include:

- Resource Requirements (Critical Materials, Energy, and Land)

Figure 23: Artist's concept of Solar Power Satellite in place. Shown is the assembly of a microwave transmission antenna. The solar power satellite was to be located in a geosynchronous orbit, 36,000 miles above the Earth's surface. NASA 1976

- Financial/Management Scenarios
- Public Acceptance
- State and Local Regulations as Applied to Satellite Power System Microwave Receiving Antenna Facilities
- Student Participation
- Potential of Laser for SBSP Power Transmission
- International Agreements
- Centralisation/Decentralisation
- Mapping of Exclusion Areas For Rectenna Sites
- Economic and Demographic Issues Related to Deployment
- Some Questions and Answers
- Meteorological Effects on Laser Beam Propagation and Direct Solar Pumped Lasers
- Public Outreach Experiment
- Power Transmission and Reception Technical Summary and Assessment
- Space Transportation

The project was not continued with the change in administrations after the 1980 US Federal elections.

The Office of Technology Assessment concluded

Too little is currently known about the technical, economic, and environmental aspects of SPS to make a sound decision whether to proceed with its development and deployment. In addition, without further research an SPS demonstration or systems-engineering verification program would be a high-risk venture.

In 1997 NASA conducted its "Fresh Look" study to examine the modern state of SBSP feasibility. In assessing "What has changed" since the DOE study, NASA asserted that:

US National Space Policy now calls for NASA to make significant investments in technology (not a particular vehicle) to drive the costs of ETO *[Earth to Orbit]* transportation down dramatically. This is, of course, an absolute requirement of space solar power.

Conversely, Dr. Pete Worden claimed that space-based solar is about five orders of magnitude more expensive than solar power from the Arizona desert, with a major cost being the transportation of materials to orbit. Dr. Worden referred to possible solutions as speculative, and that would not be available for decades at the earliest.

Figure 24: SERT sandwich concept.NASA

SERT

In 1999, NASA's Space Solar Power Exploratory Research and Technology program (SERT) (budget 22 millions $) was initiated for the following purposes:

- Perform design studies of selected flight demonstration concepts.
- Evaluate studies of the general feasibility, design, and

requirements.

- Create conceptual designs of subsystems that make use of advanced SSP technologies to benefit future space or terrestrial applications.
- Formulate a preliminary plan of action for the U.S. (working with international partners) to undertake an aggressive technology initiative.
- Construct technology development and demonstration roadmaps for critical Space Solar Power (SSP) elements.

SERT went about developing a solar power satellite (SPS) concept for a future gigawatt space power system, to provide electrical power by converting the Sun's energy and beaming it to Earth's surface, and provided a conceptual development path that would utilise current technologies. SERT proposed an inflatable photovoltaic gossamer structure with concentrator lenses or solar heat engines to convert sunlight into electricity. The program looked both at systems in sun-synchronous orbit and geosynchronous orbit.

Some of SERT's conclusions:

- The increasing global energy demand is likely to continue for many decades resulting in new power plants of all sizes being built.
- The environmental impact of those plants and their impact on world energy supplies and geopolitical relationships can be problematic.
- Renewable energy is a compelling approach, both philosophically and in engineering terms.
- Many renewable energy sources are limited in their ability to affordably provide the base load power required for global industrial development and prosperity, because of inherent land and water requirements.
- Based on their Concept Definition Study, space solar power concepts may be ready to reenter the discussion.
- Solar power satellites should no longer be envisioned as requiring unimaginably large initial investments in fixed infrastructure before the emplacement of productive power plants can begin.

- Space solar power systems appear to possess many significant environmental advantages when compared to alternative approaches.
- The economic viability of space solar power systems depends on many factors and the successful development of various new technologies (not least of which is the availability of much lower cost access to space than has been available), however, the same can be said of many other advanced power technologies options.
- Space solar power may well emerge as a serious candidate among the options for meeting the energy demands of the 21st century.
- Launch costs in the range of $100-$200 per kilogram of payload to low-Earth orbit are needed if SPS are to be economically viable.

Advantages

The SBSP concept is attractive because space has several major advantages over the Earth's surface for the collection of solar power.

- There is no air in space, so the collecting surfaces could receive much more intense sunlight, unobstructed by weather.
- A satellite could be illuminated over 99% of the time, and be in Earth's shadow on only 75 minutes per night at the spring and fall equinoxes.
- Relatively quick redirecting of power directly to areas that need it most.
- Higher collection rate: In space, transmission of solar energy is unaffected by the filtering effects of atmospheric gasses. Consequently, collection in orbit is approximately 144% of the maximum attainable on Earth's surface.
- Longer collection period: Orbiting satellites can be exposed to a consistently high degree of solar radiation, generally for 24 hours per day, whereas surface panels can collect for 12 hours per day at most.
- Elimination of weather concerns, since the collecting satellite would reside well outside of any atmospheric gasses, cloud cover, wind, and other weather events.
- Elimination of plant and wildlife interference.

- Redirectable power transmission: A collecting satellite could possibly direct power on demand to different surface locations based on geographical baseload or peak load power needs.

Disadvantages

The SBSP concept also has a number of problems.

- The space environment is hostile; panels suffer about 10 times the degradation they would on Earth. System lifetimes on the order of a decade would be expected, which makes it difficult to produce enough power to be economical.
- Space debris are a major hazard to large objects in space, and SBSP systems have been singled out as a particularly hazardous activity.
- The broadcast frequency of the microwave downlink (if used) would require isolating the SBSP systems away from other satellites. GEO space is already well used and it is considered unlikely the ITU would allow an SPS to be launched.
- Only about half the power generated by the SSP would be delivered to the grid, once all losses are factored in. These losses are on the same order as modern fossil fuel plants.

Design

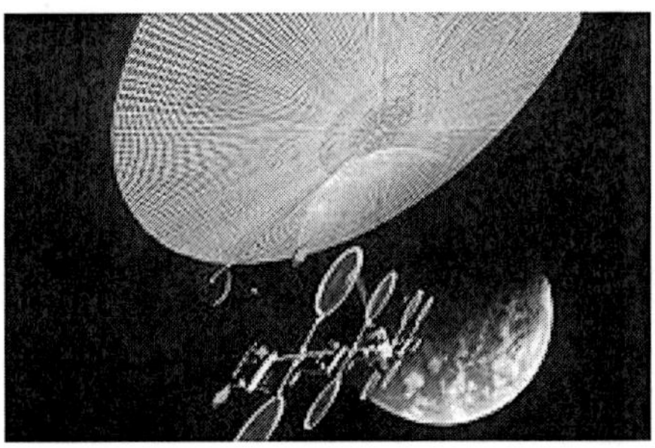

Figure 25: Artist's concept of a solar disk on top of a LEO to GEO electricaly-powered space tug.

Space-based solar power essentially consists of three elements:

- a means of collecting solar power in space, for example via solar cells or a heat engine
- a means of transmitting power to earth, for example via microwave or laser
- a means of receiving power on earth, for example via a microwave antenna (rectenna)

The space-based portion will not need to support itself against gravity (other than relatively weak tidal stresses). It needs no protection from terrestrial wind or weather, but will have to cope with space hazards such as micrometres and solar flares. Two basic methods of conversion have been studied: photovoltaic (PV) and solar dynamic (SD). Most analyses of SBSP have focused on photovoltaic conversion (commonly known as "solar cells"). Photovoltaic conversion uses semiconductor cells to directly convert photons into electrical power. Wireless power transmission was proposed early on as a means to transfer energy from collection to the Earth's surface, using either microwave or laser radiation at a variety of frequencies.

Microwave Power Transmission

William C. Brown demonstrated in 1964, during Walter Cronkite's CBS News program, a microwave-powered model helicopter that received all the power it needed for flight from a microwave beam. Between 1969 and 1975, Bill Brown was technical director of a JPL Raytheon program that beamed 30 kW of power over a distance of 1-mile (1.6 km) at 84% efficiency. Microwave power transmission of tens of kilowatts has been well proven by existing tests at Goldstone in California (1975) and Grand Bassin on Reunion Island (1997).

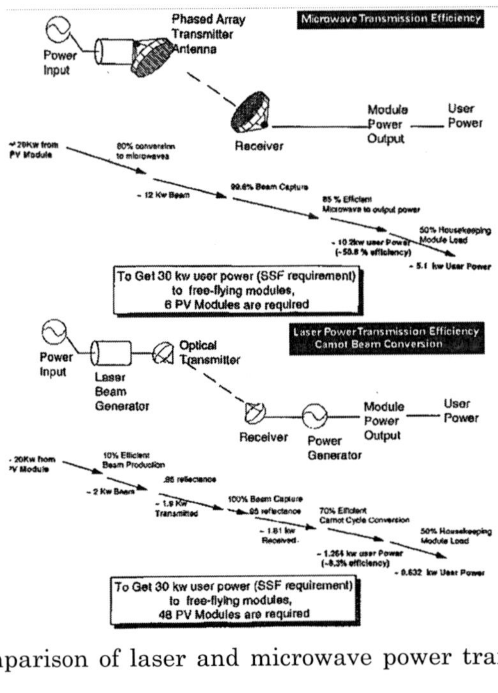

Figure 26: Comparison of laser and microwave power transmission. NASA diagram

More recently, microwave power transmission has been demonstrated, in conjunction with solar energy capture, between a mountain top in Maui and the main island of Hawaii (92 miles away), by a team under John C. Mankins. Technological challenges in terms of array layout, single radiation element design, and overall efficiency, as well as the associated theoretical limits are presently a subject of research, as it is demonstrated by the upcoming Special Session on "Analysis of Electromagnetic Wireless Systems for Solar Power Transmission" to be held in the 2010 IEEE Symposium on Antennas and Propagation.

Laser Power Beaming

Laser power beaming was envisioned by some at NASA as a stepping stone to further industrialisation of space. In the 1980s, researchers at NASA worked on the potential use of lasers for space-to-space power beaming, focusing primarily on the development of a solar-powered laser. In 1989 it was suggested that power could also be usefully beamed by laser from Earth to space. In 1991 the SELENE project (SpacE Laser ENErgy) had begun, which included the study of laser power beaming for supplying power to a lunar base. The SELENE program was a two-year research effort, but the cost of taking the concept to operational status was too high, and the official project ended in 1993 before reaching a space-based demonstration.

In 1988 the use of an Earth-based laser to power an electric thruster for space propulsion was proposed by Grant Logan, with technical details worked out in 1989. He proposed using diamond solar cells operating at 600 degrees to convert ultraviolet laser light, a technology that has yet to be demonstrated even in the laboratory.

Orbital Location

The main advantage of locating a space power station in geostationary orbit is that the antenna geometry stays constant, and so keeping the antennas lined up is simpler. Another advantage is that nearly continuous power transmission is immediately available as soon as the first space power station is placed in orbit; other space-based power stations have much longer start-up times before they are producing nearly continuous power. A collection of LEO (Low Earth Orbit) space power stations has been proposed as a precursor to GEO (Geostationary Orbit) space-based solar power.

Earth-based Receiver

The Earth-based rectenna would likely consist of many short dipole antennas connected via diodes. Microwaves broadcasts from the satellite would be received in the dipoles with about 85% efficiency. With a conventional microwave antenna, the reception efficiency is better, but its cost and complexity is also considerably greater. Rectennas would likely be multiple kilometres across.

In Space Applications

A laser SBSP could also power a base or vehicules on the surface of the moon or mars, saving on mass costs to land the power source. A spacecraft or another satellite could also be powered by the same means.

Dealing with Launch Costs

One problem for the SBSP concept is the cost of space launches and the amount of material that would need to be launched. Reusable launch systems are predicted to provide lower launch costs to low Earth orbit (LEO). Much of the material launched need not be delivered to its eventual orbit immediately, which raises the possibility that high efficiency (but slower) engines could move SPS material from LEO to GEO at an acceptable cost. Examples include ion thrusters or nuclear propulsion.

Power beaming from geostationary orbit by microwaves carries the difficulty that the required 'optical aperture' sizes are very large. For example, the 1978 NASA SPS study required a 1-km diametre transmitting antenna, and a 10 km diametre receiving rectenna, for a microwave beam at 2.45 GHz. These sizes can be somewhat decreased by using shorter wavelengths, although they have increased atmospheric absorption and even potential beam blockage by rain or water droplets. Because of the thinned array curse, it is not possible to make a narrower beam by combining the beams of several smaller satellites. The large size of the transmitting and receiving antennas means that the minimum practical power level for an SPS will necessarily be high; small SPS systems will be possible, but uneconomic.

To give an idea of the scale of the problem, assuming a solar panel mass of 20 kg per kilowatt (without considering the mass of the supporting structure, antenna, or any significant mass reduction of any focusing mirrors) a 4 GW power station would weigh about 80,000 metric tons, all of which would, in current circumstances, be launched from the Earth. Very lightweight designs could likely achieve 1 kg/kW, meaning 4,000 metric tons for the solar panels for the same 4 GW capacity station. This would be the equivalent of between 40 and 150 heavy-lift launch vehicle (HLLV) launches to send the material to low earth orbit, where it would likely be converted into subassembly solar arrays, which then could use high-efficiency ion-engine style rockets to (slowly) reach GEO (Geostationary orbit). With an estimated serial launch cost for shuttle-based HLLVs of $500 million to $800 million, and launch costs for alternative HLLVs at $78 million, total launch costs would range between $11 billion (low cost HLLV, low weight panels) and $320 billion ('expensive' HLLV, heavier panels).

For comparison, the direct cost of a new coal [1] or nuclear power plant ranges from $3 billion to $6 billion dollars per GW (not including the full cost to the environment from CO2 emissions or storage of spent nuclear fuel, respectively); another example is the Apollo missions to the Moon cost a grand total of $24 billion (1970's dollars), taking inflation into account, would cost $140 billion today, more expensive than the construction of the International Space Station.

Non-conventional Launch Methods

SBSP costs might be reduced if a means of putting the materials into orbit were developed that did not rely on rockets. Some possible technologies include ground launch systems such as mass drivers or Lofstrom loops, which would launch using electrical power, or the geosynchronous orbit space elevator. However, these require technology that is yet to be developed. John Hunter of Quicklaunch is working on commercialising the 'Hydrogen Gun', a new form of mass driver which proposes to deliver unmanned payloads to orbit for around 5% of regular launch costs (i.e. $500/lb or US$1,000/kg) and perform 5 launches per day.

Building from Space

From Lunar Materials Launched in Orbit: Gerard O'Neill, noting the problem of high launch costs in the early 1970s, proposed building the SPS's in orbit with materials from the Moon. Launch costs from the Moon are potentially much lower than from Earth, due to the lower gravity. This 1970s proposal assumed the then-advertised future launch costing of NASA's space shuttle. This approach would require substantial up front capital investment to establish mass drivers on the Moon. Nevertheless, on 30 April 1979, the Final Report ("Lunar Resources Utilisation for Space Construction") by General Dynamics' Convair Division, under NASA contract NAS9-15560, concluded that use of lunar resources would be cheaper than Earth-based materials for a system of as few as thirty Solar Power Satellites of 10GW capacity each.

In 1980, when it became obvious NASA's launch cost estimates for the space shuttle were grossly optimistic, O'Neill et al. published another route to manufacturing using lunar materials with much lower startup costs. This 1980s SPS concept relied less on human presence in space and more on partially self-replicating systems on the lunar surface under remote control of workers stationed on Earth. The high Net energy gain of this proposal derives from the Moon's much shallower gravitational well. Having a relatively cheap per pound source of raw materials from space would lessen the concern for low mass designs and result in a different sort of SPS being built. The low cost per pound of lunar materials in O'Neill's vision would be supported by using lunar material to manufacture more facilities in orbit than just solar power satellites.

Advanced techniques for launching from the Moon may reduce the cost of building a solar power satellite from lunar materials. Some proposed techniques include the lunar mass driver and the lunar space elevator, first described by Jerome Pearson. It would require establishing silicon mining and solar cell manufacturing facilities on the Moon.

On the Moon

David Criswell suggests the Moon is the optimum location for solar power stations, and promotes lunar solar power. The main advantage he envisions is construction largely from locally available lunar materials, using in-situ resource utilisation, with a teleoperated mobile factory, a crane to assemble the microwave reflectors, and rovers to assemble solar cells, which would significantly reduce launch costs compared to SBSP designs. Power relay satellites orbiting around earth and the Moon reflecting the microwave beam are also part of the project. Another design combined the rovers with the factory and directly paves the Moon with a thin film of solar cells. The Shimizu Corporation proposed using combination of lasers and microwave for the lunar ring concept, along with power relay satellites.

From an Asteroid

Asteroid mining has also been seriously considered. A NASA design study evaluated a 10,000 ton mining vehicle (to be assembled in orbit) that would return a 500,000 ton asteroid fragment to geostationary orbit. Only about 3,000 tons of the mining ship would be traditional aerospace-grade payload. The rest would be reaction mass for the mass-driver engine, which could be arranged to be the spent rocket stages used to launch the payload. Assuming that 100% of the returned asteroid was useful, and that the asteroid miner itself couldn't be reused, that represents nearly a 95% reduction in launch costs. However, the true merits of such a method would depend on a thorough mineral survey of the candidate asteroids; thus far, we have only estimates of their composition.

Counter Arguments

Safety: The use of microwave transmission of power has been the most controversial issue in considering any SPS design.

At the Earth's surface, a suggested microwave beam would have a maximum intensity at its centre, of 23 mW/cm^2 (less than 1/4 the solar irradiation constant), and an intensity of less than 1 mW/cm^2 outside of the rectenna fenceline (the receiver's perimetre).

These compare with current United States Occupational Safety and Health Act (OSHA) workplace exposure limits for microwaves, which are 10 mW/cm^2, - the limit itself being expressed in voluntary terms and ruled unenforceable for Federal OSHA enforcement purposes. A beam of this intensity is therefore at its centre, of a similar magnitude to current safe workplace levels, even for long term or indefinite exposure. Outside the receiver, it is far less than the OSHA long-term levels Over 95% of the beam energy will fall on the rectenna.

The remaining microwave energy will be absorbed and dispersed well within standards currently imposed upon microwave emissions around the world. It is important for system efficiency that as much of the microwave radiation as possible be focused on the rectenna. Outside of the rectenna, microwave intensities rapidly decrease, so nearby towns or other human activity should be completely unaffected.

Exposure to the beam is able to be minimised in other ways. On the ground, physical access is controllable (e.g., via fencing), and typical aircraft flying through the beam provide passengers with a protective metal shell (i.e., a Faraday Cage), which will intercept the microwaves. Other aircraft (balloons, ultralight, etc.) can avoid exposure by observing airflight control spaces, as is currently done for military and other controlled airspace.

The microwave beam intensity at ground level in the centre of the beam would be designed and physically built into the system; simply, the transmitter would be too far away and too small to be able to increase the intensity to unsafe levels, even in principle. In addition, a design constraint is that the microwave beam must not be so intense as to injure wildlife, particularly birds. Experiments with deliberate microwave irradiation at reasonable levels have failed to show negative effects even over multiple generations.

Some have suggested locating rectennas offshore, but this presents serious problems, including corrosion, mechanical stresses, and biological contamination. A commonly proposed approach to ensuring fail-safe beam targeting is to use a retrodirective phased array antenna/ rectenna. A "pilot" microwave beam emitted from the centre of the rectenna on the ground establishes a phase front at the transmitting antenna. There, circuits in each of the antenna's subarrays compare the pilot beam's phase front with an internal clock phase to control the phase of the outgoing signal. This forces the transmitted beam to be centred precisely on the rectenna and to have a high degree of phase uniformity; if the pilot beam is lost for any reason (if the transmitting antenna is turned away from the rectenna, for example) the phase control value fails and the microwave power beam is automatically defocused. Such a system would be physically incapable of focusing its power beam anywhere that did not have a pilot beam transmitter. The long-term effects of beaming power through the ionosphere in the form of microwaves has yet to be studied, but nothing has been suggested which might lead to any significant effect.

Atmospheric Damage Due to Launches

When hot rocket exhaust reacts with atmospheric nitrogen, it can form nitrogen compounds. These nitrogen compounds are problematic when they form in the stratosphere, as they can damage the ozone layer. However, the environmental effect of rocket launches is negligible compared to higher volume polluters, such as aeroplanes and automobiles.

Timeline

- *1968:* Dr. Peter Glaser introduces the concept of a "solar power satellite" system with square miles of solar collectors in high geosynchronous orbit for collection and conversion of sun's energy into a microwave beam to transmit usable energy to large receiving antennas (rectennas) on Earth for distribution.
- *1973:* Dr. Peter Glaser is granted United States patent number 3,781,647 for his method of transmitting power over long distances using microwaves from a large (one square kilometre) antenna on the satellite to a much larger one on the ground, now known as a rectenna.
- *1978-81:* The United States Department of Energy and NASA examine the solar power satellite (SPS) concept extensively, publishing design and feasibility studies.

- *1982:* Boeing proposal
- *1987:* Stationary High Altitude Relay Platform a canadian experiment
- *1994:* The United States Air Force conducts the Advanced Photovoltaic Experiment using a satellite launched into low Earth orbit by a Pegasus rocket.
- *1995–97:* NASA conducts a "Fresh Look" study of space solar power (SSP) concepts and technologies.
- *1998:* The Space Solar Power Concept Definition Study (CDS) identifies credible, commercially viable SSP concepts, while pointing out technical and programmatic risks.
- *1998:* Japan's space agency begins developing a Space Solar Power System (SSPS), a program that continues to the present day.
- *1999:* NASA's Space Solar Power Exploratory Research and Technology program begins.
- *2000:* John Mankins of NASA testifies in the U.S. House of Representatives, saying "Large-scale SSP is a very complex integrated system of systems that requires numerous significant advances in current technology and capabilities. A technology roadmap has been developed that lays out potential paths for achieving all needed advances — albeit over several decades.
- *2001:* Dr. Neville Marzwell of NASA states, "We now have the technology to convert the sun's energy at the rate of 42 to 56 percent... We have made tremendous progress. ...If you can concentrate the sun's rays through the use of large mirrors or lenses you get more for your money because most of the cost is in the PV arrays... There is a risk element but you can reduce it... You can put these small receivers in the desert or in the mountains away from populated areas. ...We believe that in 15 to 25 years we can lower that cost to 7 to 10 cents per kilowatt hour. ...We offer an advantage. You don't need cables, pipes, gas or copper wires. We can send it to you like a cell phone call—where you want it and when you want it, in real time."
- *2001:* NASDA (Japan's national space agency) announces plans to perform additional research and prototyping by launching an experimental satellite with 10 kilowatts and 1 megawatt of power.

- *2003:* ESA studies
- *2007:* The US Pentagon's National Security Space Office (NSSO) issues a report on October 10, 2007 stating they intend to collect solar energy from space for use on Earth to help the United States' ongoing relationship with the Middle East and the battle for oil. The International Space Station may be the first test ground for this new idea, even though it is in a low-earth orbit.
- *2007:* In May 2007 a workshop is held at the US Massachusetts Institute of Technology (MIT) to review the current state of the SBSP market and technology.
- *2009:* Several companies announce future SBSP partnerships and commitments, including Pacific Gas and Electric (PG&E) & Solaren, Mitsubishi Electric Corp. & IHI Corporation, Space Energy, Inc., and Japan Aerospace Exploration Agency.
- *2010:* Europe's EADS Astrium announces SBSP plans.
- *2010:* Professors Andrea Massa and Giorgio Franceschetti announce a special session on the "Analysis of Electromagnetic Wireless Systems for Solar Power Transmission" at the 2010 Institute of Electrical and Electronics Engineers International Symposium on Antennas and Propagation.

In fiction

- Space stations transmitting solar power have appeared in science-fiction works like Isaac Asimov's *Reason* (1941), that centres around the troubles caused by the robots operating the station. Asimov's short story "The Last Question" also features the use of SBSP to provide limitless energy for use on Earth.
- In the novel "Skyfall" (1976) by Harry Harrison an attempt to launch the core of powersat from Cape Canaveral ends in disaster when the launch vehicle fails trapping the payload in a decaying orbit.
- Solar Power Satellites have also been seen in the work of author Ben Bova's novels "Powersat" and "Colony".
- In Sid Meier's *Alpha Centauri*, an endgame 'building' that fulfills the same function as an SPS is the 'Orbital Power Transmitter' which provides every city that you own with a unit of energy per satellite launched, providing the city has

an Aerospace Command building or your faction controls the space elevator. Building multiple Orbital Power Transmitters provides massive bonuses to energy generation and soon pay for themselves many times over.

- In a 1981 storyline from the Iron Man comic book (issues #142-144), a rogue microwave transmission from a secret Solar Power Satellite is responsible for numerous deaths in Allentown, Iowa.
- In the computer games *SimCity 2000* and *3000*, plants that implemented solar satellite technology called microwave powerplants were available in the future. One disaster scenario involved the beam missing the receiver and hitting the city's infrastructure. The plant was discontinued in *SimCity 4* but several fan-made microwave powerplants were available on various *SimCity 4* fan sites.
- In the film *Die Another Day*, a satellite weapon is disguised as a solar power satellite.
- In *Mobile Suit Gundam 00*, a solar power satellite array is constructed around the Earth and is used to harness solar energy for use. They play a critical plot role in the superpowers' power balance.
- In *After War Gundam X*, a solar power station is built on the Moon, and is used to supply energy via microwave to various mobile suits, to energise their powerful "Satellite Cannons".
- In *Cadillacs and Dinosaurs (TV series)*, space solar power is one of the lost technologies of the ancients.

Bibliography

Armistead, George: *Safety in Petroleum Refining and Related Industries*, New York: John G. Simmons & Co., 1950.

Armstrong, J. A.: *Forces Shaping the U.S. Academic Engineering Research Enterprise*, National Academy Press, Washington DC, 1995.

Badiru, A.: *Handbook of Industrial and Systems Engineering*, CRC Press, Delhi, 2005.

Bamberger, Robert L.: *The National Energy Strategy: Summary and Analysis,* Congressional Research Service, Library of Congress, Washington, DC, 1991.

Barua, Lohit: *Natural Gas Measurement and Control: A Guide for Operators and Engineers*, McGraw-Hill Professional Book Group, London, 1991.

Bauer, M.: *Resistance to New Technology.* New York: Cambridge University Press, 1995.

Bennett, S.: *A History of Control Engineering: 1800-1930*, IEE Press, London, 1979.

Benvenuto, E.: *An Introduction to the History of Structural Mechanics*, Springer-Verlag, New York, 1991.

Bernard, Finn S.: *The History of Electrical Technology: An Annotated Bibliography*, Garland Publishing, Inc., New York, 1991.

Beychok, Milton R.: *Aqueous Wastes from Petroleum and Petrochemical Plants,* John Wiley and Sons, NY, 1967.

Bird, John: *Electrical and Electronic Principles and Technology,* Newnes, UK, 2007.

Bowers, Brian: *A History of Electric Light and Power*, Peter Peregrinus Ltd., Stevenage, UK, 1982.

British Electricity International: *Modern Power Station Practice: Incorporating Modern Power System Practice,* Pergamon. 1991.

Brower, Michael: *Cool Energy: Renewable Solutions to Environmental Problems*, MIT Press, Cambridge, MA, 1992.

Buchanan, R.: *Engineers and Engineering,* Beth University press, 1996.

Calahan, North: *TVA: Bridge Over Troubled Water*, Barnes, Cranbury, NJ, 1980.

Chapman, K.: *The International Petrochemical Industry*. Oxford: Blackwell, 1991.

Charles F.: *The Petroleum Industry: A Nontechnical Guide,* PennWell Pub. Co., Tulsa, OK, 1999.

Christopher Kilian: *Modern Control Technology*, Thompson Delmar Learning, UK, 2005.

Collier, Hugh: *Developing Electrical Power: Thirty Years of World Bank Experience,* The Johns Hopkins University Press, Baltimore, 1984.

Conaway, C.F.: *Petroleum Industry: A Nontechnical Guide*, Pennwell Publishing, Tulsa, USA, 1999.

Crozet, Francois: *The First Industrialists: The Problems of Origins*, Cambridge University Press, New York, 1985.

Dasgupta, S.: *Technology and Creativity*, Oxford University Press, New York, 1996.

Devereux, Steve: *Drilling for Oil & Gas: A Nontechnical Guide,* PennWell, Tulsa, OK, 1999.

Doebelin, E. O.: *Engineering Experimentation: Planning, Execution, Reporting*, McGraw Hill, Boston, 1995.

Domminghaus, H.: *Plastics for Engineers: Materials, Properties, Applications*, Hanser Publishers, Munich, 1993.

Douglas, Anderson D.: *Regulatory Politics and Electric Utilities: A Case Study in Political Economy,* Auburn House, Boston, 1981.

Duffin, W.J.: *Electricity and Magnetism,* McGraw-Hill, NY, 1980.

Dym, J.B.: *Product Design with Plastics: A Practical Manual*, Industrial Press, New York, 1983.

Edminister, Joseph: *Electric Circuits,* McGraw-Hill, NY, 1965.

Eveleigh, E.W.: *Introduction to Control Systems Design*, McGraw-Hill, 1972.

Fabrycky, W.: *Systems Engineering and Analysis*, Prentice-Hall, NY, 2005.

Finch, J. K.: *Engineering Classics*, Kensington, Cedar Press, MD, 1978.

Freeman, Neil B.: *The Politics of Power: Ontario Hydro and Its Government, 1906-1995*, University of Toronto Press, Toronto, 1996.

Ghalambor, Ali: *Petroleum Production Engineering: A Computer-assisted Approach,* Gulf Professional Pub., Burlington, MA, 2007.

Hammond, Percy: *Electromagnetism for Engineers*, Pergamon, NJ, 1981.

Harriott, P.: *Unit Operations of Chemical Engineering*, McGraw Hill, New York, 2004.

Hollands, J.: *Engineering Psychology and Human Performance*, Prentice Hall, Delhi, 1999.

Hughes, Thomas P.: *Networks of Power: Electrification in Western Society, 1880-1930*, The Johns Hopkins University Press, Baltimore, 1983.

Islam, M.A. : *Encyclopaedia of Electrochemistry*, Sonali Publications, Delhi, 2011.

James G.: *The Chemistry and Terminology of Petroleum*. New York: Marcel Dekker, 1980.

Jentleson, Bruce: *Pipeline Politics: The Complex Political Economy of East-West Energy Trade*. lthaca: Cornell University Press, 1986.

Joseph A. Brink Jr.: *The Chemical Process Industries*, McGraw Hill, New York, 1977.

Keey, R. B.: *Introduction to Industrial Drying Operations*, Pergamon, Oxford, 1978.

Keily, William E.: *Central-Station Electric Service, Its Commercial Development and Economic Significance as Set Forth in the Public Addresses (1897-1914) of Samuel Insull,* Privately printed, Chicago, 1915.

Kuo, B.C., *Automatic Control Systems*, Prentice-Hall, 1995.

Luyben, W. L. : *Process Modelling, Simulation and Control for Chemical Engineers*, McGraw-Hill, 1989

Miller, J. M.: *Separation Methods in Chemical Analysis,* John Wiley and Sons, New York, 1975.

Moorhouse, John C.: *Electric Power: Deregulation and the Public Interest,* Pacific Research Institute for Public Policy, San Francisco, 1986.

Moss, M. S, and Hume, J. R.: *Workshop of the British Empire: Engineering and Shipbuilding in the West of Scotland*, Heinemann, London, 1977.

Mugerwa, A. K. : *Fuel Cell Systems*. New York: Plenum Press, 1993.

Munroe, J. P.: *Technology and Industrial Efficiency: Proceedings of the Congress of Technology 1911*, McGraw Hill, New York, 1911.

Nivola, Pietro S.: *The Politics of Energy Conservation,* Brooking Institution, Washington, DC, 1986.

Ogata, K.: *Modern Control Engineering*, Prentice-Hall, 2002.

Parkash, S.: *Refining Processes Handbook*, Elsevier, Amsterdam, 2003.

Pope, Stephen B.: *Turbulent Flows*, Cambridge University Press, UK, 2000.

Prausnitz, J. M. and B. E. Poling: *The Properties of Gases and Liquids*, McGraw-Hill, New York, 1987.

Rathakrishnan, E : *Gas Tables*, Universities Press, Delhi, 2004.

Rathi, Rakesh : *Petroleum Refining Processes*, SBS Pub, Delhi, 2007.

Robert D.; Biblarz O.: *Fundamentals of Gas Dynamics*. John Wiley & Sons, 2002.

Salvendy, G.: *Handbook of Industrial Engineering: Technology and Operations Management*, Wiley-Interscience, Delhi, 2001.

Sears, Francis W.: *University Physics,* Addison Wesley, NY, 1982.

Seidel, A.: *Kirk-Othmer Encyclopedia of Chemical Technology*, Wiley-Interscience, New York, 2004.

Sheng, James: *Modern Chemical Enhanced Oil Recovery: Theory and Practice,* Gulf Professional Pub., Amsterdam, Boston, 2011.

Talukdar, A. K. : *Catalysis in Petroleum and Petrochemical Industries*, Alpha Science International, Delhi, 2005.

Tiwari, S.K. : *Fuel Minerals and Other Energy Resources*, Atlantic Pub, Delhi, 2010.

Treager, I.E.: *Aircraft Gas Turbine Engine Technology*. . McGraw-Hill Book Co. Glencoe Div., 1995.

Turner, W.: *Introduction to Industrial and Systems Engineering*, Prentice Hall, NY, 1992.

Vijayalakshmi, G S : *Fuels and Biofuels*, Agrobios, Delhi, 2007.

Vilim, T.: *Vapor–Liquid Equilibrium*, Pergamon, Oxford, 1967.

Waddell, J. J.: *Construction Materials Ready Reference Manual,* McGrawHill Book Co., New York, 1985.

Wayne, Beaty, H. : *Standard Handbook for Electrical Engineers,* McGraw-Hill, New York, 1978.

William Alec: *The Chemical Industry*. London: Longman Group Limited, 1971.

Wiser, Wendell H.: *Energy Resources: Occurrence, Production, Conversion*, Birkhäuser, 2000.

Zucker, Robert D.; Biblarz O.: *Fundamentals of Gas Dynamics*. John Wiley & Sons, 2002.

Index

I

J

K

L

M

N

P

S

T

V

W

❍❍❍